A CONCISE GUIDE TO COMMUNICATION IN SCIENCE AND ENGINEERING

A Concise Guide to Communication in Science and Engineering

David H. Foster

University of Manchester, UK

OXFORD
UNIVERSITY PRESS

OXFORD
UNIVERSITY PRESS

Great Clarendon Street, Oxford, OX2 6DP,
United Kingdom

Oxford University Press is a department of the University of Oxford.
It furthers the University's objective of excellence in research, scholarship,
and education by publishing worldwide. Oxford is a registered trade mark of
Oxford University Press in the UK and in certain other countries

First Edition published in 2017

Published in the United States of America by Oxford University Press
198 Madison Avenue, New York, NY 10016, United States of America

British Library Cataloguing in Publication Data

Data available

Library of Congress Control Number: 2017934656

ISBN 978–0–19–870423–2 (hbk.)
ISBN 978–0–19–870424–9 (pbk.)

To Fran

Preface

My reason for writing this guide is to help scientists and engineers communicate their research. There is plenty of advice and guidance available but the sources are widely scattered. There are books, articles, pamphlets, and web pages on writing research articles, analysing and graphing data, and formulating mathematics; on preparing dissertations, theses, literature reviews, and technical reports; on presenting posters and giving lectures and talks; on effective writing and the use of English; on data management, ethics, and malpractice; and on the publication and promotion of research. Yet these topics are rarely covered in depth and at a similar level within the one volume, and oriented towards both scientists and engineers.

Several factors have shaped the content of the guide. A direct influence has been the lectures on communication in science and engineering I give to postgraduate students and researchers at my own and other institutions. Being a journal editor and reviewer has afforded a continual reminder of what makes good scientific writing, as has interaction with other editors and reviewers and with journal publishers. And the usual academic duties, including supervising and examining theses, along with delivering support for research assessment exercises, have provided their individual insights. My own experience too as an author has had an inevitable effect. For self-preservation, however, I should add a caveat best expressed by a great expositor of mathematics, Paul Halmos [150, p. 48], thus:

> The recommendations [here] are based partly on what I do, more on what I regret not having done, and most on what I wish others had done for me. You may criticize what I've said on many grounds, but I ask that a comparison of my present advice with my past action not be one of them. Do, please, as I say, and not as I do, and you'll do better.

As expected, this guide has benefited from the contributions of many colleagues, students, and members of the larger research community. In particular, for providing technical advice, I thank Kersti Börjars, Patrick Gaydecki, Matthew P. Halsall, Robert Lucas, Eleanor Pye, Simon M. Rowland, Rob Sloan, and Daniel K. C. So. For providing abridged examples of peer reviews, I thank Fumie Costen, Mario Ferraro, Matthew P. Halsall, Alexander Lanzon, Robert Lucas, Jovica V. Milanovic, Mark Muldoon, Krikor B. Ozanyan, Adam Reeves, Dov Sagi, Roelof G. van Silfhout, and Johan Wagemans.

For allowing the use of extracts from drafts and unpublished versions of reports, dissertations, and theses, I thank six former research students, who prefer to remain anonymous.

For providing experimental data for published figures, I thank Charles Craik, Pantelis Mikropoulos, Simon M. Rowland, Alastair Sanderson, and Mario Schirmer.

For agreeing to make software available for download, I thank Richard Johnson, Iván Marín-Franch, and Kamila Żychaluk.

For reviewing individual chapters, I thank Paul Artes, Alex Baratta, Mike Barnes, Jim Boran, Kersti Börjars, Tim Echtermeyer, Gaoyang Feng, Chris Graf, Sara Grimme, Chris Isham, Dee-Ann Johnson, Alys Kay, Iván Marín-Franch, Krikor B. Ozanyan, Adam Reeves, Daniel K. C. So, Toby Stafford, Timothy Stibbs, Ann Thorson, and Kamila Żychaluk. For reviewing the whole volume, I am especially grateful to Kinjiro Amano and Sérgio M. C. Nascimento.

For support in the development and production of this guide, I thank the Oxford University Press team, particularly Sonke Adlung and Ania Wronski, Indumadhi Srinivasan, and Douglas Vipond, my copyeditor.

No guide of this kind is likely to be entirely free from errors, lapses, and other infelicities. Those that remain are entirely my own responsibility.

Contents

Permissions and Copyright Statements

Numerals in square brackets refer to entries in the Bibliography and list of Research Works.

Preface, quotation by P. R. Halmos, published in [150], copyright 1973. Used by permission of the American Mathematical Society.

Examples 2.11 and 3.24. Copyright 1997 IEEE. Reprinted, with permission, from [416].

Example 3.26. Reprinted from [674], copyright 2002, with permission from Elsevier.

Example 3.31. Reprinted from [657], with the permission of AIP Publishing.

Example 3.51. Reprinted from [585], copyright 2010, with permission from Elsevier.

Example 3.55. Copyright 2010 IEEE. Reprinted, with permission, from [588].

Example 3.56. Reprinted from [448], by permission of Oxford University Press.

Example 3.58. Reprinted from [421], copyright 2000, with permission from Elsevier.

Example 3.59. Reprinted from [421], copyright 2000, with permission from Elsevier.

Example 3.60. Reprinted from [402] with permission from PNAS.

Example 3.69. Reprinted from [285], work of the Federal Government.

Example 3.80. Reprinted from [694], copyright 2012, with permission from Elsevier.

Examples 3.85 and 3.89. Reprinted from [626] under Creative Commons Attribution Licence CC BY.

Example 3.86. Reprinted from [397] with permission from PNAS.

Example 3.87. Reprinted from [463] with permission from PNAS.

Examples 3.91 and 3.92. Reprinted from [671], copyright 2011, with permission from Elsevier.

Example 3.94. Reprinted from [719], copyright 2012, with permission from Elsevier.

Example 3.95. Reprinted from [415] under Creative Commons Attribution Licence CC BY-NC.

Example 3.97. Reproduced from [485] with permission of The Royal Society of Chemistry.

Example 3.99. Reprinted from [530].

Example 3.102. Copyright 2010 IEEE. Reprinted, with permission, from [668].

Example 3.105. Reprinted from [632] with permission from PNAS.

Example 3.106. Reprinted from [502] with permission from PNAS.

1 Introduction

Why should we care about the way we communicate research in science and engineering? After all, as long as the content of the work is reported accurately, you might reasonably assume that the style of reporting should not matter. The strength of the research itself—or the scholarship—ought to secure its recognition. Sadly, that assumption would be wrong.

Since 1665, when the world's first science journal appeared [249], the number of active refereed academic periodicals has been growing steadily, so that by 2009 the rate at which refereed articles were being published exceeded one million a year [28] [372]. Although only a very small fraction of this literature is relevant to any specific research area, the challenge for the author remains—to attract and hold the interest of readers with limited attention, pressed for time, with other articles to examine. Failure to meet this challenge may mean that your work languishes unseen and unrecognized, without influence or impact, and that the chief motive for undertaking it is lost.

Unfortunately, writing to engage the reader is not something that comes effortlessly for most of us. Effective communication requires thought, trial, reflection, and revision, summarized neatly in Dickens's journal [84, p. 189] thus:

> It is a rule that easy reading is hard writing, and to construct anything that the mind takes in without effort, and without being puzzled by it, is a triumph of art.

Nevertheless, the endeavour—the hard writing—need not be an unhappy experience. And the effort you invest can lead to your gaining a better understanding of your work, its strengths and weaknesses, and its theoretical and practical significance.

1.1 Reasons for This Guide

Communication in science and engineering ought not to be more difficult than any other kind of communication. But the nature of the subject matter, its peculiar constraints, and its emphasis on what was done rather than on who did it can lead to language that varies from the awkward to the misleading.

Tracts on the proper use of language, and in particular scientific language, have been with us since the seventeenth century, though most have been firmer on opinion than scholarship [167]. More modern guides to scientific communication, many of which are listed in the Bibliography, have a better provenance; yet they still tend to be discipline-specific and written with an individual view of what is proper. On the continuum between prescribing and describing, they generally settle on prescription.

The aim of this guide is not to offer another personal, prescriptive view, nor to advocate a passive, descriptive one. Instead, it is to bring together sound advice and accepted good practice, supported and moderated by a wide range of examples from the scientific and engineering literature. The emphasis is on a logical approach to communication [108], informed by what needs to be achieved and what difficulties get in the way.

Where rules or, more frequently, principles and conventions [167, p. 190] are invoked, I usually cite their provenance so that you can examine the evidence for and against and come to your own conclusion. Where they appear prescriptive, it is because I feel that the evidence is overwhelming.

What I hope you will gain from this guide is access to a comprehensive body of work that will enable you to prepare and deliver material that is of a professional standard, that engages the reader, and that in due course helps your research achieve its rightful place in the community.

1.2 Sources and Referencing

The advice offered here draws on many sources, including handbooks, guides, and other works, in all about 390. The Bibliography at the end of the guide contains their full publication details. As already demonstrated, a citation to a source is shown by a numeral in square brackets. For example, '[40]' refers to entry number 40 in the Bibliography, namely Burchfield's *The new Fowler's modern English usage*. No further details are given when the source is a short document, or has informative headings, an index, a dictionary structure, or if web-based a site search function. If a chapter, section, or page number is needed, then it is included in the citation; thus '[297, §2.28]' refers to Section 2.28 of entry number 297 in the Bibliography. In the interests of transparency and coverage, sources are cited more often than is usual in a work of this kind and not always by priority.

To support and motivate the advice, both good and bad examples are presented, in all about 350. They are extracts mainly from published research articles, with a few from drafts of theses, papers, and reports, along with a sampling of comments from peer reviewer reports. Examples are displayed in shaded boxes for ease of identification, accompanied by comment, analysis, or annotations.[1] To avoid uncertainty, any example that ought not to be imitated has a cross placed close to it. Where I have edited an example to illustrate a different way of phrasing, the caption contains the phrase 'edited', 'revised', or the equivalent. Ellipsis points '. . .'

mark where text is omitted from a quotation and square brackets show where original text has been replaced.

In fairness to the larger literature, I emphasize that the examples were not chosen randomly: each was cherry-picked for a purpose. The more troublesome ones should therefore be taken as representative not of any individual author's work or discipline but of the kinds of errors that we can all make from time to time.

A separate list of Research Works at the end of the guide contains the full publication details of the sources of the examples. Citations to these sources are shown in the text with the same numeric referencing system used for the Bibliography.

Comments, additional information, and statistical analyses whose inclusion would interrupt the narrative are gathered together in the Notes also at the end of the guide. Citations to the Notes are shown in the text by superscript numerals.

1.3 Using This Guide

Although the guide can be used systematically either for self-instruction or for taught courses, it need not be read from beginning to end. It can be used as an occasional reference, for browsing, or to address particular problems in communication as they occur.

To see its more ad-hoc use, suppose that you want to work on a discussion section of a paper. You could first turn to the Research Articles chapter (ch. 3), then to the Discussion and Conclusion section (§3.8), and then to the Fundamentals subsection (§3.8.1), which sets out the requirements. If you want to learn more about how to craft one of those elements, say the conclusion, you could then turn to the Conclusion subsection (§3.8.6), which explains what it should contain and gives examples from published articles. The two subsections that follow (§3.8.7 and §3.8.8) illustrate how to say too much and too little.

If you have already drafted a discussion and want merely to confirm that it contains the expected elements, you could turn straight to the Checklist section (§3.13) at the end of the chapter, where the main points are summarized.

If instead you want a less algorithmic approach, you could simply browse the examples from published articles displayed in the Discussion and Conclusion section (§3.8) and be guided by the annotations and comments in the text.

1.4 Content and Organization

Communication is here interpreted broadly and refers to both written and spoken forms. The subject coverage—albeit oriented towards the physical and life sciences and engineering—extends also to mathematics, the behavioural and clinical sciences, and medicine.

The content is organized as follows. Chapters 2–5 deal with expository matters, including general principles for scientific writing, the structure of research articles, literature reviews, technical reports, dissertations and theses, and the use and abuse of English in science and engineering. Chapters 6–8 deal with quantitative matters, including the delivery of mathematical material, describing data, statistical inference, data fitting, and plotting graphs. Chapters 9–11 deal mainly with procedural issues, including submitting papers for publication, measuring impact, presenting posters, and giving lectures and talks. Lastly, Chapter 12 deals with ethical matters, including plagiarism and data management. Each of the chapters contains a checklist at the end for easy reference and verification.

The level of technical difficulty is intended to be suited to final-year undergraduate students undertaking project work, master's students, doctoral and postdoctoral researchers, early-career academics, and beyond.

2

General Principles

This chapter sets out some commonly accepted principles and conventions for writing in science and engineering. More specific advice on writing research articles is given in Chapter 3 and on the use of English in Chapter 5.

Scientific writing is sometimes seen as being at one end of a stylistic spectrum, epitomized by Dirac thus:[2]

> In science, you want to say something nobody knew before, in words everyone can understand. In poetry, you are bound to say something that everybody knows already in words that nobody can understand.

A more nuanced view is offered by Medawar [251, p. 54] to the effect that there is poetry in science—and also a lot of bookkeeping. Some bookkeeping matters are indeed considered here, but along with many other issues, especially to do with the immediacy and impact of scientific language.

A theme in this and other chapters is the importance of the trade-off between the effort invested by the writer and that by the reader [150, p. 43]; that is, by spending more time carefully wording a phrase, sentence, or paragraph, you can save the reader very much more time and a lot of confusion. The aim should be always to write for the reader, not for yourself [161, p. 2] [389, p. 3].

Section 2.1 deals with the logical organization of a research article and how it can fail or confuse. Section 2.2 suggests ways to ease the path of the reader through the article. Section 2.3 examines how to achieve greater immediacy and impact. Section 2.4 outlines strategies for writing and Section 2.5 shows how reworking can increase clarity. Section 2.6 provides a checklist for the items in this chapter.

2.1 Logical Organization

A research article needs to have a logical organization at many levels. Most importantly, it needs to represent what the experiment or theory delivers; it needs also to give an account of that experiment or theory in an orderly fashion; it needs to recognize the subtle influences of one part of the account on another (e.g. in setting the reader's expectations about the degree of difficulty); and it needs to be consistent in how ideas and material things are identified and explained.

How should all this be achieved? Approaches vary from person to person. Some authors are systematic, first creating a structure for a paper as a sequence of headings at various levels, somewhat like a flowchart, then adding notes under each heading, accommodating any interdependencies between sections, and eventually turning the notes into sentences. Others are more spontaneous, writing what first comes to mind in no special order, concentrating on generating potentially usable material unhindered by analysis or evaluation; and only after this creative activity is complete do they edit the material into a coherent, consistent whole.

Neither method is necessarily better than the other [258, p. 41], as long as the eventual logical structure makes sense to the reader. Even so, because creating text and analysing it depend on different kinds of thinking, it is usually more efficient to treat them as separate tasks.

Specific strategies for writing are discussed in Section 2.4.

2.1.1 Representing Content

There are several ways in which a paper can fail to represent an experiment or theory correctly. Possibly the most frequent derives from a misunderstanding of logical direction. Here is an observation by Reeves,[3] an experienced reviewer (and author and editor):

> It seems to me that about half of the time I write a negative review, or get one from a reviewer, it is because of logical errors. I say 'logical' because these are the most common form of reasoning errors; straight mathematical errors are less frequent, perhaps because people who know they are weak in math get help, but few are willing to admit they are weak in logic, although, in truth, we are all weak in logic. So, what do I mean by 'logical' [error]? Often it is simple, as in A implies B, B is true, therefore A is true; or, trying yet again to prove the null hypothesis; or, the commonest, overgeneralization.

The first of the three errors in the last sentence represents a misunderstanding of the rules of inference; that is, if A implies B and if A is true, then B is true, not vice versa. For example, a chemical compound may produce a peak in the spectrum from an analyser, but a peak in the spectrum may not be due to a chemical compound. It may be due to noise.

The second error refers to an attempt to prove the null hypothesis, which typically is that a treatment has no effect (§7.5.1). The normal course experimentally is not to prove but to disprove the null hypothesis. A failure to disprove it does not mean the hypothesis is true. For example, the absence of a spectral peak does not mean the absence of a particular chemical compound. The analyser may not be sensitive enough.

The third and last error refers to overgeneralization. It is what it says: inferring more than can be supported by the data or analysis. For example, just because a particular sample of a chemical compound produces a spectral peak does not mean that other samples will.

There are many other kinds of logical errors reported by reviewers. Examples are gathered together in Section 9.3.5.

2.1.2 Sections and Headings

Depending on the length of the paper, its logical structure can be made evident by dividing it into separate sections and subsections with informative headings, in accord with any constraints imposed by the journal. Headings act as signposts to readers, helping them see the organization separate from the detail [284]. Well-chosen headings can also help with database indexing [131].

With purely experimental work, there is limited flexibility. The main headings are expected to follow a standard pattern of introduction, methods, results, and discussion, or something similar (some general interdisciplinary journals such as *Nature* have a nonstandard structure, with the methods section placed at the end of the text [261]). Within each section, however, more informative subheadings can be created to sharpen the structure, as in Example 2.1, where experimental measures, effects, and inferences are indicated.

Heading — RESULTS

Subheadings:

- Movement timing and EMG
- Motor-evoked potentials elicited by single-pulse magentic stimulation
- . . .
- Control experiment

Heading — DISCUSSION

Subheadings:

- Early triggering and EMG pattern of activation
- The early release of the prepared response and the build-up of response-related activation
- Cortical or subcortical effects of LAS in late preparation?
- . . .
- Conclusions

Example 2.1. *Headings and subheadings, from a neurosciences and neurology article [600].*

With theoretical work, there is much more flexibility. Headings and subheadings can be created that capture the content or intent of each section precisely, as in Example 2.2, where particular procedures and variables are identified.

Headings	I. INTRODUCTION II. BAYESIAN AUTOREGRESSION III. VARIATIONAL BAYES IV. VARIATIONAL BAYESIAN AUTOREGRESSION
Subheadings	A. AR Coefficients, θ B. Excitation Noise Precision, λ ... F. Summary
Heading	V. ILLUSTRATION: SYNTHETIC DATA
Subheadings	A. Synthesizing Data B. Parameter Estimation C. Model Selection D. Robustness to Outliers
Headings	VI. RESULTS: REAL DATA VII. CONCLUSION

Example 2.2. *Headings and subheadings, from an electrical and electronic engineering article [447].*

2.1.3 Scope of Text and Headings

Despite their importance for structure, headings may be passed over unnoticed by readers as they work through the text in the body of an article, much as other interruptions such as illustrations and callouts would be ignored.[4] So as to avoid confusion, the text in the body should make sense when read without the headings [263, §1.3.5], unlike the text in Example 2.3.

✗

All persons of no fixed abode who happened to be admitted from the area were excluded.

RESULTS

Misleading referent — These are presented in detail in tables I and II. It should be noted that . . .

Example 2.3. *Treating headings as text, from a general and internal medicine article [393].*

If the reader misses the heading 'RESULTS', the effect is as illustrated in Example 2.4. The underlined pronoun 'These' refers to all persons of no fixed abode. The ambiguity is easily removed by replacing 'These' by 'The results'. Ambiguous referents are discussed more fully in Section 5.4.

✗

All persons of no fixed abode who happened to be admitted from the area were excluded.

= 'All persons of no fixed abode' — These are presented in detail in tables I and II. It should be noted that . . .

Example 2.4. *Misreading of Example 2.3.*

Much as the text in the body should make sense when read without the headings, the headings should make sense when read without the text in the body. Their hierarchy should reflect the content of each section and subsection [49, §2.56], as in Examples 2.1 and 2.2. The reader can then narrow down the topics of interest and navigate through the article [258, p. 40].

Headings of different levels are usually distinguished by their typeface or font size. They may also be numbered (or lettered, or some combination of the two) to help cross-referencing, but not all journals allow this choice. Two levels normally suffice for journal articles [68, §27.7.2]; that is, headings and subheadings as in Examples 2.1 and 2.2. Three or more levels are needed only in complicated works [49, §1.54], such as technical and legal documents with a strong hierarchical structure.

Even where heading levels are obvious, the reader may sometimes get confused about their scope. Consider Example 2.5, which is from a textbook where the author uses two levels, here 2.2 and 2.2.1.

2.2 A SURVEY OF INFORMATION THEORIES

✗

Work on theories of physical information has, over the past century, arisen from three distinct, though interconnected fields. . . .

Scope of section — This section examines these three approaches and the relationship to each other in some detail.

2.2.1 Thermodynamic Information Theory

All modern physical theories of information make reference to a quantity known as *entropy*. . . .

Scope of section — In this section we examine thermodynamic entropy, Statistical Mechanical entropy.

Example 2.5. *Loss of scope, from a monograph on information [621].*

At the end of the section headed '2.2 A SURVEY OF INFORMATION THEORIES', the reader is told that the section will examine three approaches and their relationship. At the end of the later section headed '2.2.1 Thermodynamic Information Theory', the reader is then told that the section will examine thermodynamic entropy and related topics. At that point, whether thermodynamic entropy is a new topic or a subtopic of the earlier topic is unclear. The reader has to turn back and examine the hierarchy of section headings.

2.1.4 Levels of Difficulty

Subsections and finer divisions into paragraphs within subsections should have about the same level of difficulty and amount of detail. Variations in pace can be disconcerting for readers, who may assume they have failed to grasp the content.

It is therefore helpful to decide in advance on the level of technical competence and experience expected of readers and not to depart too far from it or to make unintuitive reversals [79, p. 59]. A shift in technical level occurs in Example 2.6.

Defined in following text — MicroRNAs are small (20–25 nt), noncoding RNAs that regulate gene expression by targeting nascent transcripts . . . Mounting evidence suggests that microRNAs play important roles in the regulation of mitogen-induced gene expression (6–8).

✗

. . .

In the current study, we set out to use siRNA to examine the role of Ars2 in the mitogen-stimulated proliferation of quiescent cells. — Undefined

Example 2.6. *Different levels of explanation, from a cell biology article [632].*

The term microRNA is defined at the beginning of the first extract in the example and is followed by more explanation. Yet the less intuitive term siRNA introduced in the second extract is neither defined nor explained. If readers do not know what microRNAs are, and need the definition, then they are unlikely to know what siRNAs are.

An opposite shift in technical level occurs in Example 2.7.

Reeves et al. investigated the effect of adding biochar to soil on the spectra of mixed biochar and soil in the near-infrared (NIR, 10,000 to 1000 cm^{-1}) and mid-infrared (mid-IR, 4000 to 400 cm^{-1}).[21] — Expert

✗

. . .

Elementary — Infrared (IR) light is electromagnetic radiation with a wavelength longer than that of visible light. The range of such radiation is measured from the end of visible light at 0.7 micrometers to 300 micrometers.

Example 2.7. *Different levels of explanation, from a materials science article [489].*

The first extract in the example assumes that the reader is familiar with the concepts of spectra, infrared radiation, wavenumber (wavenumber is the reciprocal of wavelength), and wavenumber ranges. The second extract is more elementary. It introduces the idea of infrared radiation and its wavelength range, which then has to be translated into wavenumber to make sense of the preceding data. Reversing the order of the material would give a more natural development, but the different units used to express the wavelength ranges might still be confusing.

2.1.5 Late Definitions

If a technical term, concept, or figure needs a definition or explanation, then delaying it can be frustrating for the reader. In Example 2.8 the term 'Mahalanobis distance' is introduced. The term is assumed to be sufficiently unfamiliar that it needs to be defined, yet the definition is withheld until the effects of applying it are discussed. Unfortunately, readers may not persevere, stopping instead at the first appearance of the term in the belief that it was defined earlier and they missed it, or that it was not defined earlier and the work requires a more expert audience.

✗

First use — The nearest-neighbour measure using the Euclidean distance can be substituted for other distance measures such as the Mahalanobis distance. This measure was shown to increase recognition accuracy . . . when compared with the Euclidean distance. The Mahalanobis distance is defined by Eq. . . . — Application — Late definition

Example 2.8. *Late definition of a technical term, adapted from the draft of a doctoral thesis in electrical and electronic engineering.*[5]

If a definition or an explanation is necessary, then it needs to be placed where it helps, namely before the term is used. If it is not necessary, it can be omitted. In Example 2.9, the opening sentence refers to 'mitogenic stimuli'. The authors believe that the term needs to be explained and they spend the next several sentences doing that. But as with Example 2.8, the reader may not continue beyond its first appearance.

✗

First use — Cellular proliferation is a highly regulated process that is driven by mitogenic stimuli and fueled by available nutrients. Mitogenic stimuli (growth factors, cytokines, etc.) reprogram cellular metabolism . . . This metabolic reprogramming causes proliferating cells to rely heavily on uptake of extracellular nutrients. — Late definition

Example 2.9. *Late explanation of a technical term, from a cell biology article [632].*

The impediment caused by a late definition or explanation in a mathematical context is called a roadblock (§6.1.4). The reader can be eased past the roadblock in Example 2.9 with some material transferred from the second sentence, as illustrated in the edited version in Example 2.10.

> Cellular proliferation is a highly regulated process driven by mitogenic stimuli such as growth factors and cytokines — *Examples from second sentence* — and is fueled by available nutrients. Mitogenic stimuli reprogram cellular metabolism . . . This metabolic reprogramming causes proliferating cells to rely heavily on uptake of extracellular nutrients.

Example 2.10. *Edited version of Example 2.9 with roadblock removed.*

Alternatively, the explanation of the term mitogenic stimuli could be placed at the beginning of the statement and then followed by the material on cellular proliferation.

2.1.6 Needless Variation

Both technical and nontechnical terms should be used consistently. Consider the text in Example 2.11. It is part of a journal abstract from a highly cited article and is used in full in Section 3.3.2 to illustrate an effective abstract structure. Here, however, it is the use of the terms 'technique', 'approach', and 'method' that is of interest.

✗

> This paper introduces a new source separation technique — *First use* — exploiting the time coherence of the source signals. In contrast with other previously reported techniques, — *Consistent repetition* — the *First variation* — proposed approach relies only on stationary second-order statistics that are based on a joint diagonalization of a set of covariance matrices. Asymptotic performance analysis *Second variation* — of this method is carried out; some numerical simulations are provided to illustrate the effectiveness of the proposed *Second variation* — method.

Example 2.11. *Unnecessary word variation, from an electrical and electronic engineering article [416].*

All three terms have slightly different meanings. In order of specificity, an approach is, figuratively, a way of addressing a task (e.g. numerical simulation); a method is a way of doing something in particular (e.g. the method of images); and a technique is a particular, concrete way of doing something, a trick (e.g. the technique of Lagrange interpolation). The distinction between a method and a technique is not precise and depends on the application.

In Example 2.11 it is not apparent whether the variations represent the same or different things. The term 'technique' is introduced in the first sentence. It is then repeated consistently in the second sentence, but then 'approach' and then 'method' and 'method' again are used. The reader is left to decide whether the different word choice is a matter of style or content.

Varying words to achieve a pretty effect rather than clarity has been called 'elegant variation' [122]. Extreme examples of elegant variation can be found in [40] [122].

In general, repetition can be avoided by recasting or by using pronouns [122] [280, p. 8], although care should be taken to avoid ambiguity (§5.4.3). The preceding Example 2.11 contains some redundancy, the removal of which reduces the need to find alternatives to 'technique' (if the authors were indeed seeking alternatives). An edited version is shown in Example 2.12.

✕

> A new source separation technique is introduced that exploits the time coherence of source signals. It relies only on stationary second-order statistics based on a joint diagonalization of a set of covariance matrices. The asymptotic performance of the technique is analysed and its effectiveness illustrated with some numerical simulations.

— Pronoun ("It")

Possessive pronoun — ("its")

Example 2.12. *Edited version of Example 2.11.*

The introductory 'This paper' has been omitted to focus attention on the technique.[6] The pronoun 'It' refers to 'A new source separation technique' and the possessive 'its' refers to 'the technique'. Both 'approach' and 'method' have been eliminated. In the last sentence, the lengthy phrase 'analysis . . . is carried out' has been replaced by 'analysed', and 'provided to illustrate' has been replaced by 'illustrated'.

The edited abstract now has only 49 words, delivering the same message as the original with 67 words. Notice the correct use of the present tense 'is introduced' here and 'introduces' in the original. The present tense should be used only if you are describing an analysis that takes place within the paper and is not a report of procedures you undertook earlier (§3.6.1).

2.2 Guiding the Reader

One of the principal difficulties of writing about your work is what makes it possible, namely your familiarity with it. Whatever words you place on the page you perceive within a larger body of private knowledge, so that what is perfectly clear to you may be incomprehensible to someone else, a bias called the curse of knowledge [7, p. 78].

The aim, then, is to put yourself in the reader's place [391, p. 84]. Changing viewpoint may require a distance from your work that only time can achieve, but there are measures you can take to anticipate and address potential problems of comprehension.

2.2.1 Reminders and Anticipations

Allow for readers forgetting key points, and remind them where necessary. One reason for a reminder may be to enable another point to be developed, as in Example 2.13, where the second statement summarizes some observations made in the first.

Key point — For example, opiates and stimulants regulate structural plasticity in the opposite direction. Opiates decrease the number and complexity of dendritic spines . . . stimulants such as cocaine, amphetamine and methylphenidate consistently increase dendritic complexity and spine density. [More detail]

. . .

The functional relevance of structural plasticity in addiction models is complicated, as noted earlier, by the fact that morphine and cocaine have opposite effects on MSN spine density. — Reminder

Example 2.13. *Reminding the reader of an earlier point, from a neuroscience article [652].*

But signalling the fact that it is a reminder and not a lapse is important. Readers expect authors to know what they wrote. Seeing material repeated without an explanation can be unsettling, as in Example 2.14.

✕

But some basis functions obtained from PCA have negative values, which may lead to approximations with negative values. — First statement

. . .

But sometimes the value reproduced by PCA is negative. — Second statement
Non-negative matrix factorization is an option.

Example 2.14 *Inadvertent repetition of the same point, adapted from the draft of a doctoral thesis in electrical and electronic engineering.*[7]

On reaching the second statement, readers cannot immediately see whether it is redundant with the first or is a new point—or whether their own understanding is faulty.

Uncertainty can be avoided by acknowledging the earlier statement, such as (the easily overworked) 'As noted earlier', or making a more tailored change, as in the edited version in Example 2.15. The reminder now recognizes the existence of the earlier statement.

> But some basis functions obtained from PCA have negative values, which may lead to approximations with negative values. — First statement
>
> . . .
>
> Acknowledges earlier statement — The problem identified earlier of PCA approximations sometimes being negative can be avoided with non-negative matrix factorization. — Repetition part of acknowledgement

Example 2.15 *Edited version of Example 2.14.*

Phrases fulfilling a similar function include 'It has already been shown that' and, say, for a specific section, 'The analysis in Section 5.1.1 demonstrated that'. Other phrases of the form 'Remember that' and 'Recall that' are also used with material that is assumed to be known generally in the field but which needs to be made explicit. 'Of course' is often introduced for the same reason, although it is best used sparingly and not abused [40] [280].

In a complementary way, introducing an anticipatory phrase can prepare the reader for a later, more detailed account, as in Example 2.16, where 'As shown later' is used.

> The receiver's password tables and other authentication data are then more vulnerable to theft than those of the transmitter (an individual user). As shown later, some techniques for protecting against this threat also protect against . . . — Foreshadowing

Example 2.16 *Foreshadowing, from a computer science article [476].*

Anticipation of this kind is called foreshadowing. It is commonly used in the introduction of a paper to indicate the outcomes of the research (§3.5.9). As with reminders, it should avoid any sense of redundant repetition [284, p. 463].

2.2.2 Useful Rephrasing

Sometimes it is worth explaining difficult or important things twice, but in a different way, as in Example 2.17.

> Because these cells both possess chemotactic ability and furthermore are known to secrete some chemokines (*SI Appendix*), this effect is most likely caused by one or several secreted chemokine(s), as evidenced by . . . which sets the time scale for chamber filling to ~40 min. [Statement >70 words]
>
> In other words, the secreted chemokines will have saturated the chamber by 40 min, effectively reducing chemokine gradient depths and the signal-to-noise ratio of chemokine receptor activity. [Rephrasing 29 words]

Example 2.17. *Saying the same thing twice, from a biophysics article [704].*

The rephrasing beginning 'In other words' is succinct and easy to grasp. It brings out an implication of the work that might not have been evident on first reading of the longer and more complicated statement that preceded it. In Example 2.18, the rephrasing beginning 'That is to say' fulfils a similar function.

> The second Sakharov criterion is required because if C and CP are exact symmetries, then one can prove that the total rate of any process that produces an excess of baryons is equal to the rate of the complementary process that produces an excess of antibaryons, so that no net baryon number can be created. [Statement 55 words]
>
> That is to say that the thermal average of the baryon-number operator B, which is odd under both C and CP, is zero unless those discrete symmetries are violated. [Rephrasing 30 words]

Example 2.18. *Saying the same thing twice, from a nuclear physics article [649].*

In Example 2.19, the rephrasing beginning 'in other words' fulfils a different function, to remind the reader of the definition of full rank.

> For a fully controllable system, the matrix C should have a full rank; in other words, $\text{rank}(C) = N$. [— Rephrasing] This ensures that all dynamical states of the system are accessible.

Example 2.19. *Saying the same thing twice, adapted from a neuroscience article [571].*

Whether or not to provide explanations of this kind may be hard to decide. The reader who has already met the idea of matrix rank will very likely have met the idea of full rank. Yet some redundancy is reassuring and aids understanding.

As with other explanatory devices, introductory phrases such as 'in other words' should not be used repeatedly, especially within a few lines of each other, as occasionally happens [460]; otherwise the effect is laboured. Alternatives are, for example, 'equivalently', 'that is to say', and, if the intention is to be more exact, 'that is', which is followed by a comma and which does bear repetition.

2.3 Immediacy and Impact

There are many ways that careless writing can impede access to your work or lessen its influence. These include the use of convoluted expressions, stereotyped text, indirect or ill-defined linking phrases, repeated qualifiers, and unnecessary terminology, along with a failure to recognize alternative interpretations. The general advice is that where possible prefer the simpler expression to the more complex [389, p. 8], the concrete to the abstract [142, pp. 78–9], and verbs to nouns [274, p. 256]. Above all, do not hide the status of your statements: when they hold and when they do not. Honesty gives clarity [150, p. 33].

Specific suggestions to do with writing for impact are made in Chapter 3 and summarized in Section 9.3.1.

2.3.1 Anglo-Saxon vs Latin

In order to avoid pomposity, choose the simpler, usually Anglo-Saxon word or phrase over the more elaborate, usually Latinate one [280, pp. 165–6] [391, pp. 67–8]. For example, use 'before' instead of 'prior to', 'show' instead of 'illustrate' (unless you mean 'make clear' or 'explain'), and 'after' instead of 'subsequent to' (unless you mean 'as a consequence').[8]

Some Latinate phrases do, however, achieve a life of their own, as in Examples 2.20, 2.21, and 2.22. Each is redundant: 'in terms of' can be replaced here by 'in', 'In case of' by 'In' or 'With', and 'in respect to' by 'in'.

✕

Antipsychotic drugs are usually the first line of treatment for schizophrenia, and clinical guidelines report clear benefits in terms of symptom reduction. — Unnecessary

Example 2.20. *Redundant phrase, from a psychiatry article [619].*

Unnecessary — In case of MOS transistors, the basic signals are derived from the threshold voltage and the mobility.

✕

Example 2.21. *Redundant phrase, from an electrical and electronic engineering article [604].*

✕

Measuring mechanical stresses is an important task in respect to various areas of application.

Unnecessary — in respect to

Example 2.22. *Redundant phrase, from an instrumentation engineering article [685].*

More elaborate Latinate constructions appear routinely in comparisons, as in Example 2.23.

✗

People with dementia may be disproportionately affected by pain and undiagnosed illnesses compared with those without cognitive impairment.

Elaborate comparison

Example 2.23. *Elaborate Latinate phrasing, from a general and internal medicine article [556].*

The phrase 'disproportionately . . . compared with' can be replaced by a simpler 'more . . . than', as in the revised version in Example 2.24.

People with dementia may be more affected by pain and undiagnosed illnesses than those without cognitive impairment.

Simple comparison

Example 2.24. *Simpler phrasing of Example 2.23.*

An even more elaborate version of this Latinate construction is shown in Example 2.25.

✗

The surface roughness of the replicated mold is relatively poor compared with that of the master mold.

Elaborate comparison

Example 2.25. *Elaborate Latinate phrasing, from an applied physics article [541].*

The phrase 'relatively poor compared with' can be replaced by 'poorer than' or, better still, 'worse than', as in the revised version in Example 2.26.

✗

The surface roughness of the replicated mold is worse than that of the master mold.

Simple comparison

Example 2.26. *Simpler phrasing of Example 2.25.*

The original and revised versions suggest that the authors wanted to make a judgement about quality [541], and so 'poor' and 'worse' seem appropriate. Otherwise the more neutral 'greater than' could be used instead of 'worse than' in Example 2.26. An alternative simplification is 'The surface of the replicated mold is rougher than that of the master mold'.

2.3.2 Stereotyped Text

Stereotyped text, also called boilerplate text, is text that is unoriginal, mechanical, or commonplace; in other words, generic text that could be used in almost any paper [151]. The uniform advice is to avoid it, along with all other clichés [79, p. 194] [293, §3.10] [391, p. 33]. Otherwise, the reader may be left with an impression of laziness or thoughtlessness [277]. As Orwell put it [274, p. 255], in a critique of English prose:

> [It] consists less and less of *words* chosen for the sake of their meaning, and more and more of *phrases* tacked together like the sections of a prefabricated hen-house.

The piece of boilerplate text in Example 2.27, taken from an abstract, is almost content-free. Statements like 'The implications are discussed' are more likely to encourage the reader to turn to another article than find out what the implications are.

✗

Multiple regression analyses indicated, however, that mechanisms other than the type II fiber atrophy might be responsible for the decline in strength performance during aging. <u>The implications of these findings are discussed</u>. — Boilerplate text

Example 2.27. *Boilerplate text, from the abstract of a physiology article [576].*

Some boilerplate text is informative yet takes a convoluted form that would be unusual in everyday speech, as in Example 2.28.

✗

Six observers including the authors participated in this experiment. The observers other than the authors were <u>naive as to the purpose of this experiment</u>. — Boilerplate text

Example 2.28. *Boilerplate text, from a multidisciplinary sciences article [724].*

The phrase 'naïve as to the purpose of this experiment' could be replaced by the more immediate and shorter 'unaware of the purpose of this experiment'.

2.3.3 Indirect Links

The phrases 'based on' and 'on the basis of' are stock linking devices, inserted almost by reflex, leaving the relationship between two things vague and indirect, as in Example 2.29.

✕

We conducted a cross-sectional study, <u>based on</u> telephone survey data from 1,012 Appalachian residents, ages 50–75, with 1 or more chronic conditions. Measures of perceived MM [multiple morbidity] burdens were developed <u>based on</u> 85 pilot interviews previously undertaken with providers and patients with MM residing in Appalachian Kentucky.

— Indirect link

Indirect link —

Example 2.29. *Stock linking phrase 'based on', adapted from a health care sciences article [690].*

The first instance of 'based on' is unobjectionable, but both can be replaced by something more direct, as in the edited version in Example 2.30.

We conducted a cross-sectional study <u>that used</u> telephone survey data from 1,012 Appalachian residents, ages 50–75, with 1 or more chronic conditions. Measures of perceived MM burdens were developed <u>from</u> 85 pilot interviews previously undertaken with providers and patients with MM residing in Appalachian Kentucky.

— Direct link

— Direct link

Example 2.30. *Edited version of Example 2.29.*

A variant of 'on the basis of' appears in Example 2.31.

✕

Data are collected and analyzed <u>on a statistical basis</u>. — Indirect link

Example 2.31. *Stock linking phrase, from an electrical and electronic engineering article [477].*

Here the phrase 'on a statistical basis' expresses indirectly the fact that the analysis (and perhaps the data collection) used statistics. The same information can be expressed more directly and compactly by 'statistically', as in the edited version in Example 2.32.

Data are collected and analyzed statistically.

Example 2.32. *Edited version of Example 2.31.*

The statement is three words shorter and is more immediate. Since the collection and analysis were performed in the past, the verb 'are' might also be replaced by 'were'.

There are several other phrases such as 'in terms of' and 'in relation to' that indirectly link one thing to another. Example 2.33 contains both.

Indirect link

✕ Melanoma in relation to skin colour is analysed in terms of exposed (facultative) and non-exposed (constitutive) skin colour measurements. — Indirect link

Example 2.33. *Stock linking phrases, from a computational biology article [439].*

Although the phrases 'in relation to' and 'in terms of' might be replaced by something more precise, the whole sentence could be recast to improve clarity. Example 2.34 shows one possibility drawn from the article. Both indirect linking phrases have been eliminated and the extract is now six words shorter.

Melanoma risk is predicted by exposed (facultative) and non-exposed (constitutive) skin colour measurements.

Example 2.34. *Edited version of Example 2.33.*

Linking devices such as 'on the basis of', 'in relation to', and 'in terms of' may serve to focus on something, to particularize. Nevertheless, they should not be used by default as 'vague all-purpose connectives' [40, p. 406], which the reader has then to decode.

2.3.4 Qualifiers

Research findings are by their nature provisional, susceptible to better measurement, more stringent tests, and new interpretations. Introducing qualifiers into statements to allow for the possibility of error or exception is a normal part of scientific writing [258, p. 39]. In Example 2.35 'may' allows the possibility that the reported blue shift in the spectrum is not due to the modulation of the band gap but to something else.

Hedge — The small blue-shift in the UV emission peak observed in the present investigation may be due to modulation of the band gap that is caused by incorporation of Sn ions in the ZnO film matrix [26, 28].

Example 2.35. *Qualifying a cause, from a materials science article [644].*

The softening of any statement with a word or a phrase in this manner is called hedging [284]. It is used so that the reader sees the provisional status of the claims and so that the author avoids having to retract them later [176].

Qualifiers may be introduced in many ways [71], with subtly different effects, as in the following, where the hedge word or phrase is underlined:

- The blue shift might be due to the band gap.
- Arguably the blue shift is due to the band gap.
- It is possible that the blue shift is due to the band gap.
- The blue shift may be due to the band gap.
- The blue shift could be due to the band gap.
- The blue shift seems to be due to the band gap.
- The blue shift appears to be due to the band gap.
- It is likely that the blue shift is due to the band gap.
- It is suggested that the blue shift is due to the band gap.

The strength of conviction generally (another hedge word) increases with descending position in the list. Thus 'The blue shift might be due to the band gap' readily allows a subsequent contradiction, e.g. 'but it is not', whereas 'It is likely that the blue shift is due to the band gap' does not invite the same contradiction.

Unfortunately, it is easy to be too tentative and to add more and more hedge words to secure more and more protection. It is then hard for the reader to see what is actually being claimed, as in Example 2.36, which has three hedge words.

First hedge Second hedge

✗ This opposite mode of regulation might possibly indicate that the ratio of ERα to ERβ might be of importance in determining the susceptibility . . . — Third hedge

Example 2.36. *Over-hedging, from an endocrinology and metabolism article [529].*

One hedge per assertion is usually enough [258, pp. 39–40]; more than one signals a lack of conviction [391, p. 22]. Without compromising the qualification, two of the three hedges in Example 2.36—'might', 'possibly', and 'might'—can be eliminated, as in the edited version in Example 2.37.

Single hedge

This opposite mode of regulation might indicate that the ratio of ERα to ERβ is of importance in determining the susceptibility . . .

Example 2.37. *Edited version of Example 2.36 with just one hedge.*

A still more extreme instance of hedging is shown in Example 2.38.

First hedge — Second hedge — Third hedge

✗ Although, on balance, testosterone might possibly be beneficial to the cardiovascular system, testosterone might also have some detrimental effects.

Fourth hedge — Fifth hedge

Example 2.38. *Extreme hedging, from an oncology article [490].*

Again without compromising the qualification, several of the five hedges—'on balance', 'might', 'possibly', 'might', and 'some'—could be eliminated, as in the edited version in Example 2.39. Whether including 'some' in 'some detrimental effects' would help understanding is unclear, for the effects are described in the following text (not shown here). The statement remains fairly non-committal.

First hedge

Although testosterone might be beneficial to the cardiovascular system it might also be detrimental.

Second hedge

Example 2.39. *Edited version of Example 2.38 with less hedging.*

Despite 'may' and 'might' (and also 'could') being used similarly as qualifiers [92, p. 384], it is 'might' that is usually seen as more tentative than 'may' [32, p. 139] [40]. In most articles 'might' is used sparingly.[9] Exceptionally in the source article for Example 2.38, 'might' occurred 54 times.

In some circumstances, however, it is tempting to under-qualify, especially in a summary statement. Example 2.40 shows where the authors' claims contain no hedge words.

No hedge

✗ We report detection of individual, isolated proton spins by a nitrogen-vacancy (NV) center in a diamond chip covered by an inorganic salt.

. . .

No hedge — The detection of single ^{1}H nuclei presents an important step toward diamond-based nano-MRI.

Example 2.40. *Under-hedging, from a multidisciplinary sciences article [592].*

The authors discovered subsequently that some of their reported detections could have other interpretations, and they retracted their report [593].

2.3.5 Terminology

As Halmos said, in a different context, the best notation is no notation [150, p. 40]. Although questions of notation are usually associated with mathematical and statistical material, the introduction of unnecessary labels and symbols can appear pretentious in any context and may impede access to your work. Example 2.41 contains the rather grand term 'photic stimulator'.

✗

Unnecessary terminology —

Scotopic electroretinograms (ERGs) were recorded from both eyes simultaneously and 10 responses to flashes of unattenuated white light (5 ms, 0.2 Hz) from a photic stimulator (light-emitting diodes) set at maximum brightness were amplified (9 cd s/m^2 without filter).

Example 2.41. *Unnecessary terminology, from a neuroscience article [481].*

A simpler way of saying the same thing is 'light stimulator' or 'light source' or, more to the point, 'light-emitting diodes', which is how the authors explain the term in the following parentheses. Unexpectedly the term 'photic stimulator' has gathered traction, with at least one manufacturer using it in a product name.

Example 2.42 introduces the term 'frictionless impact testing apparatus'. It occurs once in the title, once in the abstract, and as 'frictionless impact test apparatus' once in a figure caption. But it is never used in the text itself to refer to the apparatus. Its role is unclear.

✗

Unnecessary terminology —

This new apparatus can levitate a steel bar by using compressed air so that there is little friction between a steel bar and a guide rail against the horizontal movement. We name this apparatus as "frictionless impact testing apparatus".

Example 2.42. *Unnecessary terminology, from a civil engineering article [563].*

Naturally some terminology may be required for the sake of precision, though the choice of terms should aid the reader's understanding, not act as an impediment. Example 2.43 shows terminology that forces the reader to make a tedious letter-by-letter comparison to distinguish one acronym[10] from another.

✗

We call this class of ST-GTD MMSE transceiver the ST-GMD MMSE transceiver.

Similar acronyms

Example 2.43. *Confusing terminology, from an electrical and electronic engineering article [588].*

In fact, the acronym ST-GMD MMSE is used five times before this definition appears (§2.1.5). The readiness with which acronyms are used does depend on the discipline. In the article from which the text in Example 2.43 is taken, 30 lines of text contained about 50 acronyms.

Some terminology, especially when it is carefully planned, can help to reinforce the underpinning concepts or act as a prompt to memory.

Intuitive label

We call this set of equations the "sticky" equations because they are the Fokker-Planck equations for sticky Brownian motion (45), a stochastic process that has a probability atom on the boundary of its domain.

Example 2.44. *Intuitive labelling, from a physics article [543].*

The label 'sticky' in Example 2.44 is intuitive and accurate, and adds to the value of the exposition. Likewise with the imaginative label 'Goldilocks' in Example 2.45.

Imaginative label

We refer to this as a Goldilocks trial design, as it is constantly asking the question, "Is the sample size too big, too small, or just right?"

Example 2.45. *Imaginative labelling, from a pharmacology and pharmacy article [37].*

These thoughtful labels aside, if a description is used just once or twice, then little is gained by introducing a special term for it.

2.3.6 Limitations

Your paper may have a lifetime of five years or 50 years. If you want impact though, the advice is to write for posterity [150, p. 22]. One way to do this is to identify those aspects of the work that might cause you to revise your interpretation in the future, and explain them to the reader, typically in the closing sections of the paper and in the abstract. Rather than weakening your case, honesty about its limitations provides a service to the reader and demonstrates your objectivity.

Example 2.46 contains two caveats about experimental conditions, one to do with testing, the other with networking.

> Second, our surrogate gesture systems, particularly the pen-based system in Study 2, were tested under better than normal network conditions. — Caveat The wireless network we set up between the IP camera, Worker computer, and Helper tablet PC minimized the effects of jitter or delay — Caveat on communication and performance.

Example 2.46. *Identifying limitations, from a computer science article [511].*

Example 2.47 contains several caveats about a statistical analysis, one to do with rigour, another with error assumptions, and another with bandwidth selection.

> The simulation results show that the method proposed appears to be at least approximately correct Caveat — but a rigorous theoretical exposition would be required before these bands might be used with complete confidence.
>
> Caveat — One shortcoming of the method is that it relies on the assumption of constant variance in the errors. The bandwidth selection method might not be very sensitive. — Caveat

Example 2.47. *Identifying limitations, from a statistical computation article [497].*

Example 2.48 contains a caveat about experimental design (more caveats are listed in Example 3.95).

> One of the main limitations of the HYVET [Hypertension in the Very Elderly Trial] extension was Caveat — that only participants who were on double blind treatment were enrolled from the main trial.

Example 2.48. *Identifying limitations, from a general and internal medicine article [415].*

But too many caveats can be counterproductive. As with hedging (§2.3.4), they can make it hard for the reader to see what is actually being claimed and therefore to understand the significance of the work.

Phrases that anchor your work within a specific time, e.g. 'recently', may have a quite separate effect in limiting the perceived life of a paper.

2.4 Strategies for Writing

Writing is not easy, even for experienced authors [84, p. 189] [391, p. 4]. For less-experienced authors, there are some basic techniques that can help [258, p. 41]. Some of the suggestions made here reinforce those made elsewhere in this guide.

2.4.1 Getting Started

Techniques for getting started on your paper include the following [169, p. 153] [258, pp. 42–4]:

- Concentrate on one small writing task at a time, one paragraph or a section on a particular topic, for example, the experimental apparatus or the aims of the research.
- Consult leading research articles for examples of development and organization. Make notes only after you have finished reading the article and use your own words.
- Explain your research to a colleague. Make notes summarizing your discussion, especially which points need more explanation.

You do not have to wait until your research is complete in order to start writing. The best time is while the research is still in progress [229, p. 7]. Not only will your research be better documented, you will already have something to build on. There are several things to consider:

- For an experimental study, record details of the apparatus, procedure, any special conditions applying at the time, and details of data processing and analysis. All should be in a form that can be understood by someone unfamiliar with the work (necessary for data archiving).
- As you accumulate your results, prepare preliminary plots and tables. Decide what kind of story they tell and how they should be ordered.
- If the results depend on certain assumptions or non-routine methods of analysis, include them in the list of things to be mentioned in the introduction and described in detail in the methods section.
- Think about what constitute the core research findings. Decide whether any control experiments are needed to support (or reject) these findings.

You may later change your mind about what is important and how it should be presented, but it is less demanding, and more pleasurable, to rework material than to start afresh (§2.5).

2.4.2 Keeping Going

It is one thing to start writing, another to continue. Although there are several positive steps you can take to help the process—for example, choosing the right time and place of work—it is the negatives, the impediments to writing, that can be the greater challenge. Several are listed in [157, ch. 4.6]:

- Do not aim for perfection as you write each sentence.
- Do not stop at the end of a section. Instead write a line or two of the next section so that you can more easily resume later.
- Do not edit as you write. Leave corrections and revisions for a separate session.

In all this, keep in mind that writing is a process of experimentation and discovery [258, p. 42]. It may lead to new ideas, new ways of analysing the data, new control measurements, new theories, and so on. The implications of your research may evolve as your paper develops [169, pp. 145–6]. Re-evaluation is part of the process of reworking.

2.5 Reworking

Some rare authors can write (and speak) almost perfect prose at first pass. But they are truly rare, not least because even the most gifted authors usually have second thoughts about what they want to say. In reality, the first draft is—for almost all of us—a highly imperfect creation, whose defects are characterized by Zinsser [391, pp. 83] thus:

> The newly hatched sentence almost always has something wrong with it. It's not clear. It's not logical. It's verbose. It's klunky. It's pretentious. It's boring. It's full of clutter. It's full of clichés. It lacks rhythm.

It is in the reworking of subsequent drafts that the material achieves its clarity, authority, and accessibility. Leaving the draft on one side for a while aids that process [340, p. 326]. You can then view the work afresh. Unencumbered by immediate memory, you can more readily recognize any lapses and decide how to address them. Using a print copy rather than an on-screen version can also provide variety [340, p. 327].

An essential part of reworking is having others review your drafts. Encourage colleagues to offer frank criticism, the more direct the better. Unremitting praise may be good for morale, bad for communication.

2.5.1 Clarity and Simplicity

To achieve clarity and directness, Zinsser's recommendation is to simplify, relentlessly [391, pp. 6–7]. What does this mean in practice? Consider the sentence in Example 2.49, which opens an abstract.

Very long subject

✗ In this paper, the experimental results of a complete battery-less wirelessly-powered 2.45-GHz temperature sensor system with a microcontroller (μC) and an ISM-band transmitter on FR-4 is presented.

Example 2.49. *Opening statement that could be simplified, from an instrumentation engineering article [595].*

The sentence, with 28 words before the verb, is difficult to read. The verb 'is' fails to agree with the subject 'results' (§5.2.1). The abbreviation 'μC' for microcontroller saves no words and may be misread as a unit of charge. The abbreviations 'ISM' for 'industrial, scientific and medical' and 'FR-4' for a kind of substrate may not be familiar to nonspecialists, and although 'ISM' would be helped by 'band' in 'ISM-band', there is no corresponding referent for 'FR-4'. More generally, how can a sensor system have 'experimental results'?

A possible revision is shown in Example 2.50.

This paper describes a complete batteryless, wirelessly powered, 2.45-GHz temperature sensor system with microcontroller and ISM-band transmitter built on FR-4 laminate.

Example 2.50. *Revised version of Example 2.49.*

The top-heavy sentence has been recast; the abbreviation 'μC' has been omitted; and 'FR-4' now has a referent. The number of words has been reduced from 27 to 21.

Consider the two sentences in Example 2.51, which end the abstract begun in Example 2.49.

✗ The system is able to transmit 32-bit sensor data back to the base station with an extra 16 bits for cyclic redundancy check check at regular intervals. The battery-less μC can be wirelessly powered when located up to a distance of 2 m from a power transmitter supplying less than 50 mW of microwave power using a patch array.

Interrupts details

Duplication

Verbose

Unnecessary abbreviation

Superfluous

Example 2.51. *Closing statements that could be simplified, from an instrumentation engineering article [595].*

The phrase 'back to the base station' interrupts the technical detail; 'check' is duplicated; 'at regular intervals' is verbose; 'μC' remains an unhelpful abbreviation; and 'when located' is unnecessary.

A possible revision is shown in Example 2.52.

> The system can regularly transmit back to the base station 32-bit sensor data with an extra 16 bits for cyclic redundancy checking. It can be wirelessly powered up to 2 m from a patch-array microwave transmitter supplying less than 50 mW.

Example 2.52. *Revised version of Example 2.51.*

The phrase 'back to the base station' has been relocated; 'check' has been replaced by 'checking'; the phrase 'at regular intervals' has been replaced by 'regularly' and relocated; and 'when located' has been deleted. The three instances of 'power' have been rationalized to one. The number of words has been reduced from 60 to 43.

2.5.2 Writing in Spirals

Rewriting can be seen as more a continuing process than one that delivers a finished product [391, p. 84]. Indeed, tinkering with the text, making local improvements, may become compulsive. Yet local changes can have global consequences [340, p. 326], which need to be pursued and resolved if necessary.

Halmos explained that what is required is to write in spirals [150, p. 27]. Thus having made changes to a section, say Section 3, by introducing a new analysis, you see that Section 2 needs to be amended to prepare the reader for that new analysis, and that in turn needs Section 1 to broaden the overview of the methods used in the paper. All this entails starting again from Section 1, and working forwards through Section 2 to the local change in Section 3. And so on with each successive local change.

2.5.3 An Example of Progressive Rewriting

What does rewriting look like? Example 2.50 shows how Example 2.49 can be clarified and simplified, and Example 2.52 shows how Example 2.51 can be similarly improved. But there is more to rewriting a draft than this.

For good or ill, preliminary versions of published manuscripts are rarely available publicly, and therefore the text extracts in Examples 2.53–2.57 are taken from drafts of one of my own manuscripts. The material is introduced here only to illustrate progressive rewriting and has no further significance. Text that is deleted or replaced is shown with a line through it (not to be confused with underlining, which identifies annotated text).

Wrong term — Vague referent — Missing cites — Inaccurate description

Instead of making assumptions about the properties of the set of surface reflectances in a scene, other approaches, developed within a demarcated linear framework, have concentrated on the nature of those reflectances and illuminant spectra. The key idea is that if surface and illuminant spectra can be expressed as a weighted sum of a few spectral basis functions (Brill, 1978, Brill, 1979), and these basis functions are known to the observer, then spectral reflectances can be recovered exactly, without, for example, a gray-world assumption (cf. Buchsbaum, 1980).

Example 2.53. *Version 1 of text in author manuscript.*

More accurate — Clarifies referent — Corrects description — Corrects description — Balances cites in previous clause

Instead of making assumptions about the properties of ~~the set~~ ensembles of surface reflectances in a scene, other approaches to constancy, developed within a demarcated linear framework, have concentrated on ~~the nature of those~~ how those individual reflectances and illuminant spectra may be represented. The key idea is that if surface reflectance and illuminant spectra can be expressed as a weighted sum of a few spectral basis functions (Brill, 1978, Brill, 1979), and these basis functions are known to the observer (Dannemiller, 1991, Maloney & Wandell, 1986), then spectral reflectances can be recovered exactly, without, for example, a gray-world assumption (cf. Buchsbaum, 1980).

Example 2.54. *Version 7 of text in author manuscript.*

= 'In preference to', not 'In place of'

If reader needs clarification — More accurate — Balance — To reflect historical status — Redundant, but helpful; later removed — Further refinement — Omit as already mentioned

~~Instead of~~ Rather than making assumptions about the properties of ensembles of surface reflectances in a scene, as described in Section 2.2, other theoretical approaches to color constancy, developed within a ~~demarcated~~ well-defined linear framework, have instead concentrated on how ~~those~~ individual surface spectral reflectances and illuminant spectra may be ~~represented~~ described analytically. The key idea ~~is~~ was that if ~~surface~~ reflectance and illuminant spectra can be expressed as a weighted sum of a few spectral basis functions (Brill, 1978, Brill, 1979), and these basis functions are known to the observer (Maloney & Wandell, 1986, Dannemiller, 1991), then spectral reflectances can be recovered exactly, without, for example, a gray-world assumption (cf. Buchsbaum, 1980).

Example 2.55. *Version 9 of text in author manuscript.*

Consistency — Not all 'other' — Sharpen meaning, split long sentence — Expand earlier cryptic description

Rather than making assumptions about the properties of ensembles of surface spectral reflectances in a scene, as described in Section 2.2, some theoretical approaches to color constancy ~~developed within a well defined linear framework, have instead concentrated~~ took a more principled approach. Developed within a well-defined linear framework, the emphasis was on how individual surface spectral reflectances and illuminant spectra may be described analytically. The key idea is that if reflectance and illuminant spectra can be expressed as a weighted sum of a few spectral basis functions (Brill, 1978, 1979), and these basis functions are known to the observer (Maloney & Wandell, 1986, Dannemiller, 1991), then spectral reflectances can be recovered exactly, without, for example, assumptions about the mean such as the gray-world assumption (cf. Buchsbaum, 1980).

Example 2.56. *Version 13 of text in author manuscript.*

Clarifies 'assumptions'

> Rather than making ad hoc assumptions about the properties of ensembles of surface spectral reflectances in a scene, as described in Section 2.2, some theoretical approaches to color constancy have taken a more principled line ~~approach~~. Developed within a well-defined linear framework, the emphasis was on how individual surface spectral reflectances and illuminant spectra could be described analytically. The key idea was that if reflectance and illuminant spectra can be expressed as a weighted sum of a few spectral basis functions (Brill, 1978, 1979), and these basis functions are known to the observer (Dannemiller, 1991; Maloney & Wandell, 1986), then spectral reflectances can be recovered exactly, without, for example, assumptions about the mean such as the gray-world assumption (Section 2.2, cf. Buchsbaum, 1980).

= 'plan of action' — (line)

Repetitive, but unambiguous — (assumption)

Example 2.57. *Version 18 of text in author manuscript, subsequently published in [506, p. 677].*

To echo Zinsser [391, p. 86], these revisions are neither the best nor the only ones that could be made. Each sentence could be improved; the last in particular would benefit from being divided and the ideas expanded.

Were all 18 versions of the manuscript necessary? Some writers would certainly make fewer changes and some might make more, depending on how quickly the work converged to an acceptable form. In this respect, writers are no different from any other practitioner of a craft. The inventor and entrepreneur, James Dyson, allegedly made 5,126 prototypes before succeeding with what proved to be a revolutionary design of domestic vacuum cleaner.

2.6 Checklist for General Principles

Logical organization

1. No inferential errors or overgeneralizations?
2. Correctly ordered section headings and subheadings?
3. Comprehensible text without headings and subheadings?
4. Constant degree of difficulty throughout?
5. No late definitions or explanations?
6. Consistent use of technical terms?

Guiding the reader

7. Occasional reminders about key points?
8. No repetition without explanation?
9. Difficult or important ideas explained twice?

Immediacy and impact

10. Simple rather than elaborate phrasing?
11. No stereotyped text or clichés?
12. No vague links ('on the basis of')?
13. Not more than one hedge per assertion?
14. No unnecessary terminology?
15. Limitations on findings?

Reworking

16. Text decluttered? No superfluous words?
17. Local edits consistent with the global exposition?
18. Progressive refinement?
19. Critical reviews from colleagues?

3

Research Articles

Whether it is experimental or theoretical, a research article is neither a record of events nor an account of the processes of thought that led to those events [250, p. 228]. It rarely describes what was done: the trials, reversals, misinterpretations, reinterpretations, and insights that preceded the polished published form [258, p. 9]. Rather, it is simply an account that aims to inform the reader about the research as clearly and honestly as possible [80, ch. 1].

To that end, the modern experimental research article assumes a standardized structure, which over the years has proved remarkably resistant to change [345]. This structure allows the reader to quickly locate the germane parts of the article such as the purpose of the research or the findings; it aids comparison with other articles (methods with methods, data with data); and it helps focus ideas and increase understanding. Theoretical and review articles have fewer constraints, but they still have an underlying logical structure.

This chapter considers the organization of research articles in detail. Section 3.1 introduces the main elements. Section 3.2 deals with the title, Section 3.3 with the abstract, and Section 3.4 with keywords and other summary texts. Section 3.5 deals with the introduction, Section 3.6 with methods, Section 3.7 results, Section 3.8 discussion and conclusion, Section 3.9 acknowledgements, and Section 3.10 citations and references. Section 3.11 and Section 3.12 deal with associated documents and resources, including appendices as well as supplementary materials and data. Section 3.13 provides a checklist for the items in this chapter.

Related issues are dealt with in other chapters: mathematical writing in Chapter 6; summarizing numerical data and reporting statistical inferences in Chapter 7; graphing data in Chapter 8; and authorship and order of authors in Chapter 12.

3.1 Elements of Research Articles

An experimental research article commonly contains the following elements or their equivalents:

- title
- abstract

- introduction
- methods
- results
- discussion and conclusion
- references

The conclusion may be separate from the discussion. If several experiments are reported, each might have its own results and comment or discussion, with a general discussion section drawing them together at the end [293, §2.09]. Not all journals label the introduction as such.

A theoretical research article or a methods article may contain only a few of these elements, such as an introduction and a discussion or conclusion, with the rest determined by the content [89]. Some mathematics articles may settle for only an introduction [611] or something even less [609].

Other standard, not always mandatory elements are these:

- keywords
- highlights, lay summary, or significance statement
- acknowledgements
- appendices
- supplementary materials

Figures, tables, and captions (also known as legends) may be embedded in the document for ease of review, but some or all may need to be submitted separately with the eventual accepted version of the paper.

3.2 Title

In attracting the attention of the reader, the title is the most important part of your paper. It appears in the journal's list of contents, in electronic alerts, in the reference lists of other papers, and in bibliographic and other online databases. Crucially, the published article may be selected solely by the contents of the title [68, §27.7.1]. All the words in it should therefore be chosen with care and their relationship with each other carefully designed [78].

3.2.1 Fundamentals

An effective title generally satisfies the following criteria [78] [258], although not necessarily all:

- it accurately captures the content of the work;
- it distinguishes the work from other related studies;

- it is a phrase, i.e. does not constitute a sentence;
- it contains no symbols or abbreviations;
- it has as few words as possible.

Some journals are indifferent to whether titles form a sentence, but the wider preference is that they should not [78] [258]. The contrast is between what are called assertive sentence titles, such as 'DNA is the genetic material', and non-assertive versions, such as the phrase 'DNA as the genetic material' and the question 'Is DNA the genetic material?' [315]. One reason for avoiding an assertive sentence title is that it can overstate the more cautious conclusions in the body of your paper [68]. Another is that with the passage of time you may adopt a more nuanced view, which is better reflected in a non-assertive title. Yet another is that it may suggest an attempt to close off the field, rarely to anyone's benefit.

Symbols are normally excluded from titles except in disciplines such as mathematics and particle physics where they are necessary. The difficulty with symbols is that they are unlikely to be used as search terms by the reader and indeed, as with some abbreviations, they may not be searchable in many online databases [79, pp. 43, 211]. The number of words such as 'a', 'the', and 'very' (called database stop words) should also be minimized as they too are not searchable [68, §27.7.1].

How should you go about constructing a title? You may be lucky and find a compelling phrase that attracts the reader's attention and accurately conveys the essence of the work: it 'tells the story' [161, p. 61]. More usually, however, you arrive at a title by a process of trial and error, a process that can still be systematic. First, construct a list of words or phrases that represent the work. Then experiment with their order so that the most important is placed first and succeeded by the others in descending order of importance [68]. Then try linking the components with prepositions (e.g. 'on', 'by', 'in') or with participles e.g. ('causing', 'caused'), taking care to avoid awkward or inappropriate associations, especially with 'using'.[11] Next compare the title with the titles in the intended journal, conference, or other publication to ensure it is both compatible in style and distinguishable in content. Continue to experiment to see whether the number of words can be reduced and their meaning sharpened, without misrepresenting the work or drifting into hyperbole [68].

At some point, ensure that your use of language in the title is consistent with that in the abstract, introduction, discussion, and keywords.

3.2.2 Effective Titles

Although a long title can more comprehensively represent your work and improve indexing in online databases, it has to compete with the other titles in a journal's contents page and with all the titles returned in a database search list. If your title is too long, the reader may move to the next in the list. If it is too short, the reader may discount it. So, what is the right number of words?

The traditional advice is that the title should have around 10–12 words, and not more [24] [238]. In practice, the average falls closer to the upper limit of this range, and, as might be expected, long titles of more than 20 words and short ones of less than five words are uncommon.[12]

The recognition an article receives offers an uncertain guide to effective title length. There are suggestions that shorter titles receive more citations [225] and also that longer ones do [192]. Unsurprisingly, these kinds of analyses are complicated by several potential confounds [82].

Examples 3.1 and 3.2 show some effective titles of about average length. In each, the main topic of study is placed first [68], followed by a modifying prepositional phrase beginning 'by' or 'in' that indicates the method (or agent) or application. What constitutes the main topic depends on the context, though it is clear enough here.

Main topic

Regulation of pluripotency and cellular reprogramming by the ubiquitin-proteasome system

Method

Example 3.1. *Effective title, from a cell biology article [433].*

Main topic

Multidimensional fixed point theorems in partially ordered complete metric spaces

Application

Example 3.2. *Effective title, from a mathematics article [651].*

The modifying phrase needs to be carefully chosen. Example 3.3 shows a title in which the main topic is placed first but is followed by an incompatible modifier, 'by strong interfacial anchoring'. Nanoparticles are not an action, process, or state that can be modified [297, §I.35].

Main topic

✕ Stabilized gold nanoparticles on ceria nanorods by strong interfacial anchoring

Incompatible modifier

Example 3.3. *Awkward title, from a chemistry article [687].*

The title can be repaired by replacing 'Stabilized' by 'Stabilization of', as in the edited version in Example 3.4, where it has the same function as 'Regulation of' in Example 3.1. Stabilization is an action that can be modified in the way required.

Main topic

Stabilization of gold nanoparticles on ceria nanorods by strong interfacial anchoring

Method

Example 3.4. *Edited version of Example 3.3.*

Other ways to repair the title in Example 3.3 are to move 'Stabilized' from its position just before 'gold nanoparticles' to just after it or to just after 'nanorods', but both have the disadvantage that the keyword 'Stabilization' or 'Stabilized' no longer appears first.

Sometimes the modification of the main topic cannot be achieved with a prepositional phrase. Instead the modifier is given a separate status in a compound title. Compound titles consist of a main title and a subtitle, known also as a hanging title, which describes, for example, the nature of the approach taken [68], as in Example 3.5.

Main topic

Prototype selection for nearest neighbor classification: Taxonomy and empirical study

Aspects of approach

Example 3.5. *Compound title, from a computer science article [515]. The use of such titles is contentious.*

Compound titles like the one in Example 3.5 are uncommon,[13] and their use can be problematic. They may be difficult to index in online databases (the subtitle and the main title may each on their own be too general to be search terms); the subtitle itself may be discarded; and they are susceptible to errors in interpretation [78]. In addition, articles with compound titles may receive fewer citations than other articles [193].

Titles that have subtitles defining series numbers, e.g. 'Part 1','Part 2', can lead to further problems and are deprecated. If both parts are published, they usually entail some degree of duplication, e.g. in methods, materials, or subjects, and therefore risk the accusation of publication by salami slicing (§12.8.2). If duplication is avoided by cross-referencing, then the reader is forced to refer back and forth between the two articles. Moreover, the editor of a journal on receiving a series manuscript may have to make a decision about acceptance without knowing the status of the other parts, which may not even have been written [79, pp. 43–4]. Series titles may also distort the literature [339, §9c].

Titles that form a question are rare,[13] but they can be effective in identifying a readily understood challenge, as in Example 3.6.

How does the brain solve visual object recognition?

Example 3.6. *Question title, from a neuroscience article [475].*

The disadvantage of question titles is that they fail to signal whether the article actually gives an answer. Such titles are criticized as unintelligible [78]. An exception may be those that seem to invite either a clear 'yes' or 'no', although the answer generally proves to be 'no'.[14] Like articles with compound titles, articles with question titles may receive fewer citations than other articles [193].

3.2.3 Shifting Emphasis

Example 3.7 shows how simple variations in verb form and the ordering of phrases can affect meaning in a title.

Original — Control of star formation by supersonic turbulence

Variations:
- Controlling star formation by supersonic turbulence
- Star formation controlled by supersonic turbulence
- Supersonic turbulence control of star formation
- How supersonic turbulence controls star formation
- When supersonic turbulence controls star formation
- Does supersonic turbulence control star formation?

Example 3.7. *Original title and variations, from a physics article [596].*

The original title is first in the list and each variation that follows leads to a different interpretation, with the emphasis falling respectively on the notions of control, star formation, supersonic turbulence, mechanisms, conditions, and the question of control. The last version, forming a question, may invite the wrong answer (§3.2.2).

3.2.4 Advantages and Disadvantages of Short Titles

Short titles are uncommon (§3.2.2), and very short titles with three words or less are rare.[12] Yet some short titles can be remarkably effective with respect to the criteria of Section 3.2.1, particularly when the work constitutes a comprehensive review of a research area or offers a solution to a well-defined research question or problem.

Examples 3.8–3.10 show some successful very short titles from three different research areas. The one-word title in Example 3.8 is from a psychology article [471] and can be understood by any reader.

Intelligence

Example 3.8. *Minimal but effective title, from a psychology review article [471].*

The article provides a wide review of intelligence research in the first decade of the twenty-first century.

The three-word title in Example 3.9 is from a more specialized theoretical physics article [622].

Realistic Dirac leptogenesis

Example 3.9. *Very short but effective title, from a theoretical physics article [622].*

It presents a model that describes the generation of matter-antimatter asymmetry and which is compatible with the size of neutrino masses. For the specialist, the words 'Dirac' and 'Realistic' have useful connotations. 'Dirac' refers to the way in which the neutrino masses are generated and 'Realistic' to the generation being stable in a technical sense.[15]

The two-word title in Example 3.10, from a mathematical statistics article [532], has meaning for both the specialist and nonspecialist.

Random alms

Example 3.10. *Very short but effective title, from a mathematical statistics article [532].*

It deals with the mathematical problem of dividing a finite quantity in a repeatedly random fashion. The title comes from a concrete example used by the author on how to distribute charitable aid or relief, otherwise known as alms, to the needy (§3.5.3).

Despite their attractiveness, short titles do constitute a risk. Their effectiveness depends on the precise choice of words and their connotations, which may not be obvious to all readers, or indeed to database search engines. They may also prove a challenge for the author wanting to devise a matching title for a subsequent paper on a related topic.

Even a reasonably accessible short title may be open to misinterpretation, as illustrated in Example 3.11 from a clinical article.

Celiac disease

Example 3.11. *Possibly misleading very short title, from a general and internal medicine article [499].*

From the title alone, a reader might assume that the article contains a general review of the field. Only when the abstract of the article (or the title page in the journal) is inspected does it become evident that it represents a single case study, which then develops into a consideration of methods of investigation and treatment. With respect to these specific goals, the title does not distinguish the article from many others in what is an active research field.[16]

3.2.5 Advantages and Disadvantages of Long Titles

Like short titles, long titles are uncommon, and very long titles with more than 25 words are rare.[12] Long titles can better represent the work and improve database indexing, yet they are also inefficient and risk losing the reader (§3.2.2).

Examples 3.12, 3.15, and 3.18 show some very long titles from three different research areas.

✕

> Multiresidue determination of 375 organic contaminants including pesticides, polychlorinated biphenyls and polyaromatic hydrocarbons in fruits and vegetables by gas chromatography-triple quadrupole mass spectrometry with introduction of semi-quantification approach

Example 3.12. *Very long title, from an analytical chemistry article [412]. It has 30 words (with hyphens treated as word separators).*

Example 3.12 is from a chemistry article. The title has 30 words. The inclusion of named contaminants, 'pesticides, polychlorinated biphenyls and polyaromatic hydrocarbons', adds interest but slows the reader. The hyphen between 'gas chromatography' and 'triple quadrupole mass spectrometry' also interrupts the flow, as it creates a meaningless compound 'chromatography-triple'. It is unclear whether 'triple quadrupole' is necessary.

Example 3.13 shows an edited version without the named contaminants and with a simplified experimental description. The title now has 18 words.

> Multiresidue determination of 375 organic fruit and vegetable contaminants by gas chromatography and mass spectrometry with semiquantification

Example 3.13. *Edited version of the title in Example 3.12. It has 18 words.*

If more of the experimental method is omitted, then a still shorter and more readable version is possible. Example 3.14 shows the result, which now has just 12 words.

> Multiresidue determination of 375 organic fruit and vegetable contaminants with semi-quantification

Example 3.14. *Further edited version of the title in Example 3.12. It has 12 words.*

A decision about how much of the method to include might be settled by testing 'gas chromatography' and 'mass spectrometry' as search terms (§3.2.1).

Example 3.15 is from a biology article. This title has 27 words. The two statements 'import experiments in protoplasts' and 'import into chloroplasts' have a degree of overlap (significantly 'protoplasts' does not appear in the abstract of the article). The appearance of 'importance' shortly after 'import' may also be distracting, and the main topic of the work is placed late in the title.

✗ *In vivo* import experiments in protoplasts reveal the importance of the overall context but not specific amino acid residues of the transit peptide during import into chloroplasts

Main topic

Example 3.15. *Very long title, from a biochemistry and molecular biology article [581]. It has 27 words.*

Example 3.16 shows an edited version, with just 15 words. The fact that the experiment was in vivo has been omitted. As with the original, however, the main topic is still placed late.

✗ Importance of overall context not specific amino acid residues in transit peptide import into chloroplasts

Main topic

Example 3.16. *Edited version of the title in Example 3.15. It has 15 words.*

Example 3.17 shows a possible further revision with the main topic now placed first.

Main topic — Transit peptide import into chloroplasts influenced more by overall context than specific amino acid residues

Example 3.17. *Further revision of the title in Example 3.15. It has 15 words.*

Finally, Example 3.18 is from a chemistry article. This title has 23 words, and contains repetition (of 'performance', 'density', 'functional'). The general term 'calculations' could be omitted and the adjective 'recent', which may limit the long-term significance of the work. The cumbersome 'electronic transition energies', which does not appear in the abstract, could be replaced by the shorter 'excitations', which is used both in the abstract and in the body of the text. The fact that the authors tested 30 functionals (evident from an inspection of the article) might also be worth mentioning.

✗

> Performance of recent and high-performance approximate density functionals for time-dependent density functional theory calculations of valence and Rydberg electronic transition energies

Example 3.18. *Long title, from a physical chemistry article [549]. It has 23 words.*

Example 3.19 shows an edited version incorporating these changes. It has just 13 words.

> Performance of 30 time-dependent density functionals in predicting valence and Rydberg excitations

Example 3.19. *Edited version of the title in Example 3.18. It has 13 words.*

It could be improved further by moving the important part of the title to the front. Example 3.20 shows the result, which has 11 words.

> Valence and Rydberg excitations predicted by 30 time-dependent density functionals

Example 3.20. *Rearranged version of the title in Example 3.18. It has 11 words.*

In this version, 'performance' has been omitted, but it is likely to be understood given that 'predicted' appears in the title supported by 30 density functionals.

3.2.6 Problems with Prepositions

The problems with long titles often come from their having too many prepositional phrases. The examples of effective titles in Section 3.2.2 contain at most two such phrases, beginning 'of', 'by', 'in', 'on'. By contrast, the excessively long titles in Section 3.2.5 have from three to five prepositional phrases. As the number of prepositional phrases increases, so does the risk of confusion, as parsing becomes more and more challenging.

Example 3.21 has six prepositional phrases.

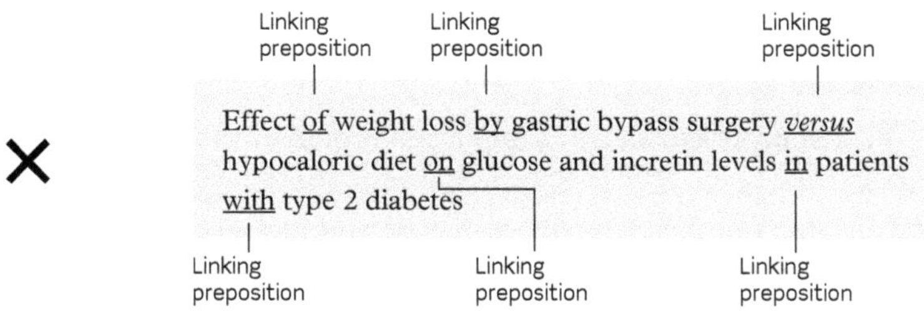

Example 3.21. *Title with multiple linking prepositions, from an endocrinology and metabolism article [573].*

The first two prepositions, 'of' and 'by' (and the last 'with'), act unambiguously. The others are confusing. Thus the third preposition, 'versus', correctly links gastric bypass surgery with a hypocaloric diet, but it could be taken to link weight loss by gastric bypass surgery with a hypocaloric diet. The fourth preposition, 'on', grammatically links a hypocaloric diet with glucose and incretin levels, whereas from the content of the article, it seems to link the effects of gastric bypass surgery and of a hypocaloric diet with glucose and incretin levels, although it could be taken to link the effects of weight loss from these procedures with glucose and incretin levels.

In fact, the authors argue in the text that the improvement of incretin levels is related not to weight loss per se but to surgery, so that the initial phrase in the title 'Effect of weight loss' may actually mislead the reader.

3.3 Abstract

After the title, the abstract is next most read section of your paper [258]. It is freely available on journal web pages without subscription, in conference proceedings, and in bibliographic and other online databases.

Its purpose is to faithfully summarize what is written in the paper. It should therefore be brief, accurate, and comprehensive: a distillation [79]. As it needs to stand alone, it should be thought of as a mini- or micro-article in its own right [179] [258].

It should not be a statement of conclusions or an advertisement for the work, or, still worse, puffery. Between 1974 and 2014 there was a more than eight-fold increase in the appearance of positive-sounding words such as 'robust,' 'novel,' 'innovative,' and 'unprecedented' [365].

3.3.1 Fundamentals

For an experimental report, the abstract generally includes the following elements [68] [258]:

- the context or background of the work
- the research question or other reason for the work, if relevant

- the approach or methods
- a key result or two
- a conclusion, if appropriate, or other implications of the work

For a report of a discovery or invention, there may be no overt research question, and the abstract then concentrates more on methods and results. For a theoretical or review paper, the emphasis is on results and implications.

To stand alone, an abstract should not normally contain references, that is, the details of other published articles, unless, for example, the article being abstracted is a critique of a particular published work. It should also not contain citations to references, that is, bracketed or superscript numbers or author names and dates, since abstracts do not have their own list of references [68] [290], unless the abstract forms part of the introduction to a research letter, as with the journal *Nature* (referencing systems are discussed in Section 3.10). Likewise, abstracts should not refer to figures or tables [79].

Although abstracts may report inferences from statistical tests, they rarely report detailed results such as *p*-values (§7.9), except in clinical studies.[17]

How should you go about constructing an abstract? As with choosing a title, it can be done systematically, albeit with trial and error. First, construct a representative sentence for each of the principal elements of the work: the context, the research question, the approach, the main result, and the conclusion. Then count the number of words. If the total is much less than the maximum allowed or what is usual for the intended journal, add another sentence or two. Do not include too many results or anything that is not stated in the body of the paper [79]. Then experiment with the sentence structure and word order so that the abstract reads as a coherent whole. As with the title, the words in the abstract may serve as search terms, and the use of language needs to be consistent with that in the title, introduction, discussion, and keywords.

As a general guide, write as if someone else had the job of abstraction. If the abstract is not part of the introduction, as with the journal *Nature*, avoid 'I' and 'we', for personal pronouns can look strangely disembodied in a database of abstracts (see Section 5.1.1 for arguments for and against using personal pronouns).

Avoid all but the most common established abbreviations [68]. If uncommon or ad-hoc abbreviations do need to be introduced to save multiple repetitions, then they should be defined on their first occurrence in the abstract, in the body of the paper, and in figures and tables so that they can be understood without reference to the text [68, §11.3.1]. They should not be used to save just one or two repetitions, for they will almost certainly force the reader to search back to where the abbreviation was first introduced.

Since the abstract is a representation of the paper, any notation and definitions should be used in a way consistent with the main text. For example, if 'frequency selectivity' appears in the abstract, then 'spectral tuning' should not be used in the main text to mean the same thing (readers may assume you are making a distinction that they have not grasped). Technical terms aside, some editors expect the abstract not to repeat sentences verbatim from the main text and especially not the opening sentence (which serves a somewhat different function). Even so, given the abstract's intimate relationship to the paper, some repetition of the phrasing is reasonable [24, p. 189], and may be essential to preserve consistency.

Some journals, usually in the life sciences or medicine, require the abstract to be structured, that is, formatted into distinct sections ('purpose', 'methods', etc.). Otherwise, the abstract is expected to form continuous text. Any paragraph breaks you try to introduce as an aid to comprehension are likely to disappear in the published article or in online databases.

3.3.2 Effective Abstracts

As with titles, an effective abstract achieves a balance between being comprehensive and not losing the reader's attention. Journals may set lower and upper word limits, for example, 150–250 words. Abstracts with fewer words—and the same content—are argued to be better than abstracts with more words [79]. Yet the evidence is ambiguous, with small increases in citations linked both to fewer (or more routine) words [226] and to more words [377]. As with titles [82], potential confounds may affect these kinds of calculations.

In practice, the average number of words in abstracts falls closer to the lower limit of the 150–250 word range. Much shorter abstracts of 50 words or less and longer ones of 300 words or more are uncommon.[18]

Example 3.22 shows an effective abstract from a neuroscience article. It has 101 words.

Context	An unresolved question in neuroscience and psychology is how the brain monitors performance to regulate behavior.
Reason for work	It has been proposed that the anterior cingulate cortex (ACC), on the medial surface of the frontal lobe, contributes to performance monitoring by detecting errors.
Methods	In this study, event-related functional magnetic resonance imaging was used to examine ACC function.
Results	Results confirm that this region shows activity during erroneous responses. However, activity was also observed in the same region during correct responses under conditions of increased response competition. This
Implication	suggests that the ACC detects conditions under which errors are likely to occur rather than errors themselves.

Example 3.22. *Effective abstract, from a neuroscience article [436]. Reprinted with permission from AAAS.*

Notice that on first use, the authors define the abbreviation 'ACC', which is later used twice.

Example 3.23 shows a successful abstract from a multidisciplinary sciences article. It has 129 words (with a little repetition).

Self-organization of liquid crystalline and crystalline-conjugated materials has been used to create, directly from solution, thin films with structures optimized for use in photodiodes. The discotic liquid crystal hexa-peri-hexabenzocoronene was used in combination with a perylene dye to produce thin films with vertically segregated perylene and hexabenzocoronene, with large interfacial surface area. When incorporated into diode structures, these films show photovoltaic response with external quantum efficiencies of more than 34 percent near 490 nanometers. These efficiencies result from efficient photoinduced charge transfer between the hexabenzocoronene and perylene, as well as from effective transport of charges through vertically segregated perylene and hexabenzocoronene π systems. This development demonstrates that complex structures can be engineered from novel materials by means of simple solution-processing steps and may enable inexpensive, high-performance, thin-film photovoltaic technology.

Context, purpose — Methods — Result — Implication — Another implication

Example 3.23. *Effective abstract, from a multidisciplinary sciences article [664]. Reprinted with permission from AAAS.*

Because this work was aimed at developing specific materials, the emphasis was less on addressing a research question and more on exploiting the properties of the materials.

Example 3.24 is more theoretical, from an article in electrical and electronic engineering (part of which forms Example 2.11). It has 121 words.

Separation of sources consists of recovering a set of signals of which only instantaneous linear mixtures are observed. In many situations, no *a priori* information on the mixing matrix is available: The linear mixture should be "blindly" processed. This typically occurs in narrowband array processing applications when the array manifold is unknown or distorted. This paper introduces a new source separation technique exploiting the time coherence of the source signals. In contrast with other previously reported techniques, the proposed approach relies only on stationary second-order statistics that are based on a joint diagonalization of a set of covariance matrices. Asymptotic performance analysis of this method is carried out; some numerical simulations are provided to illustrate the effectiveness of the proposed method.

Context, problem — Main result — Methods — Supporting results, impact

Example 3.24. *Effective abstract, from an electrical and electronic engineering article [416].*

This abstract can be improved a little as shown in Section 2.1.6, but it contains enough to let the reader decide whether to read the article or not.

3.3.3 Missing Elements

Your research findings may constitute the most important part of the abstract. Still, some contextual information is always helpful. Consider the extract in Example 3.25.

✕

> Nanocrystallyne n-ZnO and p-TiO_2 thin films were deposited on crystalline Si substrate by the simple and low-cost sol-gel method.
>
> . . . — 89 words omitted
>
> Surface modified p-TiO_2 sensors recorded higher gas response (78%), response time (3 s), and recovery time (74 s) compared with ZnO sensors under the similar conditions.

Example 3.25. *Abstract that concentrates on results, from an instrumentation engineering article [558].*

The whole abstract is devoted to methods and results. There is no background or motivation explaining, for instance, why metal oxides were deposited, why a higher gas response might be desirable, or what the implications are of the improved performance. Although the absence of any interpretation or conclusion is not surprising with an instrumentation paper, it would have been useful to have an introductory sentence or two giving the context: the importance of detecting gas (hydrogen) and the usefulness of metal oxide gas sensors.

3.3.4 Unhelpful Citations

Despite the expectation that abstracts should be able to stand alone (§3.3.1), they do sometimes contain citations, as with the extract in Example 3.26.

✕

> This paper considers the modelling and analysis of nonstationary vector ARMA processes from the theoretical point of view. Some extensions to the existing work (see, for example [1,2]) incorporating the nonstationarity of the underlying process are given using models with nonstationary innovations.
>
> — Citations without references

Example 3.26. *Abstract with unhelpful citations, from a computer science article [674].*

For the reader who has access only to the abstract, the instruction 'see, for example [1, 2]' is not useful. This abstract has other problems, which are examined in Section 3.3.6.

3.3.5 Too Long or Too Short Abstracts

Long abstracts are uncommon (§3.3.2), and those with more than 350 words are rare.[18] Example 3.27 shows a small part of an abstract that has at least 378 words (hyphens not treated as word separators). Just under one tenth of it is reproduced here.

✗

Diagenesis of the Holocene-Pleistocene volcanogenic sediments of the Mexican Basin produced, in strata of gravel and sand, 1H_2O- and 2H_2O-smectite, kaolinite, R3-2H_2O-smectite (0.75)-kaolinite, R1-2H_2O-smectite (0.75)-kaolinite, R3-kaolinite (0.75)-2H_2O-smectite and R1-1H_2O-smectite (0.75)-kaolinite . . . — 348 further words omitted

Example 3.27. *Part of a very long abstract, from a geology article [470].*

The abstract consists entirely of results. There is little to help the nonspecialist assess the significance or relevance of the work or indeed its motivation. It seems to have attracted little interest.[19]

Short abstracts too are uncommon, and those with fewer than 30 words are rare.[18] But unlike short titles, short abstracts cannot adequately deliver the required distillation of the paper. Example 3.28 is from a physics article entitled 'Unitarity constraints and role of geometrical effects'.

✗

Unitarity and geometrical effects are discussed for photon-photon scattering.

Example 3.28. *Very short abstract that adds little to the title, from a nuclear physics article [699].*

There is no motivation for the work, no methods (e.g. whether the work is purely theoretical or contains experimental results), and no implications. The abstract contains ten words, just three more than the title.

Example 3.29 is from a mathematical statistics article entitled 'Random degree and essentiality for countably condensing maps'.

✗

The aim of this paper is to provide (i) a random degree theory, and (ii) an essential map theory, for countably condensing maps.

Example 3.29. *Very short abstract that adds little to the title, from a mathematical statistics article [395].*

The abstract is a reformulation of the title. The article itself consisted of eight printed pages, from which more might have been abstracted.

3.3.6 Empty Statements

Abstracts need not be short to be uninformative. Example 3.30, introduced in Section 3.3.4, contains 78 words, but it adds little to what is already in the article title, 'A Note on the Modelling and Analysis of Vector ARMA Processes with Nonstationary Innovations'.

✗

This paper considers the modelling and analysis of nonstationary vector ARMA processes from the theoretical point of view. Some extensions to the existing work (see, for example [1,2]) incorporating the nonstationarity of the underlying process are given using models with nonstationary innovations. It is shown that this class of models provides a very strong framework for many practical situations. The estimation procedures are discussed. Various prediction results are also given. Some examples are added to illustrate the theory.

Expanded title

Uninformative detail

Example 3.30. *Abstract that adds little to the title, from a computer science article [674]. It contains multiple uninformative statements.*

The abstract tells the reader that the models provide a framework for practical situations. It also tells the reader that the paper contains results, discussion, and examples, which the reader might already have anticipated. It says nothing about the content of those results, discussion, and examples. Existence statements of this kind are of little help (§2.3.2).

Example 3.31 is more specific, yet still closes with an empty statement 'The implications of these results are discussed'.

✗

The electronic structure for graphene monolayer tubules is predicted as a function of the diameter and helicity of the constituent graphene tubules. The calculated results show that approximately 1/3 of these tubules are a one-dimensional metal which is stable against a Peierls distortion, and the other 2/3 are one-dimensional semiconductors. The implications of these results are discussed.

Uninformative detail

Example 3.31. *Abstract with an uninformative closing statement, from an applied physics article [657].*

The statement could have been expanded, for instance, by adding something about the potential applications of the work to metal-semiconductor wires, which the authors mention at the end of the article.

3.3.7 Private Conversations

Some abstracts appear to be addressed to an audience already familiar with the author's work and its significance, as with the extract in Example 3.32.

✕

In a previous study, we reported evidence for correlations between the firing of postsynaptic dorsal column (PSDC) neurons and cuneate neurons with overlapping receptive fields.

82 words omitted. — . . .

We have now re-analyzed these data using a parallel-cascade method . . .

Example 3.32. *Abstract setting the record straight, from a neurophysiology article [510].*

The abstract summarizes the authors' analysis of the results they published in an earlier article and their reanalysis in this article. The earlier article is not referenced in the abstract (one of the rare cases where a reference would have helped). Unfortunately, for readers unfamiliar with the authors' previous research, there is no clear explanation of why the reported correlations should be important or what they imply. The work seems to have generated little interest.[20]

3.3.8 Value Judgements and Aspirations

Some abstracts (and articles) go beyond a factual report of the research undertaken and drift into promotion, as with the extract in Example 3.33.

✕

Considered as a whole, the empirical and theoretical issues analyzed herein, as well as the methodologies, have sufficient quality and depth to form a generalized framework, which we hope to develop in future work.] Self-evaluation and predictions

Example 3.33. *Abstract with value judgements, from a computer science article [693].*

The abstract contains both the authors' self-evaluation of their work and their hopes for taking it further. As explained in Section 3.8.7, observations of this kind are best omitted. The article may not achieve its predicted influence[21] and the anticipated work may not take place or be published.

3.4 Keywords, Highlights, Lay Summary

In addition to the abstract, journals usually ask authors for other kinds of summary information and description.

3.4.1 Keywords

The keywords consist of a short list of words that characterize the main topics of the article [68]. They are used to help with the indexing of the article by the journal and to inform readers about its content. Some journals allow a free choice of words but others require authors to draw on a mandatory list determined by the research area.

3.4.2 Highlights

Highlights are a few short statements, each perhaps a printed line in length, describing the core findings of the research [164]. They normally exclude the aims and methods and are largely independent of each other. They are intended to answer impatient questions of the type 'What is the point of the research?' or the still more basic 'So what?' [79, p. 75].

Even if core findings are not required by the journal, identifying them at an early stage of your writing can shape the structure of your paper, sharpen the discussion, and improve the eventual impact of the published article.

3.4.3 Graphical Abstract

A graphical abstract is a single figure or image that summarizes the main outcome of the research [143]. It may be created solely for this purpose or taken from the article itself.

3.4.4 Lay Summary and Significance

A lay summary or significance statement offers a short, accessible overview that explains the relevance of the work to a nonspecialist [187]. Writing an effective lay summary can be demanding, since it needs to address two audiences at once: the informed reader outside your own field and the specialist within it. But it adds value, serving to increase interest within the community and attract attention from the larger public [211].

3.5 Introduction

The introduction of your paper performs a double function. Along with its official role in setting out the context of the work, its rationale, and the approach, it has an equally important albeit unofficial role in signalling whether the material in the rest of the document is likely to be interesting, authoritative, and professionally presented.

3.5.1 Fundamentals

Introductions can be developed in different ways [157, ch. 2.5]. But there is agreement that they should generally contain the following elements [79] [258]:

- an opening statement that sets the context
- a review of relevant published work
- the research question or other reason for the work
- the purpose or aims and objectives
- the approach or methodology, essential concepts without technical detail
- possibly an indication of the outcome of the work

Of all these elements, the most influential is the opening statement. It is one of the few parts of the paper that will convince the reader to continue reading or abandon the attempt and turn to something else. Whether or not an indication of the outcome should be included depends on the journal. Some of the arguments are considered in Section 3.5.9.

How should you go about constructing an introduction? One way is this. Start with a short general sentence that captures the reader's attention and sets the scene for the subsequent account. Then add one or more sentences that develop the topic [258]. This material should be readily comprehensible to someone who is a nonspecialist in the field. Then, over succeeding sentences and paragraphs, make the material more and more specific, preparing the specialist reader for the technical level at which the work will be reported. Develop any technical language or notation in an analogous way. Ensure that credit is given to previously published work by giving citations at suitable points, but ensure also that enough context is provided so that the reader can appreciate the material without having to consult the references [79].

After this development—and before the reader begins to ask—explain the purpose of the research, or its aims and objectives (§4.3.2). Then mention something about the experimental or theoretical approach or methodology, and any essential concepts, along with an indication of the outcome of the research if appropriate. The use of language needs to be consistent with that in the title, abstract, discussion, and keywords. Finally, consider summarizing the organization of the paper if it is a long one or one with a nonstandard structure.

Overall, a research article is expected to have an hourglass structure [79, pp. 61, 73] [258, pp. 41-2]. The introduction takes the reader from the general to the specific and the discussion reverses that process. In a long article or thesis, the hourglass structure may be repeated within the sections. Along with orienting the specialist reader, it helps the nonspecialist, who may be unable to follow the detail but can appreciate the nature of the work and what has been achieved.

3.5.2 Opening with a Generalization

By far the most common way of starting an introduction is with a generalization, an observation about the topic of general validity.[22] It works for any discipline: Examples 3.34–3.40 are drawn from pure mathematics, neuroscience, chemistry, psychology, cell biology, environmental engineering, and applied mathematics, respectively.

> Fixed point theory is one of the most powerful and fruitful tools in nonlinear analysis. Its core subject is concerned with the conditions for the existence of . . .

Example 3.34. *Opening with a general observation, from a pure mathematics article [659].*

> Retinal degeneration is a major cause of untreatable blindness. Despite very different aetiologies, degenerative retinopathies culminate in the same final common pathway, the loss of photoreceptors and vision.

Example 3.35. *Opening with a general observation, from a neuroscience article [413].*

> Gold is the quintessential noble element. By nature, it is highly unreactive and as such, historical artifacts made of gold are able to retain their brilliant luster.

Example 3.36. *Opening with a general observation, from a chemistry article [483].*

> Some people are cleverer than others. The ways in which this occurs, and the causes and consequences of these individual differences . . .

Example 3.37. *Opening with a general observation, from a psychology article [471].*

> Cancers carry somatic mutations. A small proportion are "drivers" that confer clonal advantage, are causally implicated in oncogenesis, and have been positively selected during the evolution of the cancer.

Example 3.38. *Opening with a general observation, from a cell biology article [626].*

> Solar updraft towers, also known as solar chimneys, provide a simple and reliable way of generating electricity from solar energy [1–3]. With the main principle behind these devices dating back more than 100 years . . .

Example 3.39. *Opening with a general observation, from an environmental engineering article [445].*

> Probability distributions are often observed to have power-law tails, particularly in social, economic, and biological systems. Examples include distributions of fluctuations in financial markets (1), the populations of cities (2), the distribution of Web site links (3) . . .

Example 3.40. *Opening with a general observation, from an applied mathematics article [638].*

In other contexts, general statements such as these would be seen as bland, but here they give orientation and context, priming the reader for what comes next.

3.5.3 Opening with a Research Question or Problem

The second most common way of starting an introduction is with a description of the research question or problem to be addressed.[22]

> Consider the problem of distributing one pound of gold dust at random among a countably infinite set of beggars. Let the beggars be enumerated and let the procedure for distribution be as follows: the first beggar is given a random portion of the gold . . .

Example 3.41. *Opening with a research problem, from a mathematical statistics article [532].*

In Example 3.41 the opening statement about how to divide a pound of gold dust among beggars, despite its superficial simplicity, presaged a mathematical analysis of an important general distributional problem [532]. The results proved to have relevance in many disparate research fields, including economics, materials science, signal processing, applied physics, environmental statistics, astrophysics, geology, and bacteriology.

The research problems identified in the openings in Examples 3.42–3.44 are drawn from physics, general and internal medicine, and chemistry, respectively.

> One of the most important problems in theoretical physics is the understanding of the properties of quantum systems with an infinitely large number of interacting degrees of freedom, the so-called many-body problem.

Example 3.42. *Opening with a research problem, from a physics article [565].*

> Atrial fibrillation is the most common of all sustained cardiac arrhythmias, with the prevalence increasing with age to up to 5 percent in persons more than 65 years of age, and it is a major cause of stroke.[1-3]

Example 3.43. *Opening with a research problem, from a general and internal medicine article [530].*

> The finding of clean and renewable energy is one of the major scientific and technological challenges in the 21st century. In this respect, acquiring power from the sun using photovoltaics (PV) is an attractive alternative to address global environmental issues.

Example 3.44. *Opening with a research problem, from a chemistry article [615].*

All of these openings are accessible and efficient. They not only orient the reader but motivate the account that follows.

3.5.4 Other Openings

Other less common ways of starting an introduction are by referring to recent or historic findings, by defining or explaining a key concept, by posing a question, or by explicitly stating the purpose of the work.[22] Examples 3.45–3.48 contain illustrations of each.

> Over 50 years ago, Campbell & Fiske (1959) voiced their concerns about the biasing effects that methods of measurement may have on the validity of measures.

Example 3.45. *Opening with a historical perspective, from a psychology article [640].*

A sensor network is typically composed of spatially distributed autonomous sensors that can monitor physical or environmental conditions, make computations and communicate with neighboring devices.

Example 3.46. *Opening with an explanation, from an electrical and electronic engineering article [480].*

Calculate or simulate? Predicting the evolution of a multicomponent system is often a challenge that can be solved either by direct mathematical analysis or by a device that simulates its behaviour.

Example 3.47. *Opening with a research question, from a physics article [420].*

The purpose of this paper is to establish the existence of positive solutions to the following higher order fractional differential equation.

Example 3.48. *Opening with the purpose of the work, from a mathematics article [732].*

Incidentally, although there is little risk of confusion with the compound modifier 'higher order fractional differential' in Example 3.48, a hyphen could be inserted in 'higher order'.

3.5.5 Interrupted Openings

For all its importance, the opening statement should not attempt too much, for instance, combining generalization and explanation. Consider Example 3.49.

✕

Ferroelectrics, being also pyroelectrics (thermoelectric transducers) and piezoelectric (electromechanical transducers), find applications in a wide range of sensors and transducers. Ferroelectric sensors are used to measure force, pressure, flow and motion, temperature, and IR radiation. — Nested explanations

Example 3.49. *Opening interrupted by nested explanations, from an electrical and electronic engineering article [462].*

The nested explanations 'being also pyroelectrics (thermoelectric transducers) and piezoelectric (electromechanical transducers)' are awkward and disrupt the narrative [297, §14.39]. Furthermore, they are unnecessary because the second sentence offers a more basic explanation of how ferroelectric sensors are used. In fact, the terms pyroelectrics, thermoelectric transducers, piezoelectric, and electromechanical transducers are used only later in the article (or not at all). The opening can be safely simplified, as in the edited version in Example 3.50.

> Ferrorelectrics find applications in a wide range of sensors and transducers. Ferroelectric sensors are used to measure force, pressure, flow and motion, temperature, and IR radiation.

Example 3.50. *Edited version of Example 3.49.*

Example 3.51 also has nested explanations, but in the first sentence the parenthetical definition '(where N refers to U, A, or G)' needed to make sense of the abbreviation 'CNG' is more disruptive still.

> ✗
>
> Trinucleotide repeat sequences within mRNAs, predominantly CNG sequences (where N refers to U, A, or G), are a motif prominent in neurological diseases (Box 1). Expansion of CNG- repeat sequences can cause dramatic effects on the expression and/or stability of the host mRNA, leading to the disruption of cell and protein function, resulting in neuronal degeneration and eventual cell loss.
>
> Nested explanations
>
> Ambiguous alternatives
>
> Ambiguous referents

Example 3.51. *Opening with nested explanations, from a neuroscience article [585].*

The explanation 'predominantly CNG sequences (where N refers to U, A, or G)' can be postponed. The explanations in the second sentence, however, contain some unattached participial phrases ('leading to' and 'resulting in'), which ought to be resolved (§5.4.1). Example 3.52 shows a possible revision.

> Trinucleotide repeat sequences within mRNAs are a motif prominent in neurological diseases (Box 1). These sequences are predominantly CNG repetitions, where N refers to U, A, or G, and their expansion can dramatically affect the expression and stability of the host mRNA. The consequent disruption in cell and protein function may lead to neuronal degeneration and eventual cell loss.

Example 3.52. *Revised version of Example 3.51.*

The phrase 'predominantly CNG sequences . . .' has been moved to the second sentence, where it is associated more naturally with its effects on cellular function. The unattached phrases 'leading to' and 'resulting in' have been given their own sentence. The ambiguous 'and/or' (§5.6.16) in 'expression and/or stability' has been replaced by 'and' since the action appears to be generic (some sequences affecting both expression and stability). The number of words is unaltered.

3.5.6 Repetitive Openings

Because of the central role played by some parts of an explanation in an opening statement, they may be inadvertently repeated, as in Example 3.53.

Initial size change

✕

> As a consequence of reducing the size and the dimensionality of a material, its electronic properties change drastically as the density of states and the spatial length scale of the electronic motion are reduced with decreasing size. The energy eigenstates are now determined by . . .

Dependent size change

Repeated size change?

Example 3.53. *Inadvertent repetition, from a physical chemistry article [586].*

The opening refers to three superficially similar things: 'reducing the size', 'spatial length scale . . . reduced', and 'decreasing size'. The concatenation 'reduced with decreasing size' is especially confusing. Example 3.54 shows a possible revision.

> Reducing the size and dimensionality of a material drastically changes its electronic properties. As the density of states and length scale of electronic motion decrease, the energy eigenstates are determined by . . .

Example 3.54. *Revised version of Example 3.53.*

The opening sentence has been pruned and the excised material has been moved to the second sentence where it supports the explanation originally indicated by 'now'. The number of words has decreased from 44 to 31.

3.5.7 Private Conversations

As with some abstracts (§3.3.7), some introductions seem addressed to an audience that is already engaged with the author's work.

The opening sentence of Example 3.55 starts with two self-citations '[1] and [2]', and then proceeds to review what was achieved in those studies. The review continues into a second and third paragraph (not shown here), without explaining the importance of the cited research or its relevance to the new work. Only if the reader persists to a fourth paragraph (also not shown) does the motivation become evident.

Expositions of this kind may be seen as egocentric (§9.3.5, Example 9.3).

×

> In previous papers, [1] and [2], we considered the design of zero-forcing decision feedback equalizer (ZF-DFE) transceivers over slowly time-varying multiinput multioutput (MIMO) memoryless channels. We used the block fading MIMO channel model in which the channel is constant over the coherence interval of *Nc* blocks and varies across different coherence intervals according to Jakes' model [3], [4].

Example 3.55. *Continuing story, from an electrical and electronic engineering article [588].*

3.5.8 Transitions and Purpose

The opening statement is normally followed by a description of the background or context of the work, often in the form of an abbreviated literature review. The next step in the introduction is to make the transition from previous research to the present work.

One common device is to identify the limitations of previous approaches (while giving due recognition to what has been achieved) and suggest how these limitations might be overcome [258, p. 47]. If the purpose or, if relevant, the aims and objectives of the work have not already been introduced, then the transition also provides an opportunity to describe them.

Previous limitations

> These packages do not allow editing of multiple sequence alignments. Although alignments can be edited in word processing software, such as Microsoft Word or emacs, it is often difficult to see conserved patterns without a specific colouring of the alignment that these programs do not provide. In addition, specialized multiple sequence alignment editors can provide extra features for the user including grouping and analysis of the conservation patterns in the alignment. . . .

Aims

> The Jalview program was written with the following design goals in mind. First, it should be platform independent; second, it should be fast and capable of editing of large multiple sequence alignments without significant degradation of performance; and third, it should allow multiple integrated views of the alignment and other data.

Example 3.56. *Transition from previous to present work, from a bioinformatics article [448].*

In Example 3.56 the first extract identifies the limitations of previous software packages and the second extract explains the aims of the work.

In Example 3.57 the transition proceeds similarly. Notice the restrained acknowledgement 'although valuable'.

Previous limitations	Despite this extensive research and the increasing interest in the remodeling process in pathological situations, little is known at the cellular level about the morphology of intact arteries and the structural changes associated with them. . . . Cell culture studies, although valuable, lack critical interactions with other vascular cell types and natural matrices found in the vessel wall, so the results cannot be extrapolated to the changes in intact vessels. . . .	Due recognition
Aims	The aim of the present work was to use LSCM [laser scanning confocal microscopy] to study in intact resistance arteries the cellular changes associated with remodeling in hypertension.	

Example 3.57. *Transition from previous to present work, from a vascular disease article [403]. Copyright 1997, Wolters Kluwer Health, Inc. Reprinted with permission.*

The development of the background or context, however, should not so delay the announcement of the purpose of the work that the reader begins to question whether one exists. If a prolonged development really is necessary, then the aim or aims can at least be implied and later made more explicit, as in Example 3.58. The statement 'To this end, image segmentation techniques are required' signals the aim indirectly (and the approach). The much later statement 'The purpose of

Opening statement	Multiple Sclerosis (MS) is the most common nontraumatic disabling neurological disease among young adults. It is a disease of the central nervous system that involves mainly damage to the myelin sheath that envelops neuronal fibers called axons. . . . Quantitative assessment of MR [magnetic resonance] imaging lesion load is of great interest in monitoring disease evolution and treatment trial.	
Implied aim	To this end, image segmentation techniques are required.	
	This operation consists in partitioning (clustering) the MR image into anatomical tissues, fluids, and other structures, and assigning anatomically meaningful labels to each component of the image [5].	Explanation of aim
	. . .	46 lines of background text omitted
Explicit aim	The purpose of this paper is, therefore, to develop a powerful low-level segmentation paradigm,	
	the use of which in combination with some simple knowledge-based rules gives a robust final segmentation of MS lesions in MR images.	Application

Example 3.58. *Implied general aim and explicit aim, from a computational biomedicine article [421].*

this paper is, therefore, to develop a powerful low-level segmentation paradigm' finally delivers the aim explicitly. The example continues with a few more lines explaining the paradigm and summarizing its success when applied to magnetic resonance images.

3.5.9 Anticipating the Results

Advice differs on whether the main results should be anticipated in the introduction [169]. The main argument against anticipation is logical: the exposition should proceed in linear fashion and therefore the results, even in summary form, should not precede the methods. The main argument in favour is driven by practicalities: the reader should not be kept in suspense. As Day and others emphasize, a research article is not a detective story with a surprise ending [79, p. 60]. An indication of the eventual outcome, and also the methods to be employed, helps motivate the reader to continue reading [31]. Example 3.59 (a continuation of Example 3.58) and Example 3.60 provide illustrations.

General method: The low-level segmentation paradigm is essentially based on the use of a Fuzzy C-Means (FCM) clustering algorithm, which is an unsupervised pixel-classification technique based on iterative approximation to local minima of a global objective function.

Summary result: Applied to the MR images . . . the FCM works out MS lesions with no false negative lesions and reduced false positive lesions. Furthermore, most of the remaining false positive lesions . . . can be readily eliminated.

Example 3.59. *Indication of method and experimental result, from a computational biomedicine article [421].*

General method: Finding a suitable target for drug discovery depended upon identifying the molecular mechanism regulating Msh2 protein levels.

Summary result: The results presented in this work show that unstable Msh2 variant levels are controlled by the ubiquitin/proteasome degradation pathway and that inhibition of this pathway by a clinically approved drug restores the levels, function, and chemosensitivity of unstable Msh2 missense variants, suggesting a viable adjuvant chemotherapeutic approach to treating certain mismatch-repair-defective tumors.

Example 3.60. *Indication of method and experimental result, adapted from a genetics article [402].*

Nevertheless, it may be prudent to examine articles in your intended journal to see whether they include an indication of methods and outcome, and then comply with convention [79, p. 61].

Some introductions end with this indicative material. Others end with a summary of the organization of the paper, but it should be more than a routine listing, as in Example 3.61.

> Section 2 explains the related work of the study, especially the clustering methods. . . . Section 5 presents the experimental results and explanations. Finally, Section 6 is the conclusion of the study. — Uninformative detail

Example 3.61. *Uninformative section descriptions, from a computer science article [446].*

The announcement that Section 5 contains the results and Section 6 the conclusion does not really add much to the reader's knowledge (see also Examples 3.30 and 3.31). Most experimental reports contain sections of this kind, though the advance knowledge may help some readers navigate a path through larger reports.

3.6 Methods

The introduction of a paper usually mentions something about the experimental approach or its equivalent in a theoretical study (§3.5.1). The methods section, which comes next, gives more specific information and in sufficient detail that a competent researcher can repeat the work [290] [314]. Rarely, some journals relegate the details of the methods to the end of the paper or to a supplementary document.

If the methods are similar to those in a previously published article, then it may be sufficient to give a summary, with any differences identified, and the details referred to that earlier work [293, §2.06]. If the details are reproduced, then that fact should be acknowledged; for example, with a phrase 'reproduced from . . .', to avoid what is sometimes called text recycling or self-plagiarism (§12.2).

3.6.1 Fundamentals

The methods section generally contains descriptions of the following elements, not necessarily in this order:

- apparatus (and calibrations, if appropriate)
- materials or subjects (if the research is on human beings, tissue, or animals)
- procedure
- experimental design (if sampling methods are used)
- data analysis

Depending on its length, the methods section may be subdivided according to these or equivalent subheadings [79] [293, §2.06].

Critical parts affecting the applicability of the methods or the outcome of the research should be given prominence. In this way, the reader can see immediately whether the methods are relevant. Hiding potentially damaging constraints in a mass of detail is unethical [314] and may leave the reader with a long, unhappy memory. Moreover, if there is insufficient information for the work to be repeated, then the submission may be rejected or, when published, retracted. Lapses in methodological transparency can lead to public criticism [97].

Although the description needs to allow reproduction of the work, it is not itself a recipe but an account of what was done. Conventionally [79], therefore, it is not written in the present tense (e.g. 'The GPS is mounted on top of the main cart') or as a succession of imperatives ('Mount the GPS on top of the main cart'). Rather, it is written in the past tense ('The GPS was mounted on top of the main cart' [561]). Exceptionally, the present tense or imperative may be used in the listing of an algorithm or steps in a proof or some procedure intended have relevance beyond the narrow application in the article.

3.6.2 Proprietary and Generic Products

In descriptions of commercial products, materials, or software, it is usual to give the generic name first and then, in parentheses, the proprietary details, the model or type, the manufacturer, and the place of manufacture, as in Example 3.62.

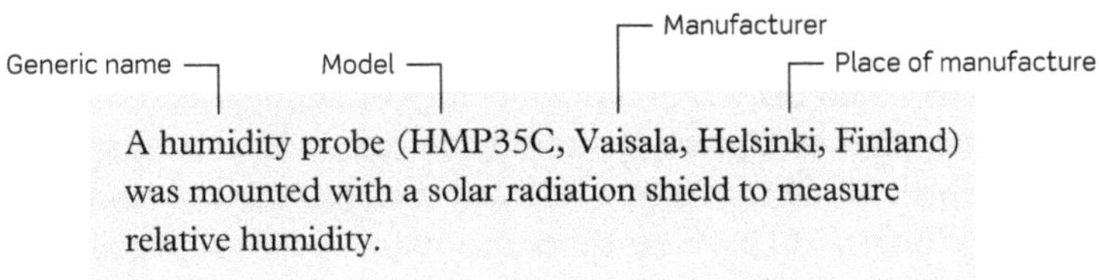

Example 3.62. *Generic and proprietary equipment descriptions, from an instrumentation engineering article [561].*

One reason for the separation is that the generic description has a meaning that lasts indefinitely or at least for the life of the research article, whereas the model and manufacturer have significance for only a limited period, terminated, for example, by the model becoming obsolete or the manufacturer ceasing to exist. Another reason is that for most readers the generic name is enough, but for researchers working in the same area, the model and manufacturer are important.

The difference in longevities of the two descriptions is evident in examples taken from articles published several decades ago. Examples 3.63 and 3.64 are both from the same issue of a psychology journal.

✕

An Automated Data Systems 1800A process control system regulated presentation of stimuli, fixation point, and postexposure mask. — Proprietary name

Example 3.63. *Uninformative proprietary equipment description, from an experimental psychology article [661].*

In Example 3.63, the phrase 'Automated Data Systems 1800A process control system' means little to the modern reader. At the time it was a well-known laboratory computer.

These stimuli were displayed on a cathode-ray tube display scope (Tektronix 604), equipped with a rapidly decaying phosphor (P-15), and controlled by a small digital computer (PDP-8/I).] Generic names

Example 3.64. *Informative generic equipment description, from an experimental psychology article [575].*

In Example 3.64, despite cathode-ray-tube display oscilloscopes being used far less widely than at the time of publication, the description is clear enough, as is 'digital computer', unlike the parenthetical model number.

3.6.3 Research on Human Beings and Animals

If the research involved the use of human beings or animals as subjects or the use of human tissue, then as a prerequisite for the research, ethical approval would normally have been obtained from a relevant research ethics committee or institutional review board (§12.4). Authors are expected to report that approval, customarily in the methods section, with details of the source of approval.

For research with human subjects, a statement is also expected indicating that informed consent was obtained from each participant. Failure to comply can lead to retraction of the published article and more serious consequences (§12.3).

For clinical research with human subjects, a statement is expected [303] that the research complied with the Code of Ethics of the World Medical Association (Declaration of Helsinki) [383] or with guidance from the equivalent national body [339, ch. 3]. For a clinical trial, a statement is also expected [303] that the trial was appropriately registered (§12.4.1).

For research with animals, a statement is expected that refers to the appropriate European Commission Directive [222], or the equivalent national directive (§12.4.2). Comprehensive advice on reporting in vivo experiments with animals is available in the ARRIVE guidelines [204].

For research with human subjects, eligibility and exclusion criteria may need to be reported, along with any other major demographic characteristics. With animals, the genus, species, and other details may be required [293, §2.06].

The journal's guide for authors usually explains which statements are required and their format.

3.6.4 Sampling Procedures

If the research entailed testing the effect of some treatment or procedure on a random sample of materials, human subjects, or animals, then the sample size may need to be justified [126] [244, §2.7]. The conventional way is to perform a statistical power calculation, which determines whether a given sample size is (or was) big enough to reveal an effect of a certain size (§7.9.3).

In practice, external factors may set a limit on sample sizes, for instance, the scarcity or cost of test materials. Still, reporting an effect size estimate allows the reader to judge whether a failure to find a reliable effect was due to the treatment itself or to the sample size being too small [244, §3.14] [270]. Care is required, since effect sizes can be misreported [227]. Other aspects of statistical reporting are considered in (§7.9).

3.6.5 Units

Quantities that have dimensions need to be expressed as the product of a number and a unit. Units are particular dimensioned quantities that act as a reference. They are normally drawn from the International System of Units [190], otherwise known as SI units (after the French *Système international d'unités*), although non-SI units may be conventional in some applications. Base units such as mass, length, and time are distinguished from derived units such as magnetic field strength and luminous flux, which can be expressed as products of powers of base units.

Importantly, the symbols for units are not abbreviations but mathematical objects [190, §5.3]. They are therefore unaltered in the plural, so that '1 cd' and '10 cd' stand for 1 and 10 candelas, respectively, and they do not take a full stop. They should always be printed in roman (upright) type, independent of the context, and in lower case unless they are derived from a person's name, when the initial letter is capitalized; for example, Pa for the unit of pressure, pascal, or Hz for the unit of frequency, hertz [190, §5.1]. The full name does not take an initial capital, except at the beginning of a sentence or in a capitalized title. Units should never be printed in italic or sloping type, as occasionally happens in documents prepared with a mathematical mark-up language (§6.4.2).

The product or quotient of two unit symbols may be indicated in any of these ways: '$\mathrm{rad\ s^{-1}}$', '$\mathrm{rad{\cdot}s^{-1}}$', 'rad/s', where rad is the symbol for radian. The space in '$\mathrm{rad\ s^{-1}}$' and the centred dot in '$\mathrm{rad{\cdot}s^{-1}}$' indicate multiplication and the solidus '/' in 'rad/s' indicates division [190]. When three or more symbols are combined, however, the solidus can lead to ambiguity, as in Example 3.65.

✗

> Achievable impedances range from 0 to 60 Nm/rad and 0 to 1 Nm/rad/s for pronation/supination.
>
> Ambiguous

Example 3.65. *Ambiguous combination of unit symbols, from a biomedical engineering article [568].*

The composition Nm/rad is unambiguous. But Nm/rad/s could represent the product of N and m divided by rad divided by s, so that N and m are actually multiplied by s, or the product of N and m divided by rad and then divided by s. Unsurprisingly the appearance of the solidus more than once in an expression is deprecated in SI units [190, §5.1]. If the solidus is still preferred, then the ambiguity can be removed by adding parentheses, as in Example 3.66.

> c_s (Nm/(rad/s)) is the inner damping coefficient of the shaft.

Example 3.66. *Unambiguous but complicated combination of unit symbols, from a control systems engineering article [627].*

Nevertheless, multiple parentheses make parsing visually difficult. The clarity achieved by avoiding the solidus and combining symbols with spaces and exponents is evident in Example 3.67 (a missing space has been inserted between N and m).

> The control value of viscous friction was set at 4.2×10^{-3} N m rad^{-1} s.

Example 3.67. *Unambiguous combination of unit symbols, from a physiology article [468].*

With just pairs of symbols there is little to choose between ‘rad s^{-1}’, ‘rad·s^{-1}’, and ‘rad/s’, unless it is to maintain consistency with usage elsewhere in the exposition.

Notice the space between the number and the unit in Example 3.65 and Example 3.67 (there should also be a space between N and m in Example 3.65 and 3.66). As with the combination of symbols, the space indicates multiplication [190, §5.3.3]; thus ‘4.2×10^{-3} N m rad^{-1}s’ means 4.2×10^{-3} times (one) N m rad^{-1} s. The only exception to this convention is where the unit symbol is for a degree, minute, or second for planar angle, that is, °, ′, and ″, respectively. Thus for planar angle, you would write 30° 22′ 8″ (no space between number and following symbol), but for temperature in degrees Celsius 30.2 °C (space between number and following symbol). It would be incorrect to write 30.2°C (no space) and 30.2° C (space before C) [190, §5.3.3].

Unlike temperature in degrees Celsius, thermodynamic temperature in kelvin, symbol K, does not take a degree sign. Example 3.68 is therefore incorrect.

✕

Incorrect degree sign

an extremely wide temperature range spanning from 4 °K to more than 1000 °K

Example 3.68. *Incorrect unit, from an instrumentation engineering article [715].*

The degree sign should be omitted from °K [190, §2.1.1.5].

3.6.6 Transforming Units

It is sometimes useful to transform numerical data. The intention may be to better represent the underlying physical variable, or to remove asymmetry in the distribution of values, or to stabilize the variance so that the standard deviation of the measurements is approximately independent of the mean [168] [244] [309]. Nonetheless, the effect of the transformation on any dimensioned quantities needs to be considered.

The unit of viscous friction in Example 3.67 has the form of a monomial with exponents that are real numbers. As explained in Section 3.6.5, units represent particular dimensioned quantities. They can be multiplied by a scalar and added to other expressions with the same dimensions and still represent dimensioned quantities [246] [337, §2.4].

By contrast, the logarithm of an expression representing a dimensioned quantity cannot itself normally represent a dimensioned quantity. Functions such as the logarithm, the exponential function, and the trigonometric functions need to take dimensionless arguments [246] [337, §2.4]. Thus the logarithmic unit for ratio quantities, the decibel, is defined with respect to a reference level of the dimensioned quantity being measured [190, §4.1]. That is, m dB means that there is a level X of the quantity and a reference level X_0 such that $\log X/X_0 = m/10$.

Yet taking the logarithm of a dimensioned quantity may be essential for the purposes of analysis. In that situation, therefore, logarithmic measures of a dimensioned quantity may be reported in logarithmic units, for example, concentration in units of $\log_2$ μg/J, on the understanding that the quantity reported is the logarithm to the base 2 of the concentration in μg/J divided by 1 μg/J [53, Fig. 4.33].

3.6.7 Combining Numbers and Units

When dimensioned quantities are brought together in the text with a mathematical operation or a linking word, unit symbols should be attached to each quantity. The proper and improper combinations of dimensioned quantities shown in Example 3.69 are from the National Institute of Standards and Technology (NIST):

Proper	Improper
35 cm × 48 cm	35 × 48 cm
1 MHz to 10 MHz	1 to 10 MHz
20 °C to 30 °C	20 to 30 °C
123 g ± 2 g	123 ± 2 g
70 % ± 5 %	70 ± 5 %

Example 3.69. *Proper and improper combinations of dimensioned quantities, from the NIST checklist [285].*

These rules may be relaxed when a range is expressed informally within a piece of text, as in Example 3.70.

> The ångström is widely used by x-ray crystallographers and structural chemists because all chemical bonds lie in the range 1 to 3 ångströms.

Example 3.70. *Informal combination of dimensioned quantities, from the SI Brochure [190, p. 127].*

In practice, authors and publishers' typesetters (who may alter the author's copy) vary in their adherence to these conventions. As with grammar, some lapses are less disruptive than others, their impact depending on the context and the extent to which a symbol is bound to the preceding number.

In Example 3.71, the expression '$0.15\ \mathrm{g{\cdot}cm^{-3}} \leq \rho \leq 0.7\ \mathrm{g{\cdot}cm^{-3}}$' follows NIST guidance exactly.

> The mass density of a range of comet nuclei ($0.15\ \mathrm{g{\cdot}cm^{-3}} \leq \rho \leq 0.7\ \mathrm{g{\cdot}cm^{-3}}$) indicate[s] that the comets described above are typical.

Example 3.71. *Proper combination of dimensioned quantities, from an applied physical sciences article [729].*

In Example 3.72, the expression '750–1500 m' is unambiguous, since the span 750–1500 can be seen as a single mathematical object and there is a space between it and the unit m.

> Towers for proposed full-scale facilities, such as the Ciudad Real Torre Solar in Spain [2,8], are much taller (750–1500 m).

Example 3.72. *Unambiguous combination of dimensioned quantities, from an environmental engineering article [445].*

In Example 3.73, the expression 'from 3,800 to 4,700 K' violates NIST guidance but is compatible with usage in the SI Brochure [190], as shown in Example 3.70. With a space before it, the unit K acts in a distributed way. Even so, adding K after 3,800 might be better.

> We performed the same calculations for a range of CMB [core–mantle boundary] temperatures from 3,800 to 4,700 K.

Example 3.73. *Informal combination of dimensioned quantities, from a geosciences article [408].*

Symbols that are not separated from the number by a space—such as the symbols for degree, minute, and second—are tightly bound to the preceding number, and therefore must be repeated [49, §9.17]. Thus in Example 3.74, the degree sign in the expression '0–80°' applies only to preceding numeral 80. The range needs to be written '0°–80°'. Likewise the range '100–180°' needs to be written '100°–180°'.

✗

> The range of epicentral distances is 0–80° for *ScS* [waves] and 100–180° for *SKKS* [waves].

Missing symbols

Example 3.74. *Improper combination of dimensioned quantities, from a geosciences article [669].*

Percentage symbols can cause a similar problem, depending on whether they are set with an intervening space or not. The SI Brochure recommends an intervening space [190, §5.3.7]; the *Chicago manual of style* [49, §9.18] does not. Most journals omit the space.

In Example 3.75, where there is no intervening space, the expression '0.5–5%' is incorrect. The percentage sign applies only to the adjacent numeral 5, with the result that the reader fleetingly interprets the preceding number 0.5 as an efficiency of 50%. The incorrect combination of 'of' and the en-dash to describe the range adds a further complication, as explained in Section 7.3.5.

✕

> The efficiency for a typical installation considered in the literature . . . is estimated to be in the range of 0.5–5% [4]. — Missing symbol

Example 3.75. *Improper combination of dimensioned quantities, from an environmental engineering article [445].*

According to both the NIST checklist [285] and the *Chicago manual of style* [49, §9.17], the expression should be written as either 'range of 0.5% to 5%' or 'range 0.5%–5%'.

3.6.8 Statistical Analyses

If the results section contains significant statistical analysis, and especially if more than elementary statistical methods or if specialist software routines are used, then a brief description of the approach to the analysis is normally included, usually at the end of the methods section [79, pp. 65–6].

Example 3.76 is from a methods section of a medical sciences article and illustrates typical phrasing.

> Data are presented as mean ± SEM. Survival statistics were calculated by the Mantel–Cox test. Serum parameters, tissue macrophage expression levels, apoptosis levels, and mRNA transcript expression were evaluated by one-way ANOVA with Bonferroni's posttest.

Example 3.76. *Brief description of statistical methods, from a medical sciences article [521].*

Other aspects of reporting statistical material are described in Chapter 7.

3.7 Results

The purpose of the results section in a paper is to summarize and provide an analysis of the data collected from an experiment or a theoretical or computational study. The results section should not be used to introduce methods omitted from the methods section [79, p. 69].

Although choices need to be made about data reduction and statistical inference, any interpretation of the results is normally reserved for the discussion [258] [293, §2.07], except in some elite journals, which take a different view [89]. The main reason for postponing the interpretation is that it allows the reader to identify what is, in principle, purely factual—the experimental outcome—independent of how it might support the author's thesis. In the long run, it may well be the results that prove more useful to the research community.

For reports of multiple experiments, however, some intermediate interpretation of the results is often introduced so that the motivation for each succeeding experiment is evident. Each individual results section needs then to be retitled as Results and Comment (or Results and Discussion). The larger implications of the work can be addressed in a general discussion at the end of the paper.

3.7.1 Fundamentals

The results section usually contains these elements:

- overview, then more detail, sufficient to support any interpretation
- other relevant findings, including unexpected ones
- statistical analysis, if appropriate
- a reference to the availability of raw data, if appropriate

Not every result from every experimental condition need be included [31] [79]. Repetitive material may be summarized or relegated to an appendix or made available separately in a supplementary document, as explained in Section 3.11. Of course, results should not be relegated or, more seriously, omitted because they run contrary to expectations [293, §2.07].

In some journals, results may be introduced by other headings, such as Performance Evaluation [478], Generalized Measurements [616], Examples [512], and Quantitative Evaluation [555]. They may also appear within sections addressed to specific research questions; for example, 'Crack Detection in a Simply Supported Beam' and 'Crack Detection in a Cantilever Beam' [552]. A decision of this kind should be guided by the practice in the journal.

The extent and type of statistical analysis varies markedly with the research discipline, most notably between the physical sciences [19], medicine [8], and social sciences [175], and, within the physical sciences, between physics and biology [96]. But whatever the discipline, some indication of the accuracy of reported values must be included. The reader should be able to judge the repeatability of the data, and, where germane, assess any differences in the data under different experimental conditions or the closeness of the data to some theoretical model. See Chapter 7 for further discussion.

3.7.2 Types of Openings

In order to orient the reader, it is helpful to start with a summary statement, after which the results can be developed in more detail, as in Examples 3.77–3.80, drawn from engineering, epidemiology, experimental psychology, and civil engineering, respectively.

> Hydraulic fracture in horizontal wells creates a network of cracks in rock that was previously impermeable, allowing gas to move. The true geometry is very complicated.

Example 3.77. *From general to particular results, from an engineering article [637].*

> Of 7·6 million children who died in the first 5 years of their life in 2010,[1] 64·0% (4·879 million) died of infectious causes (table 1). Of all infectious disorders, pneumonia, diarrhoea, and malaria were the leading causes of death worldwide (figure 2).

Example 3.78. *From general to particular results, from an epidemiology article [590]. The superscript numeral in the first sentence is a citation to a reference.*

> Table 1 provides frequencies of the most commonly used statistical analyses for each year; corresponding percentages are illustrated in Figure 1. . . . Analysis of variance (ANOVA) was reported in most articles, 83% overall, followed by *t* tests, 66% overall; these were often used . . .

Example 3.79. *From general to particular results, from an experimental psychology article [126].*

> *5.1. Uncorrected RCM simulations*
>
> There were substantial differences in the ability of RCMs [regional climate models] to reproduce temperature and precipitation data under current climate conditions, as was demonstrated earlier (Teutschbein and Seibert, 2010). The biases were found to vary substantially . . .
>
> *5.2. Statistical evaluation of bias correction procedures*
>
> According to the calculated performance measures, all temperature-bias corrections improved the raw RCM simulations. Unlike for the raw RCM temperature, a bias in mean temperature could . . .

Example 3.80. *Multi-part results, from a civil engineering article [694].*

Details of methods should not be introduced as results (§3.7, introduction), as is done in Example 3.81.

×

V. EXPERIMENTAL RESULTS

Experimental tests were conducted on a laboratory setup of a 1.5 kW DFIG system, as shown in Fig. 7. . . . A controllable three-phase power supply was built up to simulate the unbalanced grid voltage conditions [26].

Example 3.81. *Misplaced methods, from an electrical engineering article [625].*

In fact, the article contained no section with the heading 'Methods' or its equivalent. If the intention was to ensure the tight integration of methods and results, then a less restrictive heading 'Experiment' could have been used in place of 'Experimental Results'.

3.7.3 Summarizing Data

Data can be summarized numerically within the text (ch. 7) and graphically (ch. 8). Data can also be presented in tabular form. Although tables are easy to construct, they are not always easy to interpret, and graphs should generally be preferred [31] [45, §4.2], unless reporting numerical values is essential to the outcome of the work, for instance, expressing definitive values of a physical quantity.

Duplicate presentation—tabular and graphical presentation of the same data—is unlikely to be necessary [258, p. 120]. If large tables do need to be included, they can be presented in appendices or offered as supplementary materials (§3.11).

The publisher's typesetter normally formats any tables, but you do need to supply captions, usually in the same format as for figures. Abbreviations and special conditions applying to the headings or entries in a table can be defined in footnotes to the table.

3.8 Discussion and Conclusion

The purpose of the discussion in the paper is to say what the results mean [258]. It should therefore not be written in a perfunctory manner, duplicating the results section or rambling through a review of the literature. Setting out the significance of your findings for the field as a whole requires thought. As Day points out, because this section is less well defined than the others, it can be the hardest to write, and written badly, it can end in rejection [79, p. 72].

As the difference between the discussion and what may be nominated as the conclusion is not always sharp, the two are here taken together.

3.8.1 Fundamentals

The discussion generally contains the following elements [79] [258] [293, §2.08], not necessarily in this order:

- a reminder of the original research question or problem posed
- an objective review of the results
- immediate implications
- relevance to other findings in the literature
- limitations of the results and possible failings
- as a conclusion, new knowledge, research questions, or longer-term implications

It should not include any new results or methodological details that were omitted earlier. It should also not include descriptions of your work in progress or your research plans [31]. But as the preceding list implies, it can include suggestions for future research within the field. Thus in a particularly authoritative research article or literature review, the discussion might identify questions to be addressed or potential new research areas. Even so, it is rare that predictions of this kind have a major influence on research direction.[23]

3.8.2 Reminder of the Research Question

A common way to start the discussion is to remind the reader, indirectly, about the purpose or context of the work. You might refer to the original research question or problem or challenge, although not as if the reader were encountering it for the first time.

Example 3.82 is from the opening statement of an article's concluding remarks. It mentions the problem of chloride ingress into concrete, but emphasizes more its features—that is, its complexity—instead of its existence and importance, which were developed in the article's introduction.

> Chloride ingress into concrete is a complex process, which in many environments is further complicated by the temperature cycles and wet-dry cycles experienced by the reinforced concrete structures.

Example 3.82. *Opening a discussion with a reminder, from a civil engineering article [671].*

Example 3.83 is from the opening statement of another article's discussion and illustrates a more indirect approach.

> Photoreceptor transplantation has the potential to restore vision following retinal degeneration.

Example 3.83. *Opening a discussion with a reminder, from a neuroscience article [413].*

The statement concentrates on the solution rather than on the problem—the loss in vision as a consequence of retinal degeneration—yet contains enough to remind the reader of the problem described in the article's introduction (Example 3.35).

Example 3.84 is analogous to Example 3.83.

> The ability of brown fat to suppress obesity through increased energy expenditure has caused an explosion of interest in the development and function of brown adipocytes.

Example 3.84. *Opening a discussion with a reminder, from a cell biology article [719].*

In the article's introduction, the authors explain how brown fat cells or brown adipocytes produce heat, counteracting among other things obesity. In the opening statement of the discussion, the reader is reminded of that fact, but with the emphasis now on the increasing research interest.

3.8.3 Objective Review of Results

A discussion should offer an objective review of the results and a consideration of their implications, as in Example 3.85.

Reminder of approach

> Catalogues of somatic mutation from 21 breast cancers have yielded several insights into underlying mutational mechanisms. Five independent single-nucleotide substitution processes appear to have been operating . . .
>
> Signature A is likely mediated by deamination of 5-methylcytosine . . .

Findings and implications

Example 3.85. *Implications of results, from a cell biology article [626], a continuation of Example 3.38.*

The review starts with an orienting statement, a reminder of the authors' use of catalogues of somatic mutation from 21 breast cancers. The main findings and their interpretation are then developed. Importantly, the review does not constitute a recapitulation.

A similar approach is adopted in Example 3.86 and Example 3.87.

Reminder of approach

The structure described here for the human cardiac myosin filament in the normal (undiseased) state is an essential starting point from which to understand the mechanisms of various cardiac diseases . . . The current analysis shows a substantial improvement in the resolution and level of detail . . . and allows detailed modeling of myosin heads and accessory proteins and their interactions.

Findings and implications

Example 3.86. *Implications of results, from a medical sciences article [397].*

Reminder of approach

Heated amorphous carbon films display a peculiar behavior of a negative Volta potential at the place of electron beam irradiation. This behavior is responsible for the reduced performance with age of thin film ZPPs [Zernike phase plates]. The newly proposed VPP [Volta phase plate] uses this behavior constructively to generate a phase shift with the same sign and similar magnitude to that of a ZPP. . . . The VPP solves two major issues of the ZPP.

Findings and implications

Example 3.87. *Implications of results, from an applied physical sciences article [463].*

By contrast, in Example 3.88, the authors turn directly to their main findings, since the subject matter is clear: the functionality of an established database, ZINC, whose name appears in the article's title 'ZINC: a Free Tool to Discover Chemistry for Biology'. Nevertheless, the reader is quickly reminded of its meaning: a 'database of commercially available compounds for ligand discovery'.

Three main results emerge from this work. First, ZINC represents a much larger database of commercially available compounds for ligand discovery . . . The database is actively maintained and is freely available online (http://zinc.docking.org). Second, ZINC molecules . . .

Example 3.88. *Implications of results, from a chemistry article [548].*

With all these articles, the reader interested only in learning about the outcome of the research and its implications can turn directly to the discussion without having to read the introduction for context.

3.8.4 Relevance to Other Findings

In interpreting your results, due recognition should be given to the work of others, usually in the form of citations, as in Examples 3.89–3.91, which are from cell biology, physics, and civil engineering, respectively. The formatting of citations is reviewed in Section 3.10.

Relationship to other work

Regional clusters of mutations in cancer have occasionally been observed in experimental models, although not at the mutation density observed here (Wang et al., 2007). . . . Furthermore, they are closely associated with regions of rearrangement and occur on the same chromosome and chromosomal strand over long genomic distances, suggesting that they occur simultaneously or in a processive manner over a short time span (Chen et al., 2011).

Relationship to other work

Example 3.89. *Relationship with other findings, from a cell biology article [626], a continuation of Examples 3.38 and 3.85.*

Relationships to other work

Previous experimental studies of the Leggett–Garg inequality have used ensembles (15, 16), have made assumptions regarding process stationarity (17, 18), or have required weak measurements (4) to draw conclusions . . .

Example 3.90. *Relationship with other findings, from a physics article [517].*

Implications and relationships to other work

More improvements could be made to . . . better represent or simulate the field behavior of reinforced concrete structures. This could be accomplished by taking into account the time-dependency of transport properties of concrete [16,197], the repair or replacement of concrete cover [198], the corrosion propagation [120], the chloride penetration mechanisms other than diffusion (e.g., wicking) [179], the structure geometry [186], . . .

Example 3.91. *Relationship with other findings, from a civil engineering article [671], a continuation of Example 3.82.*

With these connections to other studies, their priority is recognized, albeit sometimes with caveats; new results are set in context; and support is provided for any inferences.

3.8.5 Limitations

Having identified the strengths of your work, you need also to identify its limitations and possible failings [79] [258, pp. 39–40]. Evidence of objectivity helps strengthen, not weaken, the impact of your research (§2.3.6).

Example 3.92 draws attention to theoretical modelling—the simulations in Example 3.91—being limited by the availability of reliable data, specifically of certain coefficient estimates, and it admits the need for more research.

Uncertain source data: Currently, there is a considerable amount of variability in determining chloride diffusion coefficients as an indicator of the durability of both conventional and unconventional concrete structures. In particular, the chloride diffusivity depends on the concrete pore structure and on all the factors that determine it . . .

More work needed: More testing and research are warranted.

Example 3.92. *Limitations of findings, from a civil engineering article [671], a continuation of Examples 3.82 and 3.91.*

Example 3.93 is from the closing statement of an article's conclusion. The authors suggested that with real rather than synthetic data the performance of their algorithm would have been better [689]. It seems not to have affected the impact of the work.[24]

Real data needed: One of the primary limitations of this work was the use of synthetic training data. This limited both the performance of the system and the range of algorithms

Special examples needed: . . . In addition, a set of labeled training examples would enable different types . . .

Example 3.93. *Limitations of findings, from a computer science article [689].*

Example 3.94 contains a caution about the larger health and social implications of the research.

Uncertain promise: The therapeutic potential of activating brown-fat-mediated thermogenesis in human has yet to be fulfilled.

Unsuccessful human trials: Trials of drugs that increase the β-adrenergic activation of BAT have not been successful in humans due to either lack of efficacy or to intolerable side effects resulting from activation of β-adrenergic receptors in other tissues (Collins et al., 2004; Whittle et al., 2011).

More work needed: Clearly, there is a need to develop more specific means to activate brown fat in humans in more specific ways.

Example 3.94. *Limitations of findings, from a cell biology article [719], a continuation of Example 3.84.*

Some journals formalize the reporting of strengths and weaknesses with appropriately headed subsections. Example 3.95, an extension of Example 2.48, is from a subsection headed 'Strengths and Limitations', though only limitations are reported.

> One of the main limitations HYVET extension was that only participants who were on double blind treatment were enrolled from the main trial. . . . Although participants who had received active treatment during the main trial had similar levels of diabetes, smoking, and cholesterol, they differed from those who had received placebo in that they had higher blood pressures and concentrations of uric acid at the start of the extension period. . . . Another limitation is the short length of follow-up in the extension. . . . Finally, given the small number of events in the extension, caution must be exercised in drawing firm conclusions simply from the extension.

Problems with experimental design | Comparison problem | Limited follow-up | Insufficient numbers

Example 3.95. *Limitations of findings, from a general and internal medicine article [415].*

In being careful not to claim too much, however, you can go too far in expressing caution. The effect may then be the opposite of that intended [258, p. 40]. The use of 'it is possible that' and its variants rarely (never?) needs further qualification. In Example 3.96, either of the qualifications 'it is possible that' and 'may have been' could have been omitted without increasing the authors' exposure. Hedging of this kind is discussed in Section 2.3.4.

✗

> While the review was conducted in a rigorous, systematic manner, it is possible that some studies may have been missed by not searching the 'grey' literature.

First hedge | Second hedge

Example 3.96. *Over-hedged findings, from an ophthalmology article [550].*

3.8.6 Conclusion

The closing paragraph of the discussion or a separate conclusion should complement the introduction. As Montgomery suggests [258, pp. 90–1], it is an opportunity to return to the material beginning the paper and state what new knowledge has been gained or what new research questions have been identified.

If your exposition follows an hourglass structure (§3.5.1), then having dealt with the specific in the main part of the discussion, you can expand outwards to the general in the conclusion. Although it is usually true, the phrase 'much work remains to be done' and its equivalents (e.g. Example 3.92) should not be overworked. To avoid confusion, the use of language in the conclusion (and discussion) should be consistent with that in the title, abstract, introduction, and keywords.

In Example 3.97, which comes from a review, the conclusion returns explicitly to the material in the introduction: it 'closes the loop' [219, p. 199]. The exposition is marred only slightly by the ambiguous 'the former' in 'While technology for the former is at a better developed stage' (§5.6.22).

Link back to introduction	We end where we began. Whether water splitting is achieved by electrolysis using electricity generated by photovoltaic devices or dye-sensitized solar cells, or by artificial photosynthesis remains at this point unclear.	Uncertainties remain
Implications	While technology for the former is at a better developed stage at this point, an artificial photosynthetic system for water splitting has promise of greater simplicity and efficiency. The durability, scalability, cost and efficiency will in the long run decide which approach to water splitting and hydrogen generation will win this competition.	

Example 3.97. *Concluding statements, from a chemistry article [485].*

In Example 3.98, several paradigms are linked to the authors' work on spin ice, with suggestions for new experiments.

Link back to introduction	Spin ice offers a realization of several paradigmatic concepts in nonequilibrium dynamics: dimer absorption, Coulombic reaction-diffusion physics, and kinetically constrained slow dynamics. . . .	
	The additional availability of a range of experimental probes thus allows broad and detailed experimental studies.	New work

Example 3.98. *Concluding statements, from a physics article [620].*

In Example 3.99, the conclusion focuses on a particular finding— the treatment of cardiac arrhythmias. Notwithstanding the authors' cautionary 'Additional studies are necessary', the work proved prescient and influential.

Link back to introduction — The successful ablation of foci was safely achieved, and the procedure eliminated previously frequent atrial fibrillation without a need for antiarrhythmic drugs in 84 percent of the patients during hospital monitoring and in 62 percent after discharge. . . . Additional studies are necessary to evaluate the role of ablation of triggering ectopic foci in preventing atrial fibrillation, as well as the safety of ablation in the pulmonary veins, before widespread application of ablation for the treatment of this common cardiac-rhythm disturbance can be recommended. — Cautionary note

Example 3.99. *Concluding statements, from a general and internal medicine article [530], a continuation of Example 3.43.*

All three examples complement the material in the introduction, in both style and content.

A conclusion also affords the opportunity to make an argument more forcibly than would be appropriate elsewhere, as in Example 3.100. It begins with an appeal.

Appeal — This paper is a plea for significance when using fuzzy sets as a modelling tool. As such, a fuzzy set is an abstract mathematical entity and its proper use in practice needs a proper interpretive framework. — Recommendation

Example 3.100. *Concluding statements, from an applied mathematics article [486].*

Still, as with anticipating future research, statements of this kind are most effective when they are associated with work that is recognized within the field, as this article is [486].

3.8.7 Hostages to Fortune

In focusing on new research questions or problems to address, a conclusion may go further than is necessary or desirable. In both Examples 3.101 and 3.102, the authors set out several specific goals.

First goal

✕

Future work <u>will include improving</u> the preprocessing steps such as RF inhomogeneity and partial volume effect corrections. . . . Finally, we <u>plan to investigate</u> the three-dimensional extension of the proposed method in order to measure more precisely the lesion volume. — Indirect link

Example 3.101. *Plans for further work, from a computational biomedicine article [421].*

First goal

✗

Additionally, in our future work, we will extend our model to incorporate more realistic physical effects, such as diffractions, refractions, and attenuations . . . We will expand the detection procedure over multiple range cells to consider other significant multipath reflections . . . As polarization allows identification of correlated source signals (e.g., multipath) with small angle separation [49], we will also study the detector performance employing polarized transceivers. We will integrate our detection procedure with a target tracking algorithm and explore other criteria . . . to optimally design the transmit waveform to improve the system performance. We will validate the performance of our proposed detector with real data.

Second goal — Third goal — Fourth goal — Fifth goal

Example 3.102. *Plans for further work, from an electrical and electronic engineering article [668].*

These goals may have been or may be achieved in the future. But with no other sources to draw on, it is not clear what the reader should infer at the time.

The general advice is that making a public statement about your research plans is risky and may inadvertently inhibit other researchers [31, p. 20]. If readers are also working in one or more of these areas, should they now suspend that activity or accelerate it?

3.8.8 Failed Conclusions

A conclusion can say too little. Although it may contain a restatement of the main points of the paper, it should be more than a mechanical summary of what was done [290, §VII].

Example 3.103 is from a section of an article under the heading 'Conclusions'.

Summary

✗

In this paper, we investigated the transmission completion time minimization problem in an energy harvesting communication system. We considered two different scenarios . . . We first analyzed the structural properties of the optimal transmission policy, and then developed an algorithm . . . in each scenario.

Summary — Summary — Summary

Example 3.103. *Summary not conclusion, from an electrical and electronic engineering article [725].*

In fact, it contains nothing about the new knowledge that was gained, the new research questions or problems identified, the importance of the work, or its applications and extensions [290, §VII].

Example 3.104 is taken from the final section of an article under the heading 'Conclusion and Outlook'.

✗

Experimental OSNR, EVM, and BER data were compared to analytical relations and to direct numerical simulations.	— Summary
They all showed good agreement within the specified limits.	— Summary

Example 3.104. *Summary not conclusion, from an electrical and electronic engineering article [665].*

It consists of a summary of what was done and the main experimental result 'They all showed good agreement'. Whereas the result might be regarded as a conclusion of sorts, there are no implications, no outlook for future research. Also, the multiple abbreviations are unlikely to help a reader whose first action is to inspect the conclusion for the key findings.

3.9 Acknowledgements and Declarations

At the end of the main text and before the references, it is conventional to acknowledge any funding support for your work, together with any technical help, advice, materials or other resources, intellectual contributions by colleagues, and presubmission critiques of the manuscript [68]. Some journals require a declaration by the authors about their separate contributions to the reported work (§12.6.3).

3.9.1 Acknowledgements

Acknowledging financial support is straightforward. A statement identifying grants and contracts is usually sufficient to satisfy the requirements of the funding agencies and it informs the reader of any potential conflicts of interest (§12.6.5). It can often take the simple form 'This work was supported by . . .'.

Acknowledging technical support, intellectual contributions, and criticism of the manuscript by colleagues is more complicated, since it touches on the question of authorship, which is considered in Section 12.6.

Providing that the contribution does not in itself warrant authorship, it should be acknowledged, but only if the contribution is significant. Crucially, any individuals you wish to acknowledge need to agree with being acknowledged and with the form of the acknowledgement. Although happy to offer advice, they may not want to be publicly associated with the work in case that suggests that they approve or endorse it [79, ch. 14]. Their agreement to be acknowledged should be obtained in some permanent form [303, §II.A].

Most acknowledgements are brief and free of emotion, as in Examples 3.105 and 3.106.

> The authors thank members of the C.B.T. laboratory for critical review of this manuscript and data contained herein; Nicholas Socci for analysis of deep-sequencing data; and Jidong Liu, Zissimos Mourelatos, and Gideon Dreyfuss for sharing reagents. This work was funded in part by grants from the National Cancer Institute, the Abramson Family Cancer Research Institute, and the Memorial Sloan-Kettering Cancer Center.

Example 3.105. *Acknowledgement of manuscript review, technical support, and funding, from a cell biology article [632].*

> We thank Uri Yechiali for suggesting the problem of M/M/k queues with heterogeneous servers and the solution of the M/M/2 problem. The research of E.S. was partially supported by Israel Science Foundation (ISF) Grant 212/09 and by the Google Interuniversity Center for Electronic Markets and Auctions.

Example 3.106. *Acknowledgement of ideas and funding, from an applied mathematics article [502].*

For longer works, such as master's and doctoral theses, technical reports, and books, the acknowledgements may appear before the introductory chapter. Here they can occupy a page or more and be expressed in less restrained language.

Nonstandard acknowledgements are occasionally accepted by journal editors. Example 3.107 is from an article where an author C. M. Brown, identified by his initials, poses a question at the end.

> We thank P. Hews for his many fossil discoveries in Alberta, including the holotype of *Regaliceratops*. D. Tanke, J. McCabe, and D. Lloyd assisted in the excavation and collection of the specimen. . . .
>
> C.M.B. would specifically like to highlight the ongoing and unwavering support of Lorna O'Brien. Lorna, will you marry me?

Example 3.107. *Acknowledgement with an unusual addition, from a palaeontology article [430].*

The answer was apparently yes [411].

3.9.2 Author Contributions, Competing Interests, Data Sharing

Along with details of any ethical approval for your research and sources of research funding, you may be required to report on the roles of the individual authors, any conflicts of interest, and the availability of the raw data underpinning the published results (§12.6, §12.7).

Example 3.108 is from a general and internal medicine article under the heading 'Contributors', where they are identified by their initials (the individual authors are Rachel Cooper, Bjørn Heine Strand, Rebecca Hardy, Kushang V. Patel, and Diana Kuh).

> All authors contributed to the conception and design of the study; RC and BHS undertook statistical analyses; RC drafted the article and all authors contributed to its critical revision and provided final approval of the version to be published. RC is guarantor. All authors, external and internal, had full access to all of the data (including statistical reports and tables) in the study and can take responsibility for the integrity of the data and the accuracy of the data analysis.

Example 3.108. *Author contributions, from a general and internal medicine article [455].*

Even if the journal does not require this information, you can still incorporate it in other sections of the paper.

3.10 References

Research articles almost always contain references to previously published work.[25] Indeed some journals and open access archives refuse to accept submissions without references. Among the many reasons for references, the most important and principled one is to give due credit for theories and data, ideas, procedures, results, and interpretations, and the like drawn from previously published work [262, ch. 2] [314]. That work may include your own. Other reasons for referencing are to establish the environment in which your research should be seen and which areas it agrees with and which not [258, pp. 92–3]. References should be chosen to reflect activity in the field, without bias, and with due regard to priority (§12.5).

The only exceptions to the requirement to supply references are those theories and data that have become widely known and accepted [158] [314], such as the standard model in particle physics or the method of images in electrical engineering.

In giving due recognition to the work of others (and authors can take offence about a missing citation), it is unsafe to take a scattergun approach to citing in the hope that at least some of the references will prove to be appropriate [79]. Excessive and random referencing can be misinterpreted by the reader.

Confusingly, the terms citation and reference may be used interchangeably. In this guide, citations are the indicators in the text that identify particular references, and references are the publication data gathered together in a list at or near the end of an article, usually under an obvious heading. The publication data must contain sufficient detail to allow the reader to identify and locate the original work [348]. What constitutes sufficient detail for a reference is specified by the journal, conference, or other publication in which the article appears.

The two main referencing systems—numeric and author-date—handle citations and references in fundamentally different ways, but there are also many variations within and outside these systems [68] [262]. Reference management software systems can deal efficiently with the matters of ordering and formatting citations and references in all their complexity.[26]

3.10.1 Numeric System

In the numeric system, the citation takes the form of a superscript number or a number in line with the text and set in round parentheses or square brackets. Each new citation is given a new number, in ascending order. When the same work is cited again, the original number is used. In the list of references, the references are ordered and numbered to correspond with the order of citation in the text. Examples of the numeric systems are the Vancouver system[27] and the closely related IEEE system [179] from the Institute of Electrical and Electronics Engineers.

Example 3.109 illustrates citing within the IEEE system, where square brackets are used.

> To address this issue, standard discriminative approaches . . . have been extended to the multi-label case [1]. . . . While efforts have been made to provide data sets with this information [2] . . .
>
> A preliminary version of this work was published in [17].

Example 3.109. *Citing according to the IEEE numeric system, from a computer science article [435].*

The first two citations '[1]' and '[2]' behave grammatically like parenthetical phrases, whereas the last '[17]' behaves like a noun. There is no need to prefix a citation with 'Reference' or 'Ref.' except at the beginning of a sentence [290, p. 5].

Example 3.110 shows part of the corresponding list of references in the IEEE system, in non-alphabetical order as expected.

[1] H. Wang, C. Ding, and H. Huang, "Multi-label linear discriminant analysis," in *Proc. 11th Eur. Conf. Comput. Vis.*, 2010, pp. 126–139.

[2] M. Everingham, L. Van Gool, C. K. I. Williams, J. Winn, and A. Zisserman, "The Pascal visual object classes (VOC) challenge," *Int. J. Comput. Vis.*, vol. 88, no. 2, pp. 303–338, 2010.

...

[17] R. S. Cabral, F. De la Torre, J. P. Costeira, and A. Bernardino, "Matrix completion for multi-label image classification," in *Proc. Adv. Neural Inf. Process. Syst.*, 2011, pp. 190–198.

Example 3.110. *List of references according to the IEEE system, a continuation of Example 3.109.*

In the text, if a particular part of a referenced article needs to be cited, then the extra information is inserted within the square brackets, as in Example 3.111 taken from the IEEE guide [179].

[3, Th. 1]; [3, Lemma 2]; [3, pp. 5-10]; [3, eq. (2)]; [3, Fig. 1]; [3, Appendix I]; [3, Sec. 4.5]; [3, Ch. 2, pp. 5-10]; [3, Algorithm 5].

Example 3.111. *More specific citing within the IEEE system [179, p. 34].*

In the Vancouver system, the citations of Example 3.109 would appear as superscript numbers or numbers in parentheses and the references in Example 3.110 would have the surname or family name first.

This guide, which is not a research work, uses a variant of the numeric system in which the citation number is determined by the alphabetical ordering of the references in the bibliography and the list of research works [68, §29.2.1].

3.10.2 Author-Date System

In the author-date system, the citation takes the form of a combination of the author name (or names) and the year of publication. In the list of references, the references are ordered alphabetically by surname or family name, with various rules governing multiple publications by the same authors and the same publication year. Examples of the author-date system are the Harvard system [348] and the closely similar American Psychological Association (APA) system [293]. The origins of the Harvard system are obscure [48] [68].

Example 3.112 illustrates citing within the Harvard system.

> Synaptic plasticity . . . is therefore thought to be involved in the storage of LTM (Neves et al. 2008). . . .
>
> Furthermore, Jiang et al. (2010) showed that L-LTP and LTM were blocked by an independent chemical genetic strategy . . . in vivo.

Example 3.112. *Citing within the Harvard system, from a neuroscience article [434].*

If there are two authors in the citation, then both names are included, but if there are more than two, then, in most versions of the Harvard system, all the authors except the first are abbreviated to 'et al.' for 'and others'. In Example 3.112, the first citation '(Neves et al. 2008)' behaves grammatically like a parenthetical phrase, whereas the second citation 'Jiang et al. (2010)' behaves like a noun phrase, the subject of the sentence.

Example 3.113 shows part of the corresponding list of references in the Harvard system. The references appear in alphabetical order.

> Jiang Z, Belforte JE, Lu Y, Yabe Y, Pickel J, et al. 2010. eIF2α phosphorylation-dependent translation in CA1 pyramidal cells impairs hippocampal memory consolidation without affecting general translation. *J. Neurosci.* 30:2582–94.
>
> . . .
>
> Neves G, Cooke SF, Bliss TV. 2008. Synaptic plasticity, memory and the hippocampus: a neural network approach to causality. *Nat. Rev. Neurosci.* 9:65–75.

Example 3.113. *List of references according to the Harvard system, a continuation of Example 3.112.*

This example is a particular implementation of the Harvard system. Journals usually indicate which version they require.

The APA author-date system [293] is similar to the Harvard system, but conventionally the two are treated as being distinct. The APA description is especially comprehensive.

3.10.3 Where to Cite

Citations should be inserted at the most relevant points in the sentence rather than by default all at the end, unless the end is actually the most relevant place [79, pp. 79–80]. Example 3.114 illustrates relevant citing within the numeric system. It is taken from the opening statement of an article in which citations are introduced with the mention of each new application.

> Path optimization affects many of our daily activities. Although much attention has been dedicated to routing algorithms for Internet applications, such as instant messengers and peer-to-peer systems (1, 2), many other essential routing applications have attracted less attention, ranging from water distribution networks (3) to sensor networks (4), military convoy movements (5), and journey planners (6, 7).

Internet routing

Water distribution

Wireless sensors

Transportation

Time-space network

Example 3.114. *Multiple citations within a single sentence, from an applied physical sciences article [727].*

In some circumstances the converse requirement arises where a citation has to extend in scope beyond a single sentence. If the sentences are clearly related, the citation may be placed at the end of the last sentence, as in Example 3.115 from a master's report.

> Although this is a very reliable approach, it is often expensive, and consumes already scarce resources. Arising from advances in information and communications technology, an entirely new approach to solving these problems has been developed. This approach is referred to as the smart grid (López et al. 2014).

Supports two preceding statements

Example 3.115. *Single citation with multiple sentences, adapted from an unpublished electrical and electronic engineering master's report*.[28]

The second and third sentences are closely coordinated with the repetition of 'approach'. The single citation at the end serves to identify the source of both the statement about the name of the approach and the preceding material describing its origins.

On the other hand, where attribution needs to be precise and uncertainty avoided, the citation can be repeated, as in Example 3.116.

> Estimates for the standard error of the difference in means for treatment groups in crossover studies were derived from reported P values when the standard error of the mean difference was not reported.[18] If P values for the differences were reported simply as non-significant, then P=0.2 was assumed.[18]

Precise citation

Precise citation

Example 3.116. *Repeated citation (superscript) for precision, from a general and internal medicine article [692].*

The superscript citation 18 refers to the *Cochrane handbook for systematic reviews of interventions* [162].

3.10.4 Comparison of Referencing Systems

For better or worse, the choice of referencing system for a journal article is usually settled by the publisher. With other kinds of publications, however, there may be opportunity for choice. Each of the two main referencing systems has its advantages and disadvantages [262]. In brief, for the reader, the author-date system is more informative, providing an immediate reminder of the source, but it can also be distracting when multiple citations interrupt the normal sentence flow. By contrast, the numeric system is unobtrusive, allowing the reader to concentrate on the text, but at the cost of having to repeatedly turn to the references list to identify the source.

For the author, the numeric system has the advantage that it encourages a style of writing that concentrates on the research itself rather than on who did it. Nonetheless, it can be tedious to implement without a reference management software system to automatically renumber citations and references when new ones are introduced or old ones deleted.

Compare the first sentence in Example 3.117 (reproduced from Example 3.112) with other variants in the author-date and numeric systems.

Author-date — Furthermore, Jiang et al. (2010) showed that L-LTP and LTM were blocked by an independent chemical genetic strategy . . . in vivo.

Furthermore, Jiang et al. [75] showed that L-LTP and LTM were blocked by an independent chemical genetic strategy . . . in vivo. — Numeric

Furthermore, L-LTP and LTM were blocked by an independent chemical genetic strategy . . . in vivo
Author-date — (Jiang et al., 2010).

Furthermore, L-LTP and LTM were blocked by an independent chemical genetic strategy . . . in vivo [75]. — Numeric

Example 3.117. *Differences in emphasis within the author-date and numeric systems, edited from Example 3.112.*

It is the last, numeric variant that places greatest emphasis on the research. Indeed, the IEEE *Editorial Style Manual* recommends avoiding variants with the authors' names such as 'Jiang et al. [75]', except when the names are integral to the understanding of the sentence [179, p. 16]. These recommendations parallel those concerning the use of 'we' and more objective ways of reporting [150, §12], addressed more fully in Section 5.1.

The author-date and numeric systems should not be used simultaneously, as is done in Example 3.118.

Spectral angle mapper (SAM) is . . . utilized to measure the similarity of test pixel spectra to reference spectra (Kruse, 1993) [11].

Example 3.118. *Improper merging of author-date and numeric systems, from an unpublished electrical and electronic engineering master's report.*[29]

The references in the article's list of references follow the numeric system, with the Kruse article listed as number 12. The citation should therefore be '[11] [12]', or '[11, 12]', where the ordering is conventionally ascending.

Publishers may sometimes expect authors to submit references with abbreviated journal names. Standardized abbreviations are available from several sources [43] [231] [375].

3.10.5 Reference Sources

In choosing a work to cite, you should consider its stability as a published document. The publishers of established journals, whether print and web-based or web-based alone, can usually ensure that the electronic version of an article when formally published is definitive, the article of record, which coincides with the print version if it exists. As with the print version, the only way the electronic version can be subsequently corrected is by publishing an erratum or corrigendum (§9.3.13). The article itself is retrievable through several identities, including the traditional reference format (journal title, year, volume and issue number, and pages or article number), its digital object identifier or DOI [87], and its web address in the form of a uniform resource locator or URL.[30] Cross-publisher citation linking of online journals enables content to be easily found and accessed.[31]

Other electronic sources that have only a URL may be less stable. They may be modified by the author so that the version accessed at the time of writing differs from the version accessed later by the reader. Still worse, the URL of the source may change, or the source may be deleted so that the citation becomes meaningless. If it is necessary to cite a web source (e.g. a governmental, institutional, or industrial report), then along with the URL, you should add as much other identifying information as possible: for example, the title of the page and technical report number, version, or edition, the author of the work, the owner or publisher of the website and the place of publication, the publication date, and the date of your access.

Except for reports produced by official bodies and professional and learned societies, web material with only a URL may have an unknown provenance. The quality of the work, unlike that in most journal articles, may not have been challenged: it may not have undergone independent review by experts in the field, or tested for plagiarism, or edited and prepared professionally for publication. Its usefulness as a reference is limited.

In this context, the use of online encyclopaedias such as Wikipedia [378] is controversial [269]. The central criticism is that their content can be changed, anonymously, by anyone at any time. It is not safe to assume that if an entry does contain an error then someone else will correct it. Whether a correction will take place is uncertain and whether the correction is valid or adequate is also uncertain. As a consequence, the academic integrity of such encyclopaedias as references is compromised [34]. One Wikipedian-in-Residence (a Wikipedia editor who works within institutions to facilitate their Wikipedia entries) points out that Wikipedia was never intended as an authority in its own right and 'is not itself a citable source' [300]. In practice, ignoring this advice and citing Wikipedia can damage the chances of your paper being accepted by a journal editor [34].

3.10.6 Verifying References

As emphasized elsewhere [282] [314], you should cite only articles you have read and whose reference details you can verify directly against the original source: for a journal article, the spelling of the authors' names, title of the article, title of the journal, year of publication, volume number, pages, and any further information (e.g. DOI). Each citation in the body of the text should correspond to the correct reference in the list of references.

There are good reasons not to short-circuit this process. First, it is unsafe to take a reference from another author's publication without inspecting the original. Ethical issues aside (§12.5), you may be propagating a misinterpretation, since the author of the publication may not have read the source publication, or having read it, may have misunderstood it. Second, it is unsafe to rely on the title or abstract of the article. They each provide a limited representation of the research and may occasionally misrepresent it [278]. Finally, it is unsafe to rely on the citation details captured in some online databases, which may be incomplete or inaccurate. By way of illustration, the article title in Example 3.119 is from the original printed article [723]. In one bibliographic database [326] the details are displayed correctly.

A Simple Fluorometric Assay for Lipoperoxide in
Blood Plasma

Example 3.119. *Correct article title, from a biochemistry and molecular biology article [723].*

In another bibliographic database [375], there are two small errors in the title, which is reproduced in Example 3.120.

Missing 'A'

✕ Simple Fluorometric Assay for Lipoperoxide in
Blood-Plasma

Extra hyphen

Example 3.120. *Incorrect article title, from a bibliographic database [375].*

The title has lost the initial 'A' and gained a hyphen between 'Blood' and 'Plasma'. Does it matter? Apparently it does, for if the correct title is used as the title search term, the article is not found in the second database. Although the hyphen has no effect, the initial 'A' does. Additionally, if, as some authors prefer, 'Flurometric' not 'Fluorometric' is used as the title search term, the article is found in neither database. Small differences count.

The only exception to reading the original article is when a translation is required. In that case, both the original and the translation should be cited [49, §13.76]. Where feasible, any key phrases in the translation should be verified. Even then, the citation details should be carefully checked.

Because some readers (and reviewers and examiners) use the quality of an article's references as a proxy measure for the quality of the research, one ill-chosen error may be enough to influence your audience.

3.11 Appendices and Supplementary Materials

Material underpinning your research that does not appear in the results section (§3.7.1) can be introduced in one or more appendices to your paper or, when accepted, uploaded to a separate site as supplementary materials.

3.11.1 Appendices

Appendices contain any material that readers might need and that is too long for the body of the text or would be too disruptive or repetitive, for instance, methodological details or further analyses [68]. As they form part of the article, they may be subject to length limitations. They also need to incorporate sufficient context to be able to stand alone.

3.11.2 Supplementary Materials

Supplementary materials may be large data sets, analyses, images, computer code, and animations and audio clips. They are normally posted on the journal website for electronic access or, in some cases, made available in independent data repositories. Data storage and data access are considered in detail in Section 3.12. Related ethical issues are considered in Section 12.7.

3.12 Data Storage and Access

The origin of the sources of data used in your research should be fully documented and readily accessible, with links to the original research materials [266]. Importantly, storage and access to data should be supported by information about context so that potential users can understand the significance of the data [170] [301].

3.12.1 Data Storage

Advice is usually available from your research funder, publisher, or the institution in which the research took place about intellectual property, embargo periods, confidentiality requirements, and preparation of data for storage and access. Care should be taken in dating material correctly to help demonstrate its provenance on inspection [339, §6b].

Where should data be stored? In addition to institutional data repositories, there are many external repositories, which differ in the research areas covered, the types of content, capacity limits on data, and geographic locations. There are also general repositories devoted to data for which no specialized repository exists. Fortunately, searchable registries of research data repositories are available [302].

3.12.2 Data Access

How should access to data be reported? Some journals include a statement in a footnote on the front page or near the end of an article giving details of access. Example 3.121 shows two repositories being cited.

> Data deposition: The sequences reported in this paper have been deposited in the GenBank database (accession nos. KP719983–KP720300). Datasets and accompanying files are available from the Dryad Data Repository (www.datadryad.org; accession no. 10.5061/dryad.3450r).
>
> — Genetic sequence database
>
> — Data repository

Example 3.121. *Data accessibility statement, from a multidisciplinary sciences article [560].*

The first repository is a specialized genetic sequence database and the second is a more general data repository. In the original article [560] from which the example was taken, the GenBank accession numbers are hyperlinked to a database where the sequences can be inspected. The Dryad accession number is hyperlinked to the page containing a description of the data, with links to individual data files, 'readme' text files, and other details.

3.13 Checklist for Research Articles

Title

1. Not more than 10–12 words?
2. Every word necessary?
3. Not a complete sentence?

Abstract

4. Context, research question, approach, main results, implications?
5. No unnecessary abbreviations?
6. No empty statements ('Results are discussed')?
7. Within word limit?

Keywords

8. Main points extracted?
9. Consistent with the title, abstract, introduction, discussion?

Introduction

10. Accessible opening statement, then background?
11. Relevant previous research?
12. Research question, aims and objectives of the work?
13. Nature of experimental or theoretical approach, essential concepts, no technical detail?
14. Indication of the outcome?
15. Appropriate citations?

Methods

16. Apparatus, materials, procedure, analysis: enough details for repetition?
17. Correct naming: generic first, manufacturer's details second?
18. Explanation of specialized techniques?
19. Research ethics approval?
20. Critical aspects, strengths and weaknesses?
21. Minimum notation?
22. Consistent symbols: italic, roman, and bold font? SI units?

Results

23. Overview, then detail?
24. Graphs rather than tables?
25. Statistical analysis indicating reliability or repeatability?
26. Concrete language preferred to abstractions?
27. Interpretation postponed for the discussion?

Discussion

28. Reminder of research question or problem, main outcome (not repetition of results)?
29. Immediate implications?
30. Relevance to other published work?
31. Limitations of the present findings?
32. Conclusion with new knowledge, new questions, longer-term implications?

Acknowledgements

33. Funding, technical support, and facilities identified?
34. Contributing individuals named, agreement with naming and format?

References

35. Correct and consistent use of author-date or numeric system?
36. Full reference details, journals preferred to web sources?
37. Details verified against original sources?

Data repositories

38. Data prepared for upload to institutional or external repositories?
39. Statement about where data can be accessed?

Consistency

40. Aims the same across the abstract, introduction, and discussion?
41. Main findings the same across the abstract, introduction, and discussion?

4

Literature Reviews, Technical Reports, Dissertations, and Theses

Several forms of reporting in science and engineering fall outside the research article structure found in journals and conference proceedings. These other forms include literature reviews, technical reports, dissertations, and theses.

This chapter takes each of these reporting forms in turn. Section 4.1 deals with the preparation and construction of literature reviews. Section 4.2 deals with technical reports, Section 4.3 deals with the content and preparation of dissertations and theses, and Section 4.4 deals with their assessment and examination. Sections 4.5–4.7 provide extended checklists for the items in this chapter.

4.1 Literature Reviews

A literature review represents a summary and an evaluation of previous research [258]. Literature reviews are published in dedicated review journals and sometimes in primary research journals. They may be incorporated into conference proceedings, where some reviews have had a lasting influence. They may be an essential element of dissertations and theses, and in one form or another they may also appear in technical reports.

Literature reviews published in journals, however, are usually written by experienced researchers in response to an invitation by the editor, as in [14]. The editor's decision may be guided by the timeliness of the topic, its relevance to the journal, as well as the author's international standing in the area and published research record. If, therefore, you are planning to submit an unsolicited review, perhaps from a dissertation or thesis, it is sensible to consult the editor first [79, p. 148].

As an independent document, a literature review can offer several kinds of account [157, ch. 3.3] [258, p. 101]: a history of a research field, an analysis of the main developments, a comparison of competing methodologies, or a description of outstanding research problems. In some respects, a literature review can thus act as a tutorial and critique combined [258] [293].

Within a technical report, dissertation, or thesis, a literature review can offer a frame of reference for the topic to be addressed; for example, as a critical overview or a classification of approaches into those likely to be viable, those that have failed, and those yet to be tried.

Naturally, a literature review should embody your own understanding and offer something fresh. Reproducing the content of another author's review is plagiarism. Moreover, that other review may be wrong or biased [157, ch. 3.3], and reproducing it merely continues the error. Even so, making your own survey of the field does not mean having to examine each of the hundreds, thousands, or possibly tens of thousands of potentially relevant research articles listed in a bibliographic database (it may still be necessary with smaller samples or more demanding reviews). Nor does it mean randomly sampling the database and then peppering your report with citations to simulate authenticity. Rather, it means adopting a systematic approach to the challenge.

4.1.1 Systematic Survey

Unless the field of research is exceptionally new, published articles should already exist that review the field, either as review articles in their own right or as parts of larger pieces of original research. Bibliographic databases such as Scopus [326], Web of Science [375], and Google Scholar [138] can be used to identify the most influential reviews. You can do this in two ways. The first is to enter a search term or terms (e.g. 'graphene' or 'square kilometre array'), and order the records by the number of times each is cited. Some of the more highly cited articles will be reviews or contain reviews, although you should allow for older reviews' accruing more citations. The second way is to enter your search term or terms and add the word 'review' as another search term or as the document type. Order the records by the number of times cited, and again allow for older reviews' accruing more citations. These two search methods may yield different results because not all review articles or articles containing reviews are indexed as reviews, and not all articles indexed as reviews are strictly reviews. Choose, say, three or four influential reviews, differing for example in subject emphasis, for examination:

- carefully read the reviews to gain familiarity with the topic and its diversity;
- compare what they claim with what you find in the sources they cite;
- use your own words to record the key points and the sources supporting these points.

In a complementary but separate approach, use the same databases to identify the most influential original research articles. With the same search terms, order the records by the number of times cited. Eliminate those articles that are reviews, either by inspecting them or by adding 'article' as the document type, though further inspection may be needed. If the field is mature, the leading research articles are likely be old. Give priority normally to the first or definitive report (§12.5.1), and a lower ranking to subsequent, confirmatory or incremental studies. Decide on a candidate set of influential articles for analysis:

- carefully read the articles, and note why they are influential;
- support your analysis by examining other articles that cite these leading articles;

- inspect more recent articles to see whether they significantly advance the field;
- use your own words to record the key points and the sources supporting these points.

Group key points together, experiment with their logical (or chronological) order, and then develop them into phrases, sentences, and paragraphs. Take care to support each point by citing appropriate references. Try to cover a variety of points in an even way.

4.1.2 Content

The content of a literature review depends on the aims and the context of the work, but the typical elements are as follows [79] [258]: a general introduction to the research topic, its importance or significance, its historical origins, the main theoretical and experimental findings, recent research developments, and any applications or implications of the research.

A review that was concerned with historical origins might say little or nothing about recent developments, whereas one concerned with applications might say nothing about historical origins. As with an original research article, supporting references should be chosen without bias and with due regard to priority (§12.5).

Example 4.1 shows part of the introduction to a review article in nuclear physics.

Named key historic reference

Named key historic reference, 1956

> Parity (*P*) violation was first observed by Wu et al. (1) in 1957, not long after Lee & Yang (2) made their Nobel Prize–winning suggestion that *P* may not be conserved in weak interactions (Figure 1). Atomic parity nonconservation (PNC) is caused by the weak interaction . . .

Influential experimental result

Technical calculations

Technical calculations

> The exceptionally precise measurement of the cesium 6*S*–7*S* PNC amplitude (3), in conjunction with the highly accurate calculations required for its interpretation (see References 4–9 and references therein), led to a determination of the ^{133}Cs nuclear weak charge . . . The result of this analysis differs from the SM prediction by 1.5σ (9). . . .

Application

Application

> Much of the interest in the area of atomic PNC has focused on several other important areas: measuring PNC in a chain of isotopes (10), nuclear anapole moments (AMs) (11–13), and PNC in molecules (14).

Applications and review

Example 4.1. *Use of references in an introduction, from a nuclear physics review article [650].*

The first two citations refer to original research findings in the 1950s that initiated the field. Because of their prominence, the researchers' names are included (§3.10.4). The third citation refers to a more recent influential experimental result reported in the journal *Science*. The subsequent citations refer to articles (one a review) containing technical calculations and applications.

Example 4.2 shows part of the introduction to a review article in neuroscience. The first three citations refer to a perspective piece in *Science*, a review, and a textbook. The fourth and fifth citations refer to two classic textbooks. The sixth and seventh refer to reviews, the eighth to an original analysis, and the remainder to recent findings.

> Perception and action are based on neural representations of the external world. Determining the neural code [1–3] . . . is therefore a prerequisite for understanding sensory function. The search for the neural code began many decades ago . . . [4,5]. It continues today with the exploration of different putative codes, and with systematic comparisons between the information in individual codes and behavioral performance.
>
> Two dimensions of neural representations are important for characterizing a neural code. The first is defined by space . . . [6,7]. The second dimension is defined by time . . . [8]. Over the past few years, it has become clear that. . . sensory information is multiplexed in neural responses at different timescales [9–15].

Key article, review, and textbook — [1–3]

Influential historic textbooks, 1928 and 1933 — [4,5]

Reviews — [6,7]

Original analysis — [8]

Recent findings — [9–15]

Example 4.2. *Use of references in an introduction, from a neuroscience review article [635].*

The introductions of these two review articles differ in the extent to which they draw on previous reviews and textbooks, but both describe the topic and its importance, introduce historic material, highlight the main theoretical and experimental developments, and cite recent advances.

4.1.3 Added Value

To be effective, a review needs to make an original contribution [258, p. 101]; that is, it needs to deliver more than a passive account of previously published work, something that could have been produced by a machine with a list of search terms.

You can add value in several ways: impose a fresh organization on the field, introduce a novel insight or understanding, or organize data to allow new analyses. You can also identify outstanding research problems or propose new approaches and experimental tests. These contributions are particularly germane for a review in a dissertation or thesis, where they can provide the motivation for the aims and objectives of the work.

Example 4.3 shows part of the concluding sections of a review article in nuclear physics (a continuation of Example 4.1). It contains several suggestions for experiments to test the effects described earlier in the review.

> Potential experimental test: To detect these dynamic effects, we need an experiment designed to measure small oscillations in the PNC amplitude or atomic EDM [electric dipole moment]. The frequency and amplitude of these oscillations would enable one to extract values for the relevant field parameters. Implications: For example, a determination of the frequency would provide the mass of the particle, and the amplitude of the oscillations would enable a determination of η.
>
> Potential experimental test: The high sensitivity of atomic EDM experiments makes them promising for the study of the oscillating effects considered here. Possible procedure: One can obtain improved sensitivity of the EDM measurements by tuning the experiment to a specific frequency (see, e.g., References 190, 193, and 194).

Example 4.3. *Suggestions for future experiments, from the concluding sections of a nuclear physics review article [650], a continuation of Example 4.1.*

Example 4.4 shows part of the concluding section of a review article in neuroscience (a continuation of Example 4.2). It sets out several challenges to do with the understanding of the topic addressed earlier in the review and some possible experimental approaches.

> Although the existence and the advantages of multiplexed codes are becoming increasingly evident, several challenges remain.
>
> Challenge: One is to understand the extent to which genuine temporal or multiplexed neural codes are exploited by the brain to cause or guide behavior [44,74,75,77]. Possible procedure: To this end, manipulations of neuronal activity will be needed. Such manipulations could either interfere with patterns of neural activity . . .
>
> Challenge: A final challenge will be to investigate how temporally structured responses integrate with spatial patterns of activity provided by large populations of neurons [60,82–84].

Example 4.4. *Suggestions for future experiments, from the concluding section of a neuroscience review article [635], a continuation of Example 4.2.*

If your review is of a dissertation or thesis and you prepared a preliminary version at the start of your studies, then ensure that it is brought up to date before submission [319]. A review that contains no recent work may give the impression of a field that has stagnated.

4.2 Technical Reports

A technical report is a record of research or other technical activity submitted to or contracted by a variety of organizations, including national and local government, companies, and academic and non-academic institutions, as well as individual researchers and users [68] [258]. The publication of these reports—that is, their access by the public—depends on the organization and individual. Some are never published and are classified as 'grey literature' [146].

The content and presentation of technical reports vary greatly [33] [325]. Although there are some elements in common with research articles and literature reviews, there are significant differences, too, especially in the intended audience, the amount of tutorial material, and the role of two particular elements: the executive summary and recommendations for usage or action.

4.2.1 Structure

The general requirements for technical reports are set out in [146, §4.2] [325, §4, §5]. The main elements are as follows, excluding the more formal and optional parts such as the contents list, copyright statement, and acknowledgements:

- title
- abstract
- executive summary
- introduction
- methods
- results
- discussion
- conclusions
- recommendations
- references
- appendices

The title, abstract, introduction, methods, results, discussion, conclusions, references, and appendices are much as in a journal article. The executive summary may act as an alternative to the abstract, but it differs in that it contains more detail, with an emphasis on the findings of the research and its recommendations, perhaps expressed in more accessible language than in a research article. It may be less than a page in length [514] or extend to many pages [496]. It may contain financial information or political implications that may be absent in the main report [325]. The recommendations, which may appear in the executive summary or conclusions, or more prominently as a separate section, may refer to usage, for example, of materials and systems, or for actions to be taken, for example, new experiments, feasibility studies, and market appraisals [325].

4.2.2 Audience

Unlike a research article, a technical report may be read by users with very different professional backgrounds and for different reasons [258]. Thus a technical report on developing low carbon electrical energy networks might be of interest not only to electrical engineers but also to technology developers, academics, consumers, manufacturers, and government [473].

The title, abstract, executive summary, and recommendations need therefore to be both relevant and accessible to a potentially larger audience, and each section must as far as possible be self-contained, as illustrated in Example 4.5, taken from an executive summary of an electrical energy technical report.

Opening statement	Electrical power networks will be a key enabler for securing future energy delivery and the achievement of the UK's low carbon objectives. . . .
Range of contributions	Insights have been provided by a range of contributors that include the energy storage community, technology specialists, suppliers . . .
Recommen-dations	A set of key structural principles is proposed to support solutions that are flexible and scalable, that utilise . . .

Example 4.5. *Extracts from an executive summary, from an electrical energy technical report [473].*

In this way, the potential user can recognize the significance of the report and extract useful content without having to read the whole report or consult additional sources.

4.2.3 Tutorial Aspects

Some technical reports may contain considerable explanatory material. Example 4.6 is taken from the introduction of a standards report on the stability of reference mass prototypes.

The report sets the historical context of the work, explains the choice of materials used in prototypes, the developments in the understanding of their surface properties, and their implications, all of which then lead to the work described in the body of the report.

History

The design of the prototypes was the result of careful thought and experimentation when this system was set up in late 19th Century. Platinum and iridium are noble metals, and are relatively simple to purify, so that prototypes of repeatable density could be manufactured. . . .

Traditional materials

Developments

However, two developments have renewed interest in the surface state of these prototypes. . . .

Implications

Thus the precision with which the SI unit of mass can be realised is currently limited by the surface-state of the prototypes.

Example 4.6. *Extracts from an introduction, from a physics technical report [458].*

4.2.4 Executive Summary

An executive summary emphasizes the research findings (§4.2.1). This emphasis is illustrated in Example 4.7 for a multi-team experimental research programme and in Example 4.8 for a literature review. Neither contains citations or references. The executive summary in Example 4.7 details the specific objectives and the achievements of the research.

Context

Pulsed power is a technology that is well-suited to driving electrical loads requiring very large power pulses in short bursts . . .

Aim

In 2001, the U.S. Department of Defense (DoD) launched a five-year Multidisciplinary University Research Initiative (MURI) to study fundamental issues for compact pulsed power. This research program endeavored to (1) introduce new materials for use in pulsed power systems; (2) examine alternative topologies for compact pulse generation . . .

Objectives

Results

The key achievements of this research program are as follows (i) liquid breakdown (for water and propylene carbonate) has been extensively diagnosed and new models to explain the observed behavior have been proposed; (ii) dielectric breakdown under pulsed power conditions was observed . . .

Limitations

Fundamental limitations to this approach, as well as the use of turbulent capillary flow for mitigating thermal loading were identified.

Example 4.7. *Executive summary, adapted from an electrical and electronic engineering technical report [662].*

The executive summary in Example 4.8 details the main findings from the literature and offers a recommendation.

Context	Fiber-reinforced polymer composites are increasingly being used for primary and secondary structural components in aircraft. Incineration of composites in postcrash aircraft fires generates a complex mixture of combustion products . . . including fibers. . . .	Problem
Methods and purpose	This report presents a literature review on the potential health hazards due to inhalation of carbon fibers . . .	
	The fiber retention time inside the lung is primarily dependent on the fiber dimensions. . . .	Finding
Finding	Data from fire tests and crash-site investigations suggest that only a small fraction of the fibers released in fires is of respirable size. . . .	
Recommendation	Further work is needed to assess the health implications of any synergistic interactions between the chemical and fibrous combustion products.	

Example 4.8. *Executive summary, abridged from a materials science technical report [514].*

4.2.5 Recommendations

The recommendations in a technical report may be quite specific in both objectives and methods (§4.2.1). The recommendations in Example 4.9 are taken from an extended executive summary.

1. Develop materials for hydrogen storage by weak chemisorption . . .
2. Limit development of materials in which the storage mechanism is purely physisorption to only those with optimized pore structures . . .
3. Develop substituted/heterogeneous materials with demonstrated hydrogen binding energies of 10–25 kJ/mol.

Example 4.9. *Executive summary, abridged from a chemical engineering technical report [496].*

The summary contains more than the three detailed recommendations shown here, and further details are made available separately in a full final report [496]. Nevertheless, the executive summary forms a self-contained document.

4.3 Dissertations and Theses

The size, format, and naming of dissertations and theses, usually submitted for a degree or other academic qualification, depend on the institution and country [79] [157] [322], as does the way in which they are examined. For brevity, in the remainder of this chapter the term 'theses' refers to both theses and dissertations.

The problems that occur with theses are like those with research articles, but on a larger scale. A thesis is conventionally much longer and more complicated, with more prominence given to certain elements such as its aims and objectives and future research; it is more susceptible to repetition and lapses in continuity and consistency; and it is generally more difficult to manage and prepare for submission.

4.3.1 Structure

Despite the similarity with research articles, some parts of a thesis typically require special treatment. Thus, near the beginning of the thesis, there may be a detailed list showing the aims and objectives of the research as a whole. The literature review, placed before or after the aims and objectives, may constitute a full chapter in its own right, by contrast with the few paragraphs or less it would occupy in a research article. Near the end of the thesis, there may be a section or a full chapter giving suggestions for how the research might be developed in the future.

Each individual chapter dealing with a particular research question or problem may contain its own introduction, aims, review, methods, results, and discussion, although there are many variations on this structure. Maintaining a coherent style between sections and chapters can prove a challenge.

4.3.2 Defining Aims and Objectives

The purpose of the work reported in a thesis needs to be made explicit. The division into aims and objectives fulfils a useful function. The aims of the work usually represent its general intention—what needs to be discovered or determined. The objectives usually represent the more specific tasks which, taken together, will realize those aims. Informally, the aims represent the 'what' and the objectives the 'how' of the research [352]. Not all theses observe this distinction. Some may, for instance, use the terms in the opposite way, or describe the motivation for the work or the research problem and then list research questions to be addressed or hypotheses to be tested.

Importantly, if distinct aims and objectives or their equivalents are declared, they need to be complemented by material in the subsequent methods, results, and discussion sections so that the examiner can judge how well the aims and objectives have been addressed.

Example 4.10 shows a well-defined aim and objectives.

Aim	The aim of this thesis is to quantify the safety level of the ADS-B [Automatic Dependent Surveillance Broadcast] system ... Six research objectives were formulated to achieve this aim:	— Specific action
Objectives	– Identify the deficiencies of current surveillance systems in supporting increasing air traffic.	— Specific action
	– Identify the capabilities of ADS-B to address the limitations of current surveillance systems.	— Specific action
	...	
	– Derive a mapping between ADS-B performance and the required performance of the various applications to be supported by ADS-B.	— Specific action

Example 4.10. *Aim and objectives, abridged from a civil and environmental engineering doctoral thesis [398].*

These are followed in the thesis [398] by an explanation of how each objective is to be achieved.

Notice the use of the precise words in Example 4.10 to describe the planned actions—'quantify', 'identify', and 'derive'. With less precise or altogether vague words such as 'investigate', 'explore', and 'consider', it is difficult for an examiner to decide whether what has taken place constitutes success or not. It is also difficult to see what the author is truly committed to (§2.3.4). This is not to say that broad intentions such as 'consider' and 'examine' should be avoided altogether, but they are more suited to overarching declarations, not detailed statements about what is to be delivered.

If individual research chapters contain their own aims or objectives, then, to avoid confusion, they need to be made compatible with the aims and objectives of the thesis as a whole.

4.3.3 Identifying Future Research

If a section or chapter in the thesis is devoted to possible future research (also referred to as 'further work' and similar permutations), then it should not be perfunctory [322]. Routine phrases of the kind 'more work needs to be done' are not really informative.

One way to develop a significant discussion about future research is to treat it as a response to the limitations you identify in your current research (§3.8.5). Example 4.11 shows how the two can be combined.

Making explicit the limitations of your research and how they might be addressed demonstrates both the objectivity of your analysis and your ability to take the work beyond the constraints of a thesis programme.

Limitation: The first limitation is that only the timing dimension of EV [electric vehicle] travel was assumed to be affected.
Future research: A way to overcome this limitation would be to integrate it within activity-based travel-demand modelling systems . . .

Limitation: The second limitation is the intrinsically myopic behaviour implied by the assumed independence of the charging choices.
Future research: A dynamic choice module would be necessary to overcome this limitation. The EV driver would need to maximize utility over . . .

Example 4.11. *Limitations and potential future research, abridged from a civil and environmental engineering doctoral thesis [459].*

Notice that identifying possible future research within the field is not the same as declaring your own research plans, a practice which is deprecated (§3.8.7)

4.3.4 Defendable Statements

A thesis, as with any research report, needs to cite evidence in support of the ideas and findings it contains (§3.10). But the evidence needs to be especially robust to interrogation. As part of the examination requirements, the examiner may want to ensure, among other things, that the thesis represents your own work and that you have a true grasp of the field of research and its larger context. This is not to say that there is a difference in ethical standards for a thesis and a research article; rather, there is a difference in the way the two are tested and assessed.

Thus, in the defence of your thesis, an appeal to authority—e.g. a citation of someone else's use of a method—may not in itself be enough. The citation establishes a precedent for your decision, not necessarily its wisdom. If it seems that you do not understand the reason for your decision, then the examiner may conclude that the work is not entirely your own. If you do not understand a term, do not use it [95, p. 154].

4.3.5 Automatic Writing

Because of its length and the sustained writing effort it requires, a thesis can inadvertently contain repeated text. Example 4.12 shows the same statement being made in an introduction and conclusion.

Repetition gives the impression of automatic writing, as if you were unaware of what you wrote earlier. If the same point does need to be made twice, then

		These advantages of mutual information over other measures of dependency such as linear correlation are well known.	Introduction
✗		. . .	
	Conclusion	These advantages of mutual information over other measures of dependency such as linear correlation are well known.	

Example 4.12. *Text repetition, from a draft of an electrical and electronic engineering doctoral thesis.*[32]

inserting a phrase such as 'As noted earlier' acknowledges the preceding statement (§2.2.1).

4.3.6 Time Management

It is easy to overestimate your capacity to write quickly. A doctoral thesis can take four or more months of hard, continuous work. The general advice is to start writing as soon as you have accumulated enough material [157, p. 84] [229, pp. 7–8]: the literature review at the start of your research; the results as they emerge at each stage. Details of methodology should be recorded during each experiment or analysis.

All this material will need to be updated later, but, as mentioned elsewhere, it is easier to edit than to start afresh. In any event, you will need to make changes in response to feedback on the drafts of your thesis from your thesis adviser or supervisor and your colleagues.

In order to maintain momentum, write small, manageable sections, one at a time, aiming to finish each writing session by making a start on the next section (§2.4.2).

4.3.7 Formatting

The institution to which you submit your thesis provides guidance on matters of presentation. It may specify the size and maximum number of pages (and any penalties for exceeding the maximum length); preferred fonts and sizing; pagination; special pages, such as the title page, list of contents, abstract, and declaration; details about the bound print version and number of copies; the expected electronic format; and requirements for open access.

Some institutions make available electronic thesis templates in either a word-processing application such as Microsoft Word [253] or a mark-up language such as LaTeX [217], the choice depending somewhat on the discipline (§6.4). Your adviser or supervisor may offer advice about formatting, but inspect previous theses in your research area to ensure that yours is compatible in style, content, and length [157].

4.3.8 Publication

Theses may be uploaded to institutional and national repositories for electronic access, and print copies deposited in institutional and other libraries. Yet, unlike journal articles, theses do not generally undergo even rudimentary copy-editing for language and style or professional typesetting, except for some theses that are published in abridged form. It is important, therefore, to encourage your colleagues (and indeed anyone else) to read and criticize the draft versions of your thesis, either section by section or as a whole, for both content and presentation. If your thesis is not written in your native language, ask a colleague with native-language competence to examine it for non-idiomatic constructions.

Making this effort is worthwhile. Your thesis will almost certainly be freely available for others to inspect, including potential employers, colleagues, and anyone else who might have an interest in your subsequent scientific and professional life. What might seem a semi-private conclusion to your thesis studies can have unhappy public consequences in later life (§12.3).

4.4 Examination

Depending on the degree or other academic qualification, the institution, and the country, the examination of a thesis takes place in one or two stages. The first is the reading and assessment of the thesis by an examiner or more usually examiners, a process that may take several weeks. The examiners then submit separate written reports to the institution. The second stage, if it occurs, is an oral examination or viva voce. In some countries, the oral examination is a ceremonial formality conducted in front of a large audience, which may comprise colleagues, family members, and other interested parties; in other countries, it is a prolonged, closeted interrogation by one or two experts in the field, who can and on some occasions do fail the candidate [139].

4.4.1 Thesis Assessment

The criteria for thesis assessment vary, but typically include the following:

- Is there evidence of a substantial original contribution to the field?
- Is there evidence of independent critical ability?
- Is the work placed properly in context?
- Does it demonstrate knowledge of the field as a whole?

Ensure that your thesis is crafted so that the examiners can answer each of these questions positively. Others examples of questions and concerns from examiners are included in Section 4.4.2 and in the checklist at the end of this chapter.

4.4.2 Oral Examination

The purpose of the more interrogational kind of oral examination is to enable examiners to establish that the work is the candidate's own, to allow the candidate to defend the work, to clarify any confusions, and to discuss the topic of the research more generally.

It sounds daunting, yet it need not be so. If your thesis is to be subject to an oral examination, there is much you can do to ensure that it goes well [139] [319]:

- Read your thesis thoroughly before the examination—you may have forgotten details of the work that you undertook several months or a year or more ago.
- Understand fully the concepts, procedures, and measures you describe and what their strengths and weaknesses are.
- Keep up to date with the research literature.
- Rehearse: arrange with your adviser or colleagues to test you in a mock examination.

Discuss the format of the oral examination with the examination organizer. Some practicalities to keep in mind are these:

- Take a copy of your thesis with you that is identical to the examiners' copy; otherwise it will be difficult to follow as they work their way through page by page.
- Be prepared to give a brief account of what you achieved at the beginning of the examination, as either a summary statement or a presentation lasting several minutes. Find out beforehand if the examiners want a short talk with slides.
- In the examination, note corrections in your copy of the thesis or in a notepad brought with you (not all suggestions may be included in the examiners' written reports).
- Listen to questions carefully—the important parts may be at the end. Focus on what is asked, not on what you would like to talk about.
- If you cannot answer a question completely, explain to the examiners what parts you can answer. Avoid rambling.
- Defend your choices where you believe you are right, but know when to admit defeat.

What problems do examiners identify? Here are some examples drawn from several disciplines. The distinctions between foundational, procedural, and presentational are not sharp, and depend on context:

Foundational

- unclear reason for the experiment
- misunderstanding of other authors' work
- previous work summarized but no conclusions drawn
- unexplained choice and manipulation of experimental variables
- unexplained criteria for observer judgements
- unclear purpose in fitting curves, prediction or description
- unexplained outlier points, omitted values
- meaning of correlation between variables
- results described but not explained
- missing inferences
- confusion of discussion with literature review, summary with conclusion

Procedural

- outdated reviews; wrong attribution of original findings
- missing references to earliest or most important works
- late statement of aims and objectives
- missing details of experimental apparatus, procedure
- unverified manufacturer's calibrations
- confusion of systematic and random errors
- unclear ethical approval for research with human subjects
- too few experimental participants
- ceiling effects on an experimental response range
- missing justification for a normal-theory assumption
- missing measures of goodness of fit
- post hoc tests with no adjustment for degrees of freedom
- unexplained relationship of results to the literature
- dependence of analysis on obsolete standards

Presentational

- confused or unexplained technical terms and notation
- unspecified units of measurement, unexplained symbols
- reproduction of figures from sources under copyright
- figures illegible, unlabelled axes, missing units

- missing or misunderstood error estimates, standard deviation, standard error
- spurious numerical precision
- mathematical variables not in italic; misuse of symbols
- errors in references; inconsistent citations; use of Wikipedia, student textbooks
- lapses in basic grammar, spelling, word use

No thesis is perfect. But if you know which areas are weak, you can construct your defence in advance. With solid preparation, an oral examination can be a truly satisfying experience. After all, it is work which has occupied you over a very long time and which is now the centre of attention.

4.5 Checklist for Literature Reviews

Sources

1. Influential articles identified?
2. Priority given to first reports?
3. Recent articles surveyed for advances?
4. Own words used throughout?

Structure

5. General description of the research topic?
6. Importance or significance?
7. Historical origins?
8. Main theoretical or experimental findings?
9. Recent research developments?
10. Applications?

Added value

11. Fresh organization of the field?
12. Novel insight?
13. Organization of data?
14. Outstanding research problems?
15. Suggestions for new research?

4.6 Checklist for Technical Reports

Special elements, some optional

1. Notices, e.g. copyright, distribution?
2. Contents list?
3. List of figures and tables?
4. Forward, preface?
5. Executive summary?
6. Recommendations?
7. Extended appendices?
8. List of symbols, glossary?

Executive summary

9. Research question or problem under investigation?
10. Main results and conclusions?
11. Recommendations? Separate section?

Accessibility

12. Each section accessible to the intended audience?
13. Title, abstract, executive summary, recommendations self-contained?

4.7 Checklist for Dissertations and Theses

The following checklist is similar to that for research articles, but it has additional items and differences in emphasis. The ordering is not definitive. Checklists for other aspects of the work—for example, use of English, data reduction, and graphing—can be found at the ends of the corresponding chapters.

Abstract

1. Distillation of whole study (not just e.g. motivation or main results)?
2. Context, research question, approach, main results, implications?
3. No unnecessary abbreviations?

4. No empty statements ('Results are discussed')?
5. Within word limit?
6. Style consistent with previously accepted theses?

Acknowledgements

7. Funding, technical support, and facilities identified?
8. Contributing individuals named, agreement with naming and format?
9. Restrained language?

Introduction

10. Accessible opening statement, then background?
11. Indication of previous research if separate literature review?
12. Indication of research question if separate aims and objectives?
13. Nature of experimental or theoretical approach?
14. Definition of essential concepts, no technical detail?
15. Indication of outcome?
16. Special issues, areas not addressed, technical difficulties, failures?
17. Appropriate citations?

Literature review

18. General introduction to the research field?
19. Historical origins?
20. Main theoretical and experimental findings?
21. References to influential reviews and leading journal articles?
22. Recent research findings?
23. Applications of research?
24. Critical evaluation of previous and current research?
25. Outstanding research questions or problems to be addressed?
26. New theories or experiments needed?

Aims and objectives

27. Clear aims of research?
28. Detailed objectives against which success can be evaluated?

Methods

29. Pilot work influencing the approach?
30. Apparatus, materials, procedure, analysis: enough details for repetition?
31. Correct naming: generic first, manufacturer's details second?
32. Explanation of specialized techniques? Research ethics approval?
33. Critical aspects, strengths and weaknesses?
34. Minimum notation? Consistent symbols: italic, roman, and bold font? SI units?
35. Clear evidence of informed and competent application of methods?

Results

36. Overview, then detail?
37. Graphs rather than tables?
38. Statistical analysis indicating reliability or repeatability?
39. Concrete language preferred to abstractions?
40. Interpretation postponed for the discussion?
41. Clear evidence of adequate volume and quality of results?

Discussion

42. Reminder of research question or problem and the main outcome (not repetition of results)?
43. Relationship to aims and objectives?
44. Interpretation and implications (but no new data)?
45. Relevance to other published work?
46. Longer-term implications?

Conclusions and future research

47. New knowledge, new questions?
48. Complement to the introduction?
49. Limitations of present findings and possible future research?

References

50. Correct and consistent use of author-date or numeric system?
51. Full reference details, journals preferred to web sources?
52. Details verified against original sources?

Consistency

53. Aims and objectives compatible across chapters?
54. Main findings compatible with individual chapters?

5

Use of English

The use of English in science and engineering often confuses and alienates the nonspecialist, and sometimes even the specialist. The reasons for this unsatisfactory situation are probably to do with convention, expediency, and insight, or lack of it [389, p. 3]. This situation might be seen as part of a more general and long-standing decline in the English language [274], but the use of English in science and engineering does seem to present a particular trial for both authors and readers.

This chapter identifies some of the more common problems that arise with English usage and offers some solutions. The problems can be small, such as errors in punctuation that misdirect the reader, or more fundamental, such as confounds that affect the understanding of the work itself. Most have to do with logic rather than with any dexterity with language, either as a native English speaker or as a second-language user [177]. This chapter should therefore not be taken as a wider guide to either English grammar or its standard usage. There are many authoritative sources available elsewhere, some listed in the Bibliography.

Section 5.1 considers the use of personal pronouns and the author's presence in the text. Section 5.2 deals with how the form of a verb can affect interpretation, and Section 5.3 deals with awkward passive constructions. Section 5.4 analyses the uncertainties that arise when one word or phrase refers to another. Section 5.5 describes the advantages of parallel and coordinate constructions and the use of ellipsis. Section 5.6 examines word choice, and Section 5.7 punctuation. Section 5.8 provides a checklist for the items in this chapter.

5.1 Use of Personal Pronouns

As the author of a research paper, there is a fundamental question of whether you should have a visible presence in the account. If the research refers to human beings, there is also the question of how best you can avoid sexist language. Both questions have to do with the use of personal pronouns, especially 'I' and 'we', and 'he', 'she', and 'they' [32] [44].

5.1.1 I and We

Some style guides argue strongly for use of the authorial 'I' or 'we' as the grammatical subject of a sentence [89]. One rationale is that the personal pronoun removes uncertainty about the agent of the action [80, pp. 126–7] and that it reminds the reader of a human presence, the person with whom the knowledge should be associated [258, p. 18]. Another rationale is that it avoids the construction of cumbersome passive sentences, considered separately in Section 5.3.

Yet in practice 'I' or 'we' is not allowed in large areas of writing, most notably news reporting, and for good reasons. As Zinsser points out [391, p. 21]:

> Newspapers don't want 'I' in their news stories . . . Newspaper articles should consist of news, reported objectively.

Repeated intrusions of 'I' and 'we' divert the reader's attention from the events to the person reporting them.

In a minority of special contexts, however, the use of 'we' can be entirely appropriate, as in Examples 5.1–5.4, where it refers to humankind, the research community, and the authors in identifiable roles.

> Although much of what we experience . . .

Example 5.1. *Referring to humankind, from a neuroscience article [509].*

> How can we describe turbulence driven by astrophysical processes?

Example 5.2. *Referring to the research community, from a physics article [596].*

> We argue that neurogenesis is indeed 'good' for memory . . .

Example 5.3. *Self-referral in an opinion piece, from a neuroscience article [509].*

> We surveyed the 31 most recent meta-analyses . . .

Example 5.4. *Self-referral in an analysis of published analyses, from a statistics article [396].*

Some other special uses are described in [297, §6.18]. The use of 'I' and 'we' in mathematical work is discussed in Section 6.3.5.

In addition to distracting the reader, the authorial 'I' or 'we' may be misleading when it involves contributors whose names do not appear in the author byline: for instance, a commercial programmer who wrote the code, a laboratory assistant who ran the experiment, or a consultant statistician who performed the analysis.

'We' may also be misleading when these or other special roles are fulfilled by some but not all of the authors.

The problem is that the 'we' defined by the list of authors, for which there are firm inclusion and exclusion criteria (§12.6), need not be the same as the 'we' that refers to the people involved in individual parts of the research.

Ethical issues aside, the imposition of the personal pronoun can affect the line of argument [284, p. 262] and produce less precise descriptions. In Example 5.5, the authors' original statement is explicit, making clear what process—STEM tomography—led to a particular experimental result.

> STEM tomography revealed that for this catalyst . . .

Example 5.5. *Informative impersonal subject, from a materials science article [656].*

Substituting 'I' or 'We' for 'STEM tomography' tells us less about the process.[33] Of course, the personal pronoun could be added rather than substituted, as in the edited version in Example 5.6.

Unnecessary personalization

✗ Using STEM tomography <u>we</u> revealed that for this catalyst . . .

Example 5.6. *Unnecessary introduction of first-person subject, adapted from Example 5.5.*

Yet no more information is provided to the reader than in the original, and the focus is now shifted from what was done to who did it.

Personalization can also sometimes look contrived, as in Example 5.7, taken from the introduction to an article.

Unnecessary personalization

✗ In this work, <u>we devote our attention</u> to the theoretical and experimental investigation of one-dimensional (1-D) photonic band gap (PBG) structures [1].

Example 5.7. *Unnecessary emphasis on first person, from an instrumentation engineering article [409].*

The clause 'we devote our attention' is an elaborate way of introducing what the work is about. The opening could have been expressed more simply by omitting the clause, as in the edited version in Example 5.8.

> This work is a theoretical and experimental investigation of one-dimensional(1-D) photonic band gap (PBG) structures [1].

Example 5.8. *Edited version of the statement in Example 5.7.*

For a single author, the use of 'I' can lend a slightly archaic, didactic feel to the presentation, as in Example 5.9. The formulation 'I show here' may be suitable for a review or an opinion piece, but it seems to sit less comfortably in a conventional research report.

✕

> I show here how to compute one-loop corrections to these relations for any four-dimensional model.

Example 5.9. *Unnecessary emphasis on first person, from a physics article [559].*

A more neutral formulation is shown in the edited version in Example 5.10.

> This paper shows how to compute one-loop corrections to these relations for any four-dimensional model.

Example 5.10. *Edited version of statement in Example 5.9.*

Halmos [150, p. 37] offered advice that is still timely:

> Since the best expository style is the least obtrusive one, I tend nowadays to prefer the neutral approach. That does *not* mean using 'one' often, or ever; sentences like 'one has thus proved that . . .' are awful. It does mean the complete avoidance of first person pronouns in either singular or plural. . . . The use of 'I', and especially its overuse, sometimes has a repellent effect, as arrogance or ex-cathedra preaching, and, for that reason, I like to avoid it whenever possible. In short notes, obviously in personal historical remarks, and, perhaps, in essays such as this, it has its place.

Although directed to mathematicians, these thoughts are just as relevant for scientists and engineers. The emphasis should be on what was done, not on who did it.[34] Personalization should be avoided [95, p. 155]. But care should be taken with attributing intentions. Statements of the form 'This paper shows . . .' are relatively common, whereas 'This paper hopes . . .' are not.

5.1.2 He, She, and They

Language should not appear sexist (or biased in other ways [293, ch. 3]). The usual advice is to use 'he or she' for the singular pronoun rather than a generic 'he'; likewise 'his or her' rather than 'his', and 'him or her' rather than 'him' [49]. But the repetition of 'he or she', 'his or her', and 'him or her' can seem laboured, and the plurals 'they', 'their', and 'them' offer alternatives. Although viewed by some as ungrammatical substitutes, their use is legitimate and has a long history [32, p. 54] [40].

Occasionally 'he or she' is contracted to 'he/she' or 'she/he', and even 's/he'. These constructions look clumsy and their use is deprecated [49, §5.222].

Awkward construction

✗ In this way . . . he/she would not have taken the wrong choice.

Example 5.11. *Awkward 'he/she', from a geosciences article [431].*

Example 5.11 illustrates the awkwardness of 'he/she'. Since the reader has to translate 'he/she' back into 'he or she', the contraction is unhelpful and may additionally affect credibility [49, §5.221]. Likewise with 'his/her' in Example 5.12.

✗ The answer to deciding whether to employ a PS or an ANN method when facing a large problem lies in the interest of the practitioner and his/her concrete objectives.

Awkward construction

Example 5.12. *Awkward 'his/her', from a computer science article [515].*

By recasting the text, 'his/her' and indeed 'his or her' can be avoided altogether, as in the edited version in Example 5.13.

The answer to deciding whether to employ a PS or an ANN method when facing a large problem lies in the interest and concrete objectives of the practitioner.

Example 5.13. *Edited version of Example 5.12.*

Ellipsis (§5.5.2) can also be used to avoid 'he or she', as in Example 5.14.

Omitted 'he or she'

The subject fixated the center cross and, when ready, pressed the first button to display a stimulus card.

Example 5.14. *Avoiding 'he or she' by ellipsis, from an experimental psychology article [457].*

Alternatively, the singular can be converted to the plural, as in the edited version in Example 5.15.

Plural pronoun

Plural grammatical subject — Subjects fixated the center cross and, when ready, they pressed the first button to display a stimulus card.

Example 5.15. *Edited version of Example 5.14.*

Repeated use of 'he or she' can usually be overcome by these devices, though it may take a little experimentation.

5.2 Choice of Verb

For clarity, the form a verb takes should be as simple as possible [79, p. 216]. Yet it is still easy to make errors in subject-verb agreement and in tense and other properties called mood and aspect. Subject-verb agreement is a matter of grammar,[35] but verb tense, mood, and aspect have to do with meaning and the status of the material being reported.

5.2.1 Verb Agreement

Long, complex, grammatical subjects often lead to errors in agreement between subject and verb [340, pp. 208–22]. The subject is typically a phrase, called a noun phrase, which consists of a noun as the core or head and other words or phrases that modify it [32, §4.3] [297, §2.28]. A disagreement between subject and verb is illustrated in Example 5.16, reproduced from Example 2.49.

Plural subject head

✗ In this paper, the experimental results of a complete battery-less wirelessly-powered 2.45-GHz temperature sensor system with a microcontroller (μC) and an ISM-band transmitter on FR-4 is presented.

Singular verb

Example 5.16. *Disagreement between subject and verb, from an instrumentation engineering article [595].*

The subject is the noun phrase consisting of the 23 words preceding the verb 'is presented'. The head is the plural noun 'results'. The fact that there are so many words between 'results' and the verb may have led to the assumption that the subject was instead the singular noun phrase 'sensor system', or perhaps the error was caused by the nearness of the singular noun 'transmitter'. The proximity effect, also known as attraction [40] [297], can be compelling.

Another illustration of subject-verb disagreement is given in Example 5.17.

Plural subject head — In a narrowband MIMO beamforming combining system, quantization and feedback of the optimal beamforming vector, estimated at the receiver has been considered as a means to practically obtain the benefits of instantaneous feedback [5], [6], [8], [17]. — Singular verb

✗

Example 5.17. *Disagreement between subject and verb, from an electrical and electronic engineering article [617].*

It is unclear whether the error comes from the nearness of the verb to 'receiver' (which needs a closing comma), or its nearness to the earlier 'vector', or the subject 'quantization and feedback' being interpreted as a single theme [40].

5.2.2 Tense

The most obvious variation in verb form is its tense [32] [297]. Tense refers to whether the verb is in the present or past, which is not quite the same thing as the time to which it refers.[36] Importantly, only finite verbs have tense [32, p. 58], for example, 'conduct', 'conducted'.[37] Finite clauses, mentioned elsewhere in this chapter, are ones which contain a finite verb [297, §3.52].

The present tense is used for events that happen regularly or for established knowledge, logical truths, and statements of fact ('the sun rises every day'). In theoretical, especially mathematical, papers, which also actively engage the reader, most of the exposition is in the present tense [140]. Chapter 6 contains examples.

In experimental papers, there is greater variation. Both the methods and results sections, which describe actions, are mainly in the past tense (the present tense is used to introduce figures and add inferences), but the introduction and discussion sections generally contain a mixture of tenses [79].

Illustrations of accepted usage in these different contexts are given in Examples 5.18–5.21.

Past tense, action

Samples of amorphous solid C_6H_6 were formed on the Al substrate at 10 K by the background vapor deposition of C_6H_6. The compositions of the samples were measured by in situ reflection–absorption spectroscopy with the FTIR spectrometer.

Past tense, action

Example 5.18. *Past tense in a methods section, from a chemistry article [533].*

Present tense, fact

Tunneling arises from the wave nature of matter, allowing particles to penetrate barriers that are impossible to overcome classically. . . . — Present tense, fact

In situ IR spectroscopy revealed that cyclohexane or deuterated cyclohexane [is] efficiently formed by reactions R1–R6. — Past tense, action

Present tense, fact

Example 5.19. *Present and past tenses in an introduction, from the article cited in Example 5.18.*

Present tense, fact

Fig. 1*A* shows the IR spectra of amorphous C_6H_6 and C_6H_{12} at 20 K. . . .

In the difference spectrum *B*, the C_6H_6 absorption bands . . . decreased, and new peaks for the C_6H_{12} products appeared at 2,800–3,000 and 1,453 cm^{-1}. This indicates the formation of C_6H_{12} by reactions . . .

Past tense, result

Present tense, inference

Example 5.20. *Present and past tenses in a results section, from the article cited in Example 5.18.*

Present tense, fact

Fig. 3*B* shows that the values of $\Delta C_6H_{6\_H}$ and $\Delta C_6H_{6\_D}$ strongly depend on surface temperature. They increased at low temperatures (15–25 K), until competitive adsorption of H_2 or D_2 occurred at 10–12 K (Fig. 4). This suggests that reaction R1 occurred mainly via the LH pathway.

Present tense, fact — Past tense, result — Past tense, result — Present tense, inference

Example 5.21. *Present and past tenses in a discussion, from the article cited in Example 5.18.*

Notice that although the methods section of an experimental paper is intended to allow other researchers to reproduce what was done, it remains a record of actions performed in the past, as in Example 5.18. The nonstandard description in Example 5.22 set entirely in the present tense is more a recipe than a record.

✕

> The current-controlled interfacing inverter <u>is first tested</u> in simulation with the fifth and seventh harmonics compensated. — Inappropriate present tense
>
> Inappropriate present tense — Experiments <u>are also conducted</u> on a 5-kVA grid-connected DG system to verify the proposed VCM-based compensation scheme and the circulating harmonic current phenomena.

Example 5.22. *Nonstandard use of the present tense to describe methods, from an electrical and electronic engineering article [539].*

The only exception to the expected use of the past tense in the methods section might be when listing an algorithm or describing some general procedure.

5.2.3 Aspect

The aspect of a verb expresses the duration or completion of an event or state; it is particularly useful in describing the variations in the status of research findings. Detailed accounts can be found in [32] [297]. The differences between the simple past, the present perfect, and the past perfect are as follows [224, ch. 4]:

- The simple past refers to an event that took place at a specific time in the past ('analysis showed').
- The present perfect refers to an event that took place over a nonspecific time in the past and which continues to or has relevance to the present and possibly into the future ('analysis has shown').
- The past perfect refers to an event that took place over a nonspecific time in the past and which was completed before some more recent relevant time ('analysis had shown').

Examples 5.23 and 5.24 show combinations of the simple past, the present perfect, and the past perfect in a review and a methods section.

Simple past, specific time

> This experiment <u>demonstrated</u> that a high-performance metamaterial . . .
>
> Present perfect, continuing truth — Recent work (18, 19) <u>has shown</u> that a more accurate description of a negative index material <u>is</u> negative space.

Present tense, fact

Example 5.23. *Simple past tense and present perfect, from a multidisciplinary sciences article [677].*

In the first extract in Example 5.23, the simple past is used for the result of an experiment. In the second, the present perfect is used for work that took place in the past but whose truth continues to the present and possibly beyond. The present perfect signifies its 'current relevance' [32, p. 144]. If the simple past 'Recent work (18, 19) showed . . .' had been used, it would leave open the possibility of a subsequent contradictory statement; for instance, 'but it is now known that . . .'. The present perfect eliminates that uncertainty.

Simple past, specific time — Treatment with CAMPATH-1H (with or without anti-CD4) was given electively to patients who had shown at — Past perfect, completed earlier
least one Gd-DTPA enhancing lesion during screening.

Example 5.24. *Simple past tense and past perfect, from a clinical neurology article [618].*

In Example 5.24, the simple past 'was given' is used for a treatment that took place at a specific time in the past. The past perfect 'had shown' is used for the evidence being present over some nonspecific period preceding the treatment. Without the past perfect, it would have been necessary to make explicit the fact that the evidence preceded the treatment rather than being obtained at, say, the same time.

The choice of the simple past or perfective forms in Examples 5.23 and 5.24 was determined by the facts. It can also be determined by interpretation, as in Example 5.25.

Present perfect, continuing utility — This additional filter has been useful for distinguishing
authentic plant miRNAs from false positives and recently
Present perfect, continuing utility — has been extended to . . .

Simple past, specific time — Two new target candidates were found, but evidence for
miRNA-guided cleavage of these targets was not detected — Simple past, specific time
by 5' RACE in . . .

Example 5.25. *Present perfect and simple past tenses used to support interpretation, from a plant biology article [554].*

In the first extract of the example, the present perfect is used to make a judgement about the continuing usefulness of a device, a filter, as well as for a factual comment about its continuing applicability. If instead the simple past had been used for the first statement, that is, 'This additional filter was useful', it would suggest—or at least allow the possibility—that it might not be useful now. In the second extract, the simple past is used to report a specific result and the specific failure to find a property related to that result. The second statement identifies a limitation on the first left open by the simple past.

In the same way, the present perfect can be used to moderate inferences drawn from results, as in Example 5.26.

Present perfect, continuing importance

> The simulation results <u>have shown</u> that, when the training set is sufficiently large, the classification decision of the distributed algorithm is equivalent to that of the centralized algorithm.

Example 5.26. *Present perfect to support a conclusion, from an electrical and electronic engineering article [564].*

If instead of 'have shown' the simple past 'showed' had been used, it could suggest some uncertainty about the continuing truth of the equivalence of the authors' algorithms; that is, the possibility of the equivalence not holding now. Alternatively, if the present 'show' had been used, it could appear too much of rhetorical flourish, and suggest a certainty that goes beyond the data.

5.2.4 Mood

Mood is a grammatical category into which verbs are classified as indicative, imperative, or subjunctive according to whether they express a fact, a command, or hypothesis, respectively [44] [297]. The indicative is used routinely for reporting facts, as in Example 5.27.

Indicative

> Printed circuit board failures <u>range</u> from circuit malfunctions to violent propagating PCB faults. . . .

Indicative — The arcing <u>damaged</u> the insulation of the PCB below the part.

Example 5.27. *Use of the indicative, from an electrical and electronic engineering article [676].*

The imperative expresses a command [297, §11.24] and is mostly found in mathematical expositions and algorithm listings, as in Example 5.28.

Imperative — <u>Let</u> $A = (a_{ij})$ be a symmetrizable generalized Cartan matrix in the sense of [4] . . .

Imperative — <u>Put</u> $x = e^u$ and <u>write</u> $\check{R}(x)$ for $\check{R}(u)$.

Example 5.28. *Use of the imperative in a mathematical argument, from a mathematical physics article [553].*

The imperative is also used in non-mathematical texts to secure the reader's attention. It is strikingly effective at the beginning of a paragraph, where it can set the tone for what follows, as in Example 5.29.

Imperative — First, consider anchoring. . . .

Imperative — Suppose that an individual experiences loud music for two hours.

Example 5.29. *Use of the imperative in a non-mathematical argument, from an economics article [545].*

The subjunctive is used for hypotheses, particularly those expected not to hold [297, §3.62], as in Example 5.30.

Subjunctive — If this were the only way to implement such transformations then scalable linear-optical quantum computing would be hopeless.

Example 5.30. *Use of the subjunctive, from a physics article [429].*

Its appearance in a conditional such as 'If this were' may be more a matter of style than grammar, and 'were' may be replaced by the indicative 'was' [284], but opinions differ. In spite of predictions of its demise, the subjunctive continues to survive [44].

5.3 Choice of Voice and Subject

Voice is a grammatical category that provides two ways of viewing a situation, namely as active or passive, according to the manner in which the participants in the activity are reported by the verb [32] [297]. The voice is active if the subject performs the action and passive if the subject is the recipient of the action, as illustrated in Example 5.31.

Active, subject performs action — Intercalation weakened the interface bonding. — Object of action

Passive, subject recipient of action — The interface bonding was weakened.

Passive, subject recipient of action — The interface bonding was weakened by intercalation. — Passive, agent of action

Example 5.31. *Active voice, passive voice, and passive voice with agent of action; adaptations from an applied physics article [654].*

The object of the active verb in the first sentence, 'interface bonding', is the subject of the passive verb in the second and third sentences. The only difference between the second and third sentences is that the agent of the action is made explicit in the third, introduced by 'by' [32, §9.3] [346, §412–413].

Problems with voice and agent are ubiquitous in scientific writing. They are commonly revealed in unbalanced, awkward, or roundabout constructions that are difficult to read or even parse. With some thought, most can be avoided.

5.3.1 Active versus Passive Voice

There is general agreement that the active voice should be used in preference to the passive [79] [340], yet recognition also that the choice depends on circumstances, on the point of view [49] [80], on whether the agent of the action is known or relevant [224], and on what is being described [68].

The active voice has the virtue of clarity, vigour, and immediacy [391], and it typically produces shorter, tighter sentences. But the beginning of a sentence can have a special role in providing focus [92, §30.3] [340, pp. 290–1]. The passive voice allows the grammatical object that would otherwise be nearer the end of a sentence in the active voice to be manoeuvred to the beginning.

Consider Example 5.31 again. If the intended emphasis is on intercalation, then the active construction of the first sentence is appropriate; that is, 'Intercalation weakened the interface bonding'. If the intended emphasis is on interface bonding alone, the passive construction of the second sentence is appropriate; that is, 'The interface bonding was weakened'. If the agent of the action, intercalation, is relevant, then the passive voice allows it to be included, as in the third sentence.

In reality it is the third formulation in Example 5.31 that the authors used [654]. It appeared in the discussion section where the emphasis was on interface bonding rather than on intercalation. The passive voice thus delivered the required focus, without loss in information [340].

Revealingly, the passive voice is used to great effect in journalism. Example 5.32 is from an edition of the *Times Higher Education* magazine[38] in which several summary reports begin in the passive voice.

Focus — An investigation has been launched into claims that the president of a south Texan university plagiarised portions of her dissertation.

Focus — A $2.1 million (£1.3 million) payout has been agreed by Texas A&M University over a tragedy in which 12 students lost their lives.

Example 5.32. *Two reports beginning in the passive voice, from a magazine article [695].*

In each of the two statements, the object of interest strikes the reader immediately, namely the investigation and the size of the payout. For comparison, in Example 5.33 the reports have been edited to begin in the active voice. The effect is to focus instead on the agent of the action.

Focus — A south Texan university has launched an investigation into claims that the president plagiarised portions of her dissertation.

Focus — Texas A&M University has agreed a $2.1 million (£1.3 million) payout over a tragedy in which 12 students lost their lives.

Example 5.33. *Reports in Example 5.32 edited to begin in the active voice.*

With this formulation, an impatient reader who does not have a special interest in Texas universities may never complete the rest of the sentence or indeed the report.

The recipe for choosing between active and passive constructions in these kinds of constructions is simple [340]. If the agent of the action is known and is more important than the object of the action, use the active voice, as in 'Intercalation weakened the interface bonding'. If the object of the action is more important than the agent, use the passive voice, as in 'The interface bonding was weakened' or 'The interface bonding was weakened by intercalation'.

Nonetheless, judgements of this type do not take place in isolation. The choice of voice may depend on what has already been said as well as on the properties of the agent and object within the sentence itself. Techniques for shifting the focus of interest are described comprehensively in [32, pp. 233–6] [297, §18.9–18.31].

5.3.2 Repairing Top-Heavy Passive Constructions

One rationale for the use of the authorial 'we' (§5.1.1) is that it avoids the unbalanced, top-heavy sentences that passive constructions can sometimes generate, with a long subject preceding a late-placed verb.[39] An example of a top-heavy sentence is given in Section 5.2.1. Another is given in Example 5.34.

Subject head

✗ A new approach toward target representation and localization, the central component in visual tracking of nonrigid objects, is proposed.

Late-placed verb

Example 5.34. *A top-heavy sentence in the passive voice, from a computer science article [453].*

The subject is the noun phrase consisting of the 17 words preceding the verb 'is proposed'. The head of the noun phrase is 'approach'. The sentence is clearly top-heavy.

There are several ways to repair or replace top-heavy passive constructions without introducing personal pronouns:

- Move the verb nearer the head of the noun phrase defining the subject.
- Split the sentence into two: one introducing the head, the other the detail modifying it.
- Convert the passive voice into the active voice.

Thus, in Example 5.34 the sentence can be rebalanced by moving 'is proposed' to just after the subject head 'approach', as in the edited version in Example 5.35.

Subject head Verb

> A new approach is proposed toward target representation and localization, the central component in visual tracking of nonrigid objects.

Example 5.35. *Edited version of Example 5.34.*

The sentence is no longer top-heavy, with now only three words making up the subject before the verb 'is proposed'. The following 14 words deliver the expected end weight[39], so that the extra information—the detail about the proposed approach—appears late in the sentence [32, pp. 225–6] [297, §18.40]. Readability might be improved a little further by changing 'proposed toward' to 'proposed for'. In either case, there has been a cost: the long prepositional phrase 'toward target representation . . .' is now separated from its noun 'approach' by two words 'is proposed'. Still, the discontinuity is small and conventionally acceptable [92, §30.6] [297, §18.39].

Example 5.36 is more extreme, with 40 words in the noun phrase making up the subject before the verb 'is described'.

Subject head

✗

> The application of new a segmentation software, Anatomatic™ in the evaluation of volumetric measurements of ovarian tumours and the new Medimag™ three-dimensional (3D) software in the evaluation of 3D image representation of ovarian tumours with 1.5 T magnetic resonance imaging (MRI) is described.

Late-placed verb

Example 5.36. *Top-heavy sentence in the passive voice, from a computer science article [465].*

The sentence can be rebalanced in a similar way to Example 5.35, that is, by moving 'is described' to just after the subject head 'application'. Here, however, the discontinuity is more awkward, and the remaining prepositional phrases beginning 'of', 'in', and so on remain difficult to parse. Instead, it is better to split the sentence into an introductory statement and a slightly longer explanation, as in the edited version in Example 5.37.

Subject head

The application of new software is described for 1.5 T magnetic resonance imaging (MRI) of ovarian tumours. Segmentation software, Anatomatic^{TM}, evaluates volumetric measurements, and three-dimensional (3-D) software, Medimag^{TM}, evaluates 3D image representations of the tumours. — Verb

Example 5.37. *Edited version of Example 5.36.*

The first sentence is still passive, but the number of words preceding the verb has been reduced to five. The second sentence, which is now in the active voice, complements the first by giving details of the software.

Example 5.38 is also extreme, with 39 words or symbols in the noun phrase making up the subject before the verb 'is shown':

Subject head

✗ In Fig. 7 a detailed view of the PV array voltage and of the duty cycle for the P&O controlled boost battery charger with $\varepsilon = 0.1$, $T_a = 0.01$ s, $\Delta d = 0.005$ and $S = 1000$ W/m^2 is shown.

Late-placed verb

Example 5.38. *Top-heavy sentence in the passive voice, from an electrical and electronic engineering article [500].*

The head of the noun phrase is 'view'. Because it refers to a figure, the sentence can be straightforwardly rebalanced by making the figure the subject of the sentence, as in the edited version in Example 5.39.

Verb

Subject head — Fig. 7 shows a detailed view of the PV array voltage and of the duty cycle for the P&O controlled boost battery charger with $\varepsilon = 0.1$, $T_a = 0.01$ s, $\Delta d = 0.005$ and $S = 1000$ W/m^2.

Example 5.39. *Edited version of Example 5.38.*

Another less common way of dealing with top-heavy passive sentence is by inverting the normal order of subject and verb, as in Example 5.40.

Verb | Subject head

In the following sections are presented lake-level data obtained from the sediment profiles of sites San Carlo, Ponale, and core Ledro II-3.

Example 5.40. *Subject-verb inversion, from a physical geography article [597].*

This device is called balance inversion [123]. In Example 5.40 it seems slightly artificial and may not have much advantage over a non-inverted version, as in the edited version in Example 5.41. The sentence has the normal order of subject and verb but is now somewhat top-heavy.

Subject head

✗ In the following sections lake-level data obtained from the sediment profiles of sites San Carlo, Ponale, and core Ledro II-3 are presented.

Verb

Example 5.41. *Edited version of Example 5.40.*

A stronger case for balance inversion might be made with the extremely top-heavy sentence of Example 5.38, which, when inverted, reads as in Example 5.42.

Verb | Subject head

In Fig. 7 is shown a detailed view of the PV array voltage and of the duty cycle for the P&O controlled boost battery charger with $\varepsilon = 0.1$, $T_a = 0.01$ s, $\Delta d = 0.005$ and $S = 1000$ W/m^2.

Example 5.42. *Edited version of Example 5.38 with subject-verb inversion.*

Inversion seems more natural here than in Example 5.40, although its success remains a matter of perception [123, p. 190].

A side effect of inversion can be to cause a change in emphasis [123]. In Example 5.43, 'Missing' is the focus of attention at the beginning of the sentence. If subject and verb are put in the normal order, as in the edited version in Example 5.44, then the sentence is not only unbalanced, but the focus is no longer on 'missing'.

First subject head

Verb

Focus — Missing from (5.4) is the notion of identity morphism (as an arrow from $\ulcorner$an object of $C\urcorner$ to $\ulcorner$a morphism in

Second subject head — $C\urcorner$) and the associated path equivalences, as well as the identity and associativity laws.

Example 5.43. *Subject-verb inversion, from a mathematical sciences textbook [679].*

First subject head

✗ The notion of identity morphism (as an arrow from $\ulcorner$an object of $C\urcorner$ to $\ulcorner$a morphism in $C\urcorner$) and the associated path equivalences, as well as the identity and associativity — Second subject head

laws, are missing from (5.4).

Verb

Example 5.44. *Edited version of Example 5.43 with normal order of subject and verb.*

Despite its effectiveness, subject-verb inversion of this kind represents an unnatural word order. It should not be used carelessly [123].

5.3.3 'It' and 'There'

Constructions of the form 'It is' and 'There are' are very common in scientific writing.[40] The pronouns 'It' and 'There' are called dummy subjects[41] as they allow the understood subject to appear later in the sentence [32], where its content can be developed. Examples 5.45 and 5.46 show typical instances, the first to report an observation, the second to introduce a new topic.

Dummy subject — It was found that the surfaces touched when their average separation was 32.7 ± 0.8 nm.

Example 5.45. *Dummy subject 'It', from a nuclear physics article [394].*

Dummy subject — There are three types of errors that influence the descriptors computation: the region error, the localization error, and the error of the estimated orientation angle.

Example 5.46. *Dummy subject 'There', from a computer science article [605].*

Because dummy subjects delay access to the content of the sentence and add wordiness, they are generally deprecated, although there are counterarguments [284, p. 295].

They can often be eliminated without affecting the sense too much. The dummy subject in Example 5.45 is absent in the edited versions in Examples 5.47 and 5.48.

> The surfaces were found to touch when their average separation was 32.7 ± 0.8 nm.

Example 5.47. *Edited version of Example 5.45.*

> The surfaces touched when their average separation was 32.7 ± 0.8 nm.

Example 5.48. *Further edited version of Example 5.45.*

Likewise the dummy subject in Example 5.46 is absent in the edited version in Example 5.49.

> Three types of errors influence the descriptors computation: the region error, the localization error, and the error of the estimated orientation angle.

Example 5.49. *Edited version of Example 5.46.*

All three edited versions are more immediate and slightly shorter than the originals. Nevertheless the originals may be preferred for rhetorical reasons [297, §18.48].

Not all dummy subjects can be eliminated as easily as in the preceding examples, especially when they express the existence of something [32, pp. 231–3] [297, §18.47]. Example 5.50 shows such a case.

Dummy subject —

> There are few corresponding regions for large transformations and the recall-precision curves are not smooth.

Example 5.50. *Bare existential 'There', from a computer science article [605].*

For 'There are' to be eliminated, 'exist' or the equivalent would need to be introduced, as in 'Few corresponding regions exist . . .'.

5.4 Uncertain Referents

Participles such as 'using' and 'based' and pronouns such as 'this' and 'which' cause confusion when they begin phrases or clauses that refer to the wrong grammatical subject or to no subject at all. The author may have a subject in mind, but the reader cannot tell from the phrase or clause what it is.

Unsurprisingly, not being able to identify the intended subject of a phrase or clause is viewed as a grammatical error [297, §15.59]. Burchfield devotes three columns [40, pp. 804–5] and Partridge two [280, pp. 77–8] to the problem of unattached participles, also known as dangling, hanging, misrelated, confused, or disconnected participles. In scientific writing, the use of unattached constructions is institutionalized [297, §15.59].

5.4.1 Unattached or Misattached Participial Phrases

Examples 5.51 and 5.52 illustrate instances of probably the most common kind of unattached participial phrase, namely that beginning with 'using' [79, p. 42].

Unattached participle

✗ Using the time-dependent Bogoliubov-de Gennes equation a reduced mean value of the order parameter oscillation compared to the obtained equilibrium value of the order parameter has been found.

Example 5.51. *Unattached participial phrase, from a physics article [535].*

In Example 5.51, who or what is 'using the time-dependent Bogoliubov-de Gennes equation'? Grammatically, the present participle 'Using' refers to the following noun phrase 'a reduced mean value', which interpretation is nonsensical. In fact, no noun phrase in the sentence can sensibly act as subject of this participial phrase. The only possibility is the unexpressed authors of the article.

✗

The light pulses are triggered and detected synchronously

Unattached participle — using a fast (5 MHz) DAC/ADC board which integrates the detector outputs for each pulse.

Example 5.52. *Unattached participial phrase, from a physics article [623].*

In Example 5.52, who or what is the subject of 'using a fast (5 MHz) DAC/ADC board'? It is neither the grammatical referent 'light pulses' nor, by contrast with Example 5.51, the unexpressed authors of the article.

Example 5.51 can be repaired by replacing the unattached 'Using' by an instrumental preposition 'With' or 'With the aid of' [297, §9.50], as in the edited version in Example 5.53.

Instrumental prepositional phrase — With the aid of the time-dependent Bogoliubov-de Gennes equation, a reduced mean value of the order parameter oscillation compared to the obtained equilibrium value of the order parameter has been found.

Example 5.53. *Edited version of Example 5.51.*

Incidentally, the sentence can also be simplified, as in Example 5.54. The Latinate 'reduced . . . compared to' has been replaced by 'smaller than' (§2.3.1) and ellipsis used to eliminate repetition of 'of the order parameter' (§5.5.2).

> With the aid of the time-dependent Bogoliubov-de Gennes equation, a mean value of the order parameter oscillation has been found that is <u>smaller than</u> the obtained equilibrium value.

Replaces 'reduced . . . compared to' —

Ellipsis of 'of the order parameter'

Example 5.54. *Further edited version of Example 5.51.*

Example 5.52 can be repaired by replacing the unattached 'using' with 'by' [297, §9.50], as in Example 5.55. The phrase 'a fast (5 MHz) DAC/ADC board' becomes the agent of the action. The preposition 'by' is said to be agentive [297, §9.7].

> The light pulses are triggered and detected synchronously <u>by</u> a fast (5 MHz) DAC/ADC board which integrates the detector outputs for each pulse.

Preposition —

Example 5.55. *Edited version of Example 5.52.*

The unattached participle 'providing' in Example 5.56 is harder to deal with because of its vagueness.

✗

> Hippocampal place cells fire when the animal visits a specific area in a familiar environment (1), <u>providing a population representation of self-location (2–4)</u>.

— Unattached participial phrase

Example 5.56. *Unattached participial phrase, from a systems biology article [442].*

Who or what is 'providing a population representation of self-location'? It is unlikely to be the grammatical referent 'a familiar environment' nor the preceding 'area' or 'animal'. It might be 'Hippocampal place cells', which is a long way from the participial phrase, or it might be the whole of the preceding clause 'Hippocampal place cells . . . in a familiar environment'. Depending on the authors' intention, the sentence might be recast as in Example 5.57.

> Hippocampal place cells fire when the animal visits a specific area in a familiar environment (1). <u>The effect</u> is to provide a population representation of self-location (2–4).

— Inserted noun phrase

Example 5.57. *Edited version of Example 5.56.*

The unattached participial phrase ‘providing a population representation of self-location’ has been incorporated into a sentence whose subject ‘The effect’ represents the whole of the preceding sentence. Other versions are possible.

The foregoing examples have present participles, ‘using’, ‘providing’, as the head of the unattached phrase. Example 5.58 shows an instance with a past participle, ‘known’, as the head.

✗

> The propulsion of intraluminal content throughout the gastrointestinal (GI) tract relies on the coordinated movements of the longitudinal and circumferential muscle layers in the gut wall, commonly known as peristalsis [1], [2]. — Unattached participial phrase

Example 5.58. *Unattached participial phrase, from an instrumentation engineering article [401].*

The subject of ‘commonly known as peristalsis’ is grammatically the preceding noun phrase ‘the gut wall’. The intended referent was probably ‘the coordinated movements’, although it might also have been the whole of the preceding clause. It might be argued that the referent is obvious, but only if the reader already knows what is meant by peristalsis, and in that case a definition is unnecessary. Depending on the authors’ intention, the sentence might be recast as in Example 5.59.

> The propulsion of intraluminal content throughout the gastrointestinal (GI) tract relies on the coordinated movements of the longitudinal and circumferential muscle layers in the gut wall. This process is commonly known as peristalsis [1], [2]. — Inserted noun phrase

Example 5.59. *Edited version of Example 5.58.*

The unattached participial phrase ‘commonly known as peristalsis’ has been incorporated into a sentence whose subject ‘This process’ represents the whole of the preceding sentence, in the same way that ‘The effect’ does in Example 5.57.

Unattached ‘to’ infinitive phrases, for example ‘to show . . .’, can produce similar ambiguities [297, §15.59].

Fortunately, not all words that look like participles, particularly those ending in ‘ing’, need be treated as participles. Some are prepositions or quasi-conjunctions. The following list is drawn from [40, pp. 804–5] [280, pp. 78–9], with the slightly more formal ‘that’ after some entries being optional:

- according to
- allowing for
- assuming (that)

- barring
- concerning
- considering
- depending on
- excluding
- excepting
- given (that)
- owing to
- provided (that) and providing (that)
- regarding
- supposing (that)

Examples 5.60 and 5.61 illustrate usages from this list.

Preposition — Considering the conditions of this heat treatment, it is expected that the aligned structure can also be formed in austenite to ferrite transformation.

Example 5.60. *Prepositional '-ing' word, from a materials science article [631].*

The algorithms are general and can be applied to any
Preposition — robot providing that its kinematics is known.

Example 5.61. *Quasi-conjunctive '-ing' word, from a robotics article [633].*

Whereas 'providing' in Example 5.61 has a quite different role from that in Example 5.56, to avoid misdirection it might still be better replaced here by 'provided'. The ubiquitous 'using' is yet to be accepted as a preposition.

5.4.2 Unattached or Misattached Relative Clauses

Examples 5.62 and 5.63 illustrate instances of unattached relative clauses beginning with 'which'.

✗ By west European standards, Russian adults, particularly men, have a very high risk of premature death, which has fluctuated sharply in recent decades. — Unattached relative clause

Example 5.62. *Misattached relative clause, from an epidemiology article [730].*

✕

The combination of the micro- and nano-structures increase[s] the roughness of the surface, and air can be trapped into the interspaces of the rough surface, which can enhance the WCA according to the Cassie's theory [28]. — Unattached relative clause

Example 5.63. *Unattached relative clause, from a materials science article [474].*

In Example 5.62, the relative pronoun 'which' refers grammatically to the preceding noun 'premature death', not the intended referent, namely the risk of death. In reality, the subsequent text, not shown here, concentrates on the high mortality rate and not its fluctuations per se. The relative clause could therefore have been omitted or placed rather later in the exposition.

In Example 5.63, the relative pronoun 'which' refers grammatically to the preceding noun 'surface', but from the content it seems more likely to refer to 'air' or 'interspaces' or the whole of the preceding sentence ('WCA' is water contact angle). The sentence is better recast. One possibility is shown in Example 5.64.

The combination of the micro- and nano-structures increases the roughness of the surface, and air can be trapped in the interspaces of the rough surface. Applies to preceding sentence — As a result, the WCA can be enhanced, according to Cassie's theory [28].

Example 5.64. *Edited version of Example 5.63.*

In this edited version, the relative clause has been replaced by a sentence with a passive verb so that there is no agent needed for 'enhance'. Instead the adverbial 'As a result' (or 'Consequently') refers to all of the actions described in the preceding sentence.

5.4.3 Ambiguous Pronouns

For demonstrative pronouns such as 'this' and 'these' to do their job, what they refer to needs to be obvious [40, p. 780]. Unfortunately, the opportunity for ambiguity exceeds even that with relative pronouns. Examples 5.65–5.69 illustrate several instances.

In Example 5.65 the pronoun 'This' could refer to the setups not being reported or to the setups being different or to both. If it were both, then 'This' could be replaced by something like 'Identifying these differences'.

In addition, reaction setups differ and sometimes they Ambiguous pronoun — were not even reported. This is critical since the power introduced into the reaction system will strongly depend on the detailed setup.

✕

Example 5.65. *Pronoun with uncertain referent, from a chemistry article [680].*

In Example 5.66 there is no plural referent for the pronoun 'These'. The authors may have had in mind the unexpressed parameters of the P&O (perturb and observe) method. If this interpretation is correct, 'These' could be replaced by 'The parameters of the P&O method'. Incidentally, the adverb 'merely' could also be moved from its position just before 'chosen' to just after it, so that the slightly misleading 'merely chosen' is replaced by 'merely on the basis'.

✕

> In [12] it is shown that the P&O method, when properly optimized, leads to an efficiency which is equal to that obtainable by the INC method. These are often merely chosen on the basis of trial and error tests. — Ambiguous pronoun

Example 5.66. *Pronoun with no referent, from an electrical and electronic engineering article [500].*

In Example 5.67 the pronoun 'This' could refer to pathophysiological mechanisms not being mutually exclusive or to their number or to ICU-acquired weakness. Further reading of the article suggests that the most probable referent is ICU-acquired weakness. If so, then 'This' could be replaced by an explicit noun phrase; for instance, 'This weakness' or 'The development of weakness'.

✕

> Weakness acquired in the intensive care unit (ICU) is caused by many different pathophysiological mechanisms that are not mutually exclusive. This is not surprising, given the diverse diseases that precipitate critical illness. — Ambiguous pronoun

Example 5.67. *Pronoun with uncertain referent, from a general and internal medicine article [569].*

In Example 5.68 the pronoun 'this' could refer to differentiation, regulation, or the importance of the Th2 locus control region. If it were regulation, 'this' could be replaced by 'this regulation'.

✕

> The Th2 locus control region (LCR) has been shown to be important in efficient and coordinated cytokine gene regulation during Th2 cell differentiation. However, the molecular mechanism for this is poorly understood.

Ambiguous pronoun

Example 5.68. *Pronoun with uncertain referent, from an immunology article [546].*

In Example 5.69 the pronoun 'This' could refer to 'application' or 'interest'. If it were interest, 'This' could be replaced by 'This attention'.

✕

In recent years, there has been considerable interest in the application of fiber Bragg grating (FBG) sensors. This is because they have many practical advantages.

Ambiguous pronoun

Example 5.69. *Pronoun with uncertain referent, from an instrumentation engineering article [709].*

As the suggested corrections to these examples illustrate, the ambiguity with 'this' and 'these' can often be eliminated by adding a noun or noun phrase, thereby converting the pronoun into an adjective [40, p. 780], as in Example 5.70.

In that case, of course, the same argument could apply to the other intertwined electronic phases. These ideas motivate . . . — Unambiguous noun phrase

These two gap functions are approximately described by cos k_x − cos k_y and cos k_x + cos k_y, respectively.

Unambiguous noun phrase — This result indicates that . . .

However, due to the intertwining of these instabilities, the order parameter may fluctuate from one type to

Unambiguous noun phrase — another. This fluctuation would prevent . . .

Example 5.70. *Elimination of ambiguity, from a physics article [466].*

These extracts are taken from an article [466] in which there are 35 sentences beginning with 'This' or 'These'. Yet in every occurrence, 'This' or 'These' functions as an adjective, not as a pronoun. The reader is never left with uncertainty.

5.4.4 Other Misattributions

In a kind of mental shorthand, the property of a thing is sometimes confounded with the thing itself. The misattribution can lead to nonsense.

In Example 5.71, it is the difference in the two oscillating signals that was amplified, not the difference frequency.

✕

By biasing the two oscillators (a few microvolt apart) on their velocity matching steps, the two resulting oscillations (280-330 GHz) were mixed in the SIS mixer.

= 'signal'? — The difference frequency was amplified and detected on a spectrum analyzer.

Example 5.71. *Attribute substitution, from an applied physics article [734].*

In Example 5.72, it is not the temperatures that were freezing but the weather. Actually 'temperatures' could be omitted, to read 'survive freezing', as in the second sentence.

✕

Over the course of the year, plants from temperate regions vary dramatically in their ability to survive freezing temperatures. In the warm growing seasons, such plants display little capacity to withstand freezing.

= 'weather'? — (temperatures)

Example 5.72. *Attribute substitution, from a plant sciences article [697].*

In Example 5.73, it is not results that were measured but electrical properties, namely current and voltage [534]. The results were obtained as a consequence of making those measurements.

= 'Current and voltage'? (Results)

✕

Results were measured at 25 °C with an aperture area of 0.231 cm^2 . . .

Example 5.73. *Attribute substitution, from a chemistry article [534].*

In Example 5.74, it is not the parameters that were sensed but physical properties such as strain, temperature, and refractive index [705].

✕

The fiber-optic sensor industry is becoming increasingly important, employing presently different techniques to sense a large variety of parameters in demanding applications [1]–[3].

= 'physical properties'? (parameters)

Example 5.74. *Attribute substitution, from an instrumentation engineering article [705].*

In Example 5.75, it is not the wavelengths that were red but the light. In this context there seems little to be gained in using 'red' rather than 'long'.

✕

The topographic effect has been characterised precisely from measurements of nadir-viewed radiance at red wavelengths.

— = 'long'? (red)

Example 5.75. *Attribute substitution, from an imaging science article [700].*

Misattributions such as these are known in rhetoric as hypallage or more specifically as transferred epithets. The rhetorical effect can be compelling ('he smoked a nervous cigarette') [121]. In the preceding examples, however, the transpositions seem not to have been deliberate. This is not to say that transferred epithets should be avoided. They can provide a useful shorthand; for instance, 'red shift' to refer to the displacement of spectral lines towards longer wavelengths in the radiation from distant receding stars. Nonetheless, transferred epithets should be used sparingly in case they suggest confused thinking.

5.5 Parallelism, Coordination, and Ellipsis

There are several ways to structure sentences and parts of sentences to take advantage of similarities, differences, and repetition.

5.5.1 Parallelism and Coordination

Parallelism and coordination refer to two or more neighbouring phrases, clauses, or sentences sharing grammatical features, namely word class, tense, aspect, word order, or clause structure [224] [297]. The effect is to focus attention on the similarities and differences [284], and make reading easier. Examples 5.76–5.79 illustrate several instances of parallelism. For clarity, only one instance is marked in each example, with others identified in the text.

In Example 5.76, the two clauses following 'is' on the first line are identical in form. The clause 'to show that certain systems satisfy them' is parallel with the clause 'to show that certain systems do not satisfy them'. Additionally, the clause '. . . therefore are spin glasses' is parallel with the clause '. . . therefore are not spin glasses'.

> In fact, the main use of conditions (i) and (ii) is not to show that certain systems satisfy them and therefore are spin glasses; but to show that certain systems do not satisfy them and therefore are not spin glasses.
>
> Parallel

Example 5.76. *Parallel clauses, from a physics article [708].*

In Example 5.77, the three phrases 'higher . . .', 'larger . . .', 'smaller . . .' are parallel with one another, the last phrase receiving greater emphasis with the insertion of 'consequently'.[42]

> Parallel
>
> The higher the central concentration, the larger the central value of *W*, and consequently the smaller the central value of λ.

Example 5.77. *Parallel phrases, from an astronomy and astrophysics article [562].*

In Example 5.78, the clause 'high-cholesterol diets have been shown . . .' is parallel with the clause 'cholesterol-lowering drugs have been shown . . .'. Also the clause 'to increase Aβ pathology in animals' is parallel with the clause 'to reduce pathology in APP transgenic mice'.

> Concurrently, high-cholesterol diets have been shown to increase Aβ pathology in animals (79, 80), and cholesterol-lowering drugs have been shown to reduce pathology in APP transgenic mice (81). — Parallel

Example 5.78. *Parallel clauses, from a multidisciplinary sciences article [536].*

Parallelism can be particularly helpful in setting out successive steps in a mathematical exposition, as in Example 5.79.

> Parallel — If r is even then $K^{[r]}_{\text{even}} = r + 1$ and $K^{[r]}_{\text{odd}} = r$, whereas if r is odd then $K^{[r]}_{\text{even}} = r$ and $K^{[r]}_{\text{odd}} = r + 1$.

Example 5.79. *Parallel clauses, from a computer science article [628].*

The clause 'If r is even' is parallel with the clause 'if r is odd', and the clause '$K^{[r]}_{\text{even}} = r + 1$ and $K^{[r]}_{\text{odd}} = r$' is parallel with the clause '$K^{[r]}_{\text{even}} = r$ and $K^{[r]}_{\text{odd}} = r + 1$'.

Yet precisely because of the effectiveness of parallelism, its failures can be disconcerting. Example 5.80 shows a mispairing of 'not only . . . but also' clauses.

> ✕ To dissect the mechanism by which a signaling pathway is regulated it is necessary to not only measure the localization of the key signaling proteins, but also their activation at different subcellular locations. — Non-parallel

Example 5.80. *Non-parallel clauses, from a biophysics article [540].*

Here 'measure the localization of the key signaling proteins' is paired with 'their activation at different subcellular locations', two grammatically different elements. When as a test they are interchanged, the clause 'it is necessary . . .' becomes 'it is necessary to not only their activation at different subcellular locations'. Either 'not only' needs to be moved so that it falls just after 'measure', or 'measure' needs to be inserted just after 'but also'.

✕

> A low frame rate is accompanied by a small noise bandwidth, a high dwell time, and the eye integrates only a small number of frames.

Non-parallel

Example 5.81. *Non-parallel clauses, from an instrumentation engineering article [646].*

In Example 5.81, there are problems with both parallelism and coordination.

Parallelism fails with the change in verb voice. The first clause has the passive verb 'is accompanied' and the second clause the active verb 'integrates'. Coordination fails with the sequence beginning with the noun phrases 'a small noise bandwidth' and 'a high dwell time'. The 'and' should be followed by a third noun phrase like 'noise bandwidth' and 'dwell time'. Instead, the third item is a finite clause 'the eye integrates . . .'.

The coordination can be improved by replacing the first comma by 'and', so that 'the eye integrates' is no longer a member of the sequence. Verb voice can then be made parallel as in the edited version in Example 5.82. The wordy 'small number of' has been replaced by 'few'.

> A low frame rate is accompanied by a small noise bandwidth and a high dwell time, and only a few frames are integrated by the eye.

Parallel

Example 5.82. *Edited version of Example 5.81.*

Failures in parallelism can also occur in describing numerical quantities, as in Example 5.83.

✕

> On a genetic basis approximately one in every 300 subjects lacks TPMT activity, and 11% of subjects have intermediate activities.

Non-parallel

Example 5.83. *Non-parallel measures, from a pharmacology and pharmacy article [583].*

The reader has to compare a frequency on a fractional scale that emphasizes the denominator with a frequency on a percentage scale that emphasizes the numerator. Either one can be converted to the other. In the edited version in Example 5.84, the percentage has been replaced.

> On a genetic basis approximately one in every 300 subjects lacks TPMT activity, and one in every ten subjects has intermediate activity.

Parallel

Example 5.84. *Edited version of Example 5.83.*

The foregoing examples illustrate the problems with failures in parallelism and occasionally coordination. Nevertheless Stilman [340, p. 223] has some cautionary advice that observing parallelism does not mean constructing every sentence in the same way; rather it means that

> within a sentence, a list, or a passage that contains elements related in purpose or structure, these elements should be presented in a consistent manner.

Other aspects of consistency are discussed in Section 2.1.

5.5.2 Ellipsis

Ellipsis refers to the omission of a word or words predictable from the context or grammar [40] [44]. The degree to which the missing elements can be recovered from the context or grammar varies according to the precision of the ellipsis, which can occur in many forms [297] [346]. Ellipsis in this sense should not be confused with the related punctuation mark '. . .'.

In Example 5.85, the preposition 'to' is omitted with the second infinitive but could be included for emphasis [40, p. 245].

Omitted 'to'

The database should be quick to search and download.

Example 5.85. *Ellipsis of 'to', from a chemistry article [548].*

Likewise in Example 5.86, the subject 'Jets in radio-loud AGNs' is omitted with the second verb 'show'.

Jets in radio-loud AGNs emit strong nonthermal radiation at all wavelengths and show evidence of apparent superluminal motion.

Omitted subject

Example 5.86. *Ellipsis of subject, from a nuclear physics article [577].*

Still, there are grammatical constraints on what can be omitted [297]. For example, the subject cannot be omitted in a subordinate clause, for instance, one that begins with a subordinating conjunction 'although', 'because', or 'since'. In Example 5.87, it begins with 'whereas'.

✗ However, note that the transmitting weights influence the OFDM measurements through the matrix A whereas CE-OFDM measurements through the matrix Φ.

Incorrect ellipsis

Example 5.87. *Incorrect ellipsis, from an electrical and electronic engineering article [668].*

Because 'whereas' is a subordinating conjunction, it forces 'CE-OFDM measurements' to be the subject of the clause that follows. The grammatically recovered text is then 'influence the OFDM measurements', as in the edited version in Example 5.88. The result is nonsense.

✗ However, note that the transmitting weights influence the OFDM measurements through the matrix $\boldsymbol{A}$ whereas CE-OFDM measurements <u>influence the OFDM measurements</u> through the matrix $\boldsymbol{\Phi}$. — Grammatically recovered text

Example 5.88. *Incorrect recovery, recast from Example 5.87.*

Example 5.87 was taken from [668]. An inspection of the equations preceding the extract in the article suggests that the text omitted after 'whereas' is likely to be 'the transmitting weights influence', or, with less repetition, 'the weights influence' or, in light of the parallelism, simply 'they influence'. The recovered text might then read as in Example 5.89.

Correctly recovered text = 'the transmitting weights influence' — However, note that the transmitting weights influence the OFDM measurements through the matrix A whereas <u>they influence</u> CE-OFDM measurements through the matrix Φ.

Example 5.89. *Correct recovery, recast from Example 5.87.*

If the conjunction in Example 5.87 had not been subordinating but coordinating—for example, 'and' or 'but'—the ellipsis would have worked, as in the edited version in Example 5.90.

However, note that the transmitting weights influence the OFDM measurements through the matrix A but CE-OFDM measurements through the matrix Φ.

Correct ellipsis

Example 5.90. *Correct ellipsis, recast from Example 5.87.*

In Example 5.91, the ellipsis is ambiguous owing to a change in verb form.

Ambiguous ellipsis

✗

These cells are already present when we investigate young adults (seven to eight segments), but exclude the first two pairs of appendages that are known to correspond to the last two pairs of larval parapodia (27).

Example 5.91. *Incorrect ellipsis, from a developmental biology article [407].*

Given the parallelism of the transitive verbs 'investigate' and 'exclude', the omitted subject of the second clause 'exclude the first two pairs . . .' is 'we'. The result is again nonsense.

Example 5.91 is taken from [407]. An inspection of the text in the article suggests that the subject is really 'cells' and that the transitive verb 'exclude' needs to be replaced by the non-transitive verb phrase[43] 'are absent', which parallels 'are already present' in the first clause, as in the edited version in Example 5.92.

Unambiguous ellipsis

Parallel verb phrases

These cells are already present when we investigate young adults (seven to eight segments), but are absent from the first two pairs of appendages that are known to correspond to the last two pairs of larval parapodia (27).

Example 5.92. *Unambiguous ellipsis, recast from Example 5.91.*

The recovered text might then read as in Example 5.93.

These cells are already present when we investigate young adults (seven to eight segments), but they are absent from the first two pairs of appendages that are known to correspond to the last two pairs of larval parapodia (27).

Correctly recovered text = 'these cells'

Example 5.93. *Correct recovery, recast from Example 5.91.*

Effective ellipsis depends on effective parallelism.

5.6 Word Choice

As a consequence of its mixed heritage, English contains many words that have alternatives with closely similar meanings. This variation offers great scope for precision but also for confusion, obscuration, and misdirection.

5.6.1 Consistency and Differentiation

It is sometimes tempting to substitute one word for another to avoid repetition. The general advice, however, is that for clarity the same word should be used for the same thing or, where appropriate, an unambiguous pronoun substituted [79, p. 216] [122, p. 149].

In Example 5.94, 'very small', 'little', 'minimal', and 'negligible' are all used to describe the size of an effect.

✕

Subsequent studies at the single-channel level found that external Cl^- significantly increases the opening rate but only has a very small effect on the closing rate (49, 71, 99). — Minute?

Intracellular Cl^- ions, however, have little effect on the opening rate of the fast gate (71, 112). — No effect?

On the other hand, the size or shape of the side chain of the introduced residues had only a minimal effect. — Smallest possible?

For example, S(-)-CPP significantly inhibited CLC-1 at a concentration of 50–300 μM, whereas it had negligible effect on CLC-2. — Ignorable?

Example 5.94. *Variations in meaning, from a physiology article [444].*

All four words have different meanings. In turn, 'a very small effect' is a neutral statement about the size of an effect that exists, whereas 'little effect' could include having no effect (by contrast with 'a little effect', which implies an effect exists).[44] A minimal effect can mean an extremely small effect or a negligible effect [38] or, in a more formal context, the least possible effect [44], coinciding with the usual scientific or mathematical interpretation. A negligible effect is unambiguously one that can be neglected or ignored.

In Example 2.11 of Chapter 2 (§2.1.6), 'technique', 'approach', and 'method' are all used for what may have been the same thing. Since these terms are often confused, it is worth repeating their definitions and the definitions of procedure and methodology,[45] together with examples to make the distinctions clearer.

A procedure is a particular mode or course of action, whereas an approach is a way of addressing a task or dealing with a problem, as in Example 5.95.

The efficacy of such a contraction procedure was first noted by Shavitt,[7] . . .

In all of these papers the approach to the problem tended to be mathematical, obscuring any physical basis for contraction.

Example 5.95. *Differential use of 'procedure' and 'approach', from a physics article [487].*

A method is a way of doing something in particular and a technique is a specific, concrete way of doing it, especially the formal or mechanical aspects of it [38] [280], as in Example 5.96.

> The resonant cavity perturbation technique described here provides a quick, simple and accurate method of measuring the dielectric properties of relatively small, easily prepared tissue samples.

Example 5.96. *Differential use of 'technique' and 'method', from a biomedical engineering article [574].*

Finally, a methodology means a body of methods [142]. In Example 5.97, 'methodology', 'methods', and 'techniques' are all used with distinct meanings.

> The length scale . . . depends on the fabrication methodology. Although most conventional gradient-forming methods reported in the literature result in materials gradients spanning . . . millimeters to centimeters, certain fabrication techniques facilitate the generation of materials on much smaller length scales.

Example 5.97. *Differential use of 'methodology', 'methods', and 'techniques', from a materials science article [516].*

Needless word variation is discussed Section 2.1.6.

5.6.2 Failed Conjunctions

Some adverbs are also classified as conjunctions, for instance, 'yet', but most are not; particularly 'however', 'therefore', and 'thus' [38] are not.

In Example 5.98, the adverb 'however' should be replaced by a conjunction—for instance, 'but' or 'although'—or the comma preceding it should be replaced either by a semicolon or by a full stop and a new sentence started.

✗

> Several classification algorithms that cope with concept drift have been put forward, however, most of them specialize in one type of change.

Example 5.98. *Nonstandard use of 'however', from a computer science article [432].*

In Example 5.99, the adverb 'therefore' should be preceded either by a semicolon or by a full stop and a new sentence started.

✗ There are no significant signal changes in the center of the blob therefore descriptors perform better on larger neighborhoods.

Example 5.99. *Nonstandard use of 'therefore', from a computer science article [605].*

In Example 5.100, the comma preceding 'thus' should be replaced either by a semicolon or by a full stop and a new sentence started.

✗ Note that these quantities are scalars irrespective of N, thus the DL pilot signaling need not scale with N.

Example 5.100. *Nonstandard use of 'thus', from an electrical and electronic engineering article [419].*

The role of the comma and other stops in punctuation is discussed in Section 5.7.

5.6.3 'Not' and 'Un-'

It is occasionally argued that using 'not' to modify an adjective when the negative version already exists lessens impact. Compare the closely similar Examples 5.101 and 5.102.

It is not clear how to reconcile these observations.

Example 5.101. *Moderate 'not clear', from a geochemistry and geophysics article [443].*

It is unclear how to reconcile these observations.

Example 5.102. *Strong 'unclear', from a neurosciences article [603].*

The negation 'not clear' in Example 5.101 is slightly less effective than the 'unclear' in Example 5.102. Like the prefix 'non-', 'not' expresses the logical complement rather than the opposite. The difference is more obvious in the sequence 'scientific', 'not scientific', and 'unscientific' [297, §I.21].

5.6.4 'By', 'With', and 'Via'

The prepositions 'by', 'with', and 'via' signify different roles for what follows. Deciding which is not always as straightforward as in Example 5.103.

> One of the initial observations on mirror neurons was that they fire . . . but not when the same piece of food is grasped <u>by</u> the experimenter <u>with</u> a tool.
>
> Signifies agent — Signifies instrument

Example 5.103. *Agentive 'by' and instrumental 'with', from a neuroscience article [547].*

The agent of the action, the experimenter in Example 5.103, is introduced by 'by', and the instrument, the tool, is introduced by 'with' [297, §9.50]. But when the agent of the action is unknown or unimportant it may be omitted (§5.3). The choice of 'by', 'with', or 'by means of' then depends on the intended interpretation.

In Example 5.104, the word 'filter' is first introduced by 'with', as the instrument supporting the operation of convolution, and then by 'by', as the agent of the process of wavelet generation.

> Note that multiplication by $\boldsymbol{H}_0^{\star}$ corresponds to upsampling by 2 followed by convolution <u>with</u> a filter . . . — Signifies instrument
>
> For comparison purposes we show in Fig. 16 the corresponding plots for the orthonormal wavelet generated <u>by</u> a filter of the same length.
>
> Signifies agent

Example 5.104. *Instrumental 'with' and agentive 'by', from an electrical and electronic engineering article [706].*

The preposition 'via' signifies something quite different. It refers to the path taken on a journey. In Example 5.105, 'via' is correctly used to mean 'by way of', 'by the route of'.

> = 'along the route of'
>
> [Nanoparticles] have been reported to travel <u>via</u> the nasal nerves to the brain.

Example 5.105. *Correct use of 'via', from an environmental and occupational health article [479].*

There is general agreement that it should not be used for 'by means of'; that is, the means of travel [40] [122]. It is 'wrong, very wrong, misleadingly wrong' [280]. Accordingly, 'via' in Example 5.106 is incorrect and should be replaced by 'by' or 'with'.

= 'along the route of'

✗ This is most often accomplished <u>via</u> the use of genetically encoded biosensors (1, 2).

Example 5.106. *Nonstandard use of 'via', from a biophysics article [540].*

Since the means are evident, that is, 'the use of genetically encoded biosensors', the longer substitute 'by means of' is unnecessary.

5.6.5 'Fewer' or 'Less'

The traditional view is that 'fewer' is used with countable things and 'less' with uncountable things [40].

✗ The regions are more accurate since there are <u>less</u> parameters to estimate.

Example 5.107. *Nonstandard use of 'less', from a computer science article [605].*

In Example 5.107, the use of 'less' would be regarded as nonstandard [38], and it should be replaced by 'fewer'. Even so, the use of 'less' with count nouns is established with quantities that have a collective entity such as time and length [284, p. 205].

5.6.6 'Compare To' and 'Compare With'

A distinction is traditionally drawn between the roles of 'to' and 'with' with some verbs. For example, 'compare to' means to liken, and 'compare with' to set up a comparison [122] [280]. Example 5.108 contains both.

= 'likened to'

The whole investigation may be <u>compared to</u> the analytical treatment of the Theory of Gases. . . .

Now as I– J, J– K, the mean differences in any trait due to a single factor, are small <u>compared with</u> the whole variation within the population.

= 'by contrast with'

Example 5.108. *Differentiated use of 'compared to' and 'compared with', from a statistics article [504].*

In place of 'compare', the abbreviation 'cf.' may be used. It stands for the Latin 'confer', which means 'compare', but more in the sense of 'compare to' [417], as in Example 5.109.

> A consequence of the symmetric form of the initial data
> = 'compare to' — set (cf. remark 2.4) is that all but one of the actions
> (definition 3.2) are permutations.

Example 5.109. *Use of 'cf.', from a mathematical physics article [404].*

The instruction 'cf. remark 2.4' refers to an earlier statement about the initial data set being highly symmetric. In other words, the author is pointing out the similarity, not the contrast.

5.6.7 'Above' and 'Below'

The locations 'above' and 'below' are sometimes used to refer to something mentioned earlier or later [280], as in Example 5.110. The different grammatical roles of the two words are indicated.

✗

> Noun — The above is totally dependent on . . .
> Adjective — The above procedure leads . . .
> Adverb — It was shown above that . . .
> Adverb — Everything discussed below can easily be translated . . .

Example 5.110. *Use of 'above' and 'below', from a theoretical physics article [531].*

When 'above' appears at the top of the printed page or 'below' at the bottom, the effect is incongruous. More natural alternatives are available, depending on proximity and grammatical role,[46] as illustrated in the edited version in Example 5.111.

> Or 'earlier' or 'foregoing' — The preceding is totally dependent on . . .
> The foregoing procedure leads . . . — Or 'earlier' or 'preceding' or 'previous'
> Or 'previously' — It was shown earlier that . . .
> Everything discussed later can easily be translated . . . — Or 'in the following' or 'in what follows'

Example 5.111. *Alternatives to 'above' and 'below', from Example 5.110.*

The artificiality of 'above' and 'below' is obvious when used in speech [280].

5.6.8 'Approximately', 'About', and 'Roughly'

The general advice is that 'approximately' should be used in more formal contexts and 'about' and 'roughly' in less formal ones. Also, 'approximately' suggests a more careful calculation [122], as in Example 5.112.

> The root from which the profile was obtained is reproduced above, roughly to scale . . .
>
> In approximately 15% of the profiles, velocity is constant . . .
>
> The number of marks that can be applied to the root, typically around 10 per growth zone . . .

Example 5.112. *Differential use of 'roughly', 'approximately', and 'around', from a plant physiology article [703].*

Furthermore, there are some technical phrases that require 'approximately' rather than the alternatives, as in Example 5.113.

> However, at each step it does guarantee that the likelihood increases (under the CAT approximation). Thus, FastTree 2 is an approximately maximum-likelihood method.

Example 5.113. *Special use of 'approximately', adapted from a computational biology article [645].*

Here 'approximately' refers to the fact that the likelihood, a statistical quantity, is not necessarily maximum but that it increases with each step in the method [645].

5.6.9 'First', 'Firstly'

In enumerating successive pieces of text, 'First', 'Secondly, 'Thirdly' was once advocated [122] [280]. Yet 'Firstly', 'Secondly', 'Thirdly' is more logical, and 'First', 'Second', 'Third' is both more logical and succinct [142] [284], as illustrated in Example 5.114.

> Three important points emerged from this analysis. First, the resulting force-displacement curve made by . . . Second, the simulations demonstrated that the force-displacement curve is insensitive to . . . Third, the force-displacement response is insensitive to . . .

Example 5.114. *Simple enumeration, from a materials sciences article [580].*

In longer enumerations, the last term would then be 'Last', not 'Lastly' [142] [280].

5.6.10 'Reason Why' and 'Because'

The combinations 'reason why', 'reason . . . because', and 'why . . . because' contain redundant elements [297, §15.46]. In Example 5.115, 'why' is redundant and can be omitted.

Redundant

✕ The reason why is the existence of the magnetic helicity integral of motion for ideal MHD [magnetohydrodynamics].

Example 5.115. *Redundancy, from a physics article [524].*

In Example 5.116, the redundant 'why' can be omitted and the redundant 'because' could be replaced by 'that'.

Redundant — Redundant

✕ The reason why is because these CRC5 implementations associate each slice of a stream with a separate table.

Example 5.116. *Redundancy, from a computer engineering article [566].*

The edited version is shown in Example 5.117. The explanation is now more direct.

The reason is that these CRC5 implementations associate each slice of a stream with a separate table.

Example 5.117. *Redundancy eliminated from Example 5.116.*

In Example 5.118, the redundant 'why' could be replaced by 'that', but 'reason why' heads a subordinate clause and in this context represents a long-standing, idiomatic construction [40].

Redundant
but idiomatic

The reason why late degenerative arthritis developed in some patients . . . was investigated.

Example 5.118. *Idiomatic redundancy, from an orthopaedics article [722].*

5.6.11 'Because' and 'For'

Traditionally, 'because' and 'for' introduce different kinds of explanation [280]. 'Because' is used when the explanation is direct, objective, or causal, and 'for' when it is indirect, subjective, or noncausal. In Example 5.119, 'because' is causal.

Causal

This search probe failed because the training data for this class did not include any 3D object rotation and the search probe did.

Example 5.119. *Objective 'because', from a computer science article [686].*

The reported failure of the search probe was a direct consequence of the properties of the training set and the search probe. In Example 5.120, 'for' is noncausal.

Subjective, equivalent to 'since'

Debris flows occurred in pre-vegetation times, for they have been recognised in Precambrian fan deposits by Selley (1965, p. 375) and by Maycock (cited by Bull, 1972, p. 79).

Example 5.120. *Subjective 'for', from a geology article [653].*

The reported occurrence of the debris flows in pre-vegetation times was not a consequence of their being recognized by the authors Selley and Maycock. As an alternative, 'for' could be replaced by 'since', 'seeing that', or 'given that' [40].

Nevertheless, 'because' is used routinely with noncausal explanations [297, §15.45] and 'for' can sometimes occur with causal explanations, usually in mathematics, as in Example 5.121. The 'for' statement implies the truth of the statement that precedes it, 'The relation . . .'.

Objective —

The relation $y \equiv x$ (mod. R) is equivalent to $x^{-1}y \equiv e$ (mod. R), for $y \equiv x$ implies $x^{-1}y \equiv x^{-1}x = e$ and conversely $x^{-1}y \equiv e$ implies $y = x(x^{-1}y) \equiv x$.

Example 5.121. *Objective 'for', from a mathematics textbook [423, p. 34].*

Corpus data suggest that the explanatory use of 'for' is declining [284]. Yet it can still be found in popular journalistic contexts [391, p. 74] [519].

5.6.12 'Since' and 'As'

'Since' can be ambiguous, expressing a relationship to an earlier time or event or to cause and effect [280] [284]. In Example 5.122, 'Since' is obviously temporal, for 'Since then' could be replaced by 'Subsequently' or the like.

> The MRC National Survey of Health and Development . . . was nationally representative of the population born in England, Scotland, and Wales in 1946. Since then there have been losses to follow-up from death, emigration, loss of contact, and permanent refusal.
>
> Temporal — Since

Example 5.122. *Temporal 'Since', from a general and internal medicine article [455].*

But in Example 5.123, 'since' is more ambiguous, for it could mean 'after' as well as 'because'.

> ✗ Fluorometric assay was tried, since the reaction product was found to be fluorescent.
>
> Ambiguous — since

Example 5.123. *Ambiguous 'since', from a biochemistry and molecular biology article [723].*

In Example 5.124 'since' is obviously causal and could be replaced by 'because', although 'since' may still be preferred for rhetorical reasons.

> Theorem 4.19 does not imply Theorem 4.13 even in B^n, since there are countable closed sets C in S^{n-1}.
>
> Causal — since

Example 5.124. *Causal 'since', from a mathematics article [522].*

'As' can produce similar ambiguities, expressing a relationship to a current event or to cause and effect (or something else) [280] [284]. In Example 5.125, 'As' seems to be temporal from the context, and could therefore be replaced by 'When', albeit not with the same sense of progression.

> As the γ value increases, the channel passbands shift to longer wavelengths.
>
> Temporal — As

Example 5.125. *Possible temporal 'as', from an optics article [538].*

In Example 5.126, 'As' seems to be more temporal than causal from the context, but the sentence needs recasting to remove all ambiguity.

Temporal or causal? — As α increases without any odd harmonic ($n = 1$), class B, AB, and A PAs [power amplifiers] are determined sequentially.

Example 5.126. *Ambiguous 'as', from an electrical and electronic engineering article [557].*

Although any of the four, 'because', 'for', 'since', 'as', can be used to introduce indirect causal or explanatory material [297, §15.45], it is 'as' that is most likely to cause the reader difficulty and force a rereading of the text. It should therefore be used only when there is truly no risk of ambiguity.

5.6.13 'Which' and 'That'

The general advice is that 'which' with a preceding comma should be used for non-defining or nonrestrictive clauses (with a matching closing comma) and 'that' without a preceding comma (unless it ends a parenthetical phrase) for defining or restrictive clauses [49, §5.220] [122].

In Example 5.127, the clause 'which originated . . .' is non-defining and can be omitted from the sentence without changing the identity of the samples that were said to be *T. cruzi*–positive.

Serum samples . . . were obtained from David A. Leiby (American Red Cross, Rockville, MD). These samples, which originated from US blood banks, were *T. cruzi*–positive. — Non-defining

Example 5.127. *Non-defining clause, from an immunology article [544].*

By contrast, in Example 5.128, the clause 'that were furnace cooled' is defining. It cannot be omitted without introducing uncertainty into which samples were said to be annealed and quenched.

Defining

Some of the samples that were furnace cooled were later annealed and quenched.

Example 5.128. *Defining clause, from a physics article [666].*

Traditionally, a 'which'-clause without commas is also a defining clause [122], although this usage is deprecated in some guides [49, §6.22].

The distinction between defining and non-defining clauses is not always clear [40]. Even so, there is a case for consistent usage, not observed in Example 5.129.

✗

Ambiguous relative pronoun — Non-defining relative pronoun — Defining relative pronoun

> The signal generator (see Fig. 1) is composed of a 16 kbyte EPROM, which contains the samples . . ., a 16 bit counter, that is used to scan the EPROM, and an 8 bit digital-to-analog converter (DAC) that feeds the transmitting piezoelectric transducer.

Example 5.129. *Inconsistent non-defining and defining clauses, from an instrumentation engineering article [438].*

The first non-defining clause 'which contains . . .' is correct, for it is evident from the cited figure, not shown here, that there is only one 16 kbyte EPROM and that the clause provides only incidental information. It is correctly preceded by a comma. The second clause 'that is used . . .' is ambiguous. The preceding comma suggests that it is non-defining, whereas 'that' suggests that it is defining (since there is only one 16-bit counter in the figure, it ought to be non-defining). The third clause 'that feeds . . .' is defining, for it begins with 'that' and is not preceded by a comma. Since there is only one 8-bit digital-to-analog converter in the figure, it ought to be a non-defining clause.

5.6.14 'While', 'Whereas', and 'Although'

As a conjunction, 'while' is open to ambiguity, for it can mean 'during the time that' and 'although' [284]. When there is a risk of confusion between these temporal and concessive roles, its use is discouraged by some [49, §5.220] [280], but not all [40].

The instances in Example 5.130 are all taken from a single article.

= 'Although' or 'Whereas'? — While freestanding silicene could not be produced up to now, experimental evidence even for the existence of epitaxial silicene is rather limited but does exist.

= 'despite the fact that'? — The result indicates that Si_C atoms are located at a height different from Si_A atoms while these two types of atoms form one of the two sublattices of the honeycomb structure . . .

= 'Although' or 'Whereas'? — While partial sp^3 hybridization caused by the buckling and interactions with the underlying metallic ZrB_2 substrate [is] expected to occur, the existence of a π-electronic band structure itself proves . . .

✗

= 'while at the same time'? — It seems that in order to accommodate the compressive epitaxial strain, a relatively large buckling . . . is induced while the Si bond lengths are maintained to similar value.

Example 5.130. *Possible interpretations of 'while', from a physics article [505].*

The first 'While' is almost certainly concessive, equivalent to 'Although' or 'Whereas' (the later 'up to now' suggests transiently a temporal interpretation). The second 'while' is probably also concessive, equivalent to 'despite the fact that'. The third is unambiguously concessive, equivalent to 'Although' or 'Whereas'. The last is probably temporal, which, to avoid ambiguity, might be better emphasized by 'while at the same time' [297, §8.34].

5.6.15 'Due To' and 'Owing To'

'Due to' and 'owing to' have a contentious history. Partridge [280] and others [122] argue that there is a useful distinction to be maintained, namely that 'due to' means 'caused by' and 'owing to' means 'because of'. Thus 'due to' is adjectival and 'owing to' is adverbial. The traditional distinction is still recognized [49, §5.220], though there is a shift in acceptance allowing 'due to' to function in both roles [40] [284], as in Example 5.131.

Adverbial

They typically occur due to nonuniform thermal expansions . . .

The hydraulic fluid temperature change due to a 14 MPa pressure change is on the order of 0.8 °C. — Adjectival

Example 5.131. *Different roles of 'due to', from an instrumentation engineering article [681].*

The first 'due to' is adverbial, modifying the verb 'occur'. The second 'due to' is adjectival, modifying the noun 'change'. Neither is open to misinterpretation and each reads naturally.

But Example 5.132 shows the potential ambiguity of 'due to' when its grammatical role is unclear.

✗

Dimensions $m \times n$ are not the same for experimental and simulation results due to simulation process optimization.

Ambiguous

Example 5.132. *Potential ambiguity of 'due to', from an instrumentation engineering article [707].*

One interpretation is that here 'due to' functions adjectivally and modifies the preceding noun 'results'; that is, it restricts the results to those attributable to simulation process optimization. The other interpretation is that it functions adverbially, explaining why the dimensions are not the same for experimental and simulation results. From the larger context, the adverbial interpretation seems more plausible. The use of 'owing to' would have eliminated any uncertainty, as in the edited version in Example 5.133.

> Dimensions $m \times n$ are not the same for experimental and simulation results <u>owing to</u> simulation process optimization.

Example 5.133. *Unambiguous 'owing to', edited from Example 5.132.*

Alternatively, the sentence could be recast so that 'due' is unambiguously adjectival, as in Example 5.134.

> The difference in dimensions $m \times n$ for experimental and simulation results is <u>due to</u> simulation process optimization.
>
> Adjectival

Example 5.134. *Unambiguous 'due to', edited from Example 5.132.*

'Due to' here modifies the noun 'difference'. As another alternative, the adverbial 'owing to' in Example 5.133 could be replaced by 'because of'.

In some instances, 'due to' can be confusingly duplicated, as in Example 5.135.

✗

> These frequency shifts, <u>due to</u> a static stress bias, occur <u>due to</u> nonlinear effects acting through the third-order elastic stiffness coefficients.
>
> Duplication

Example 5.135. *Duplication of 'due to', from an instrumentation engineering article [492].*

The effect is that the reported frequency shifts appear to have dual causes, namely static stress and nonlinear effects. One possibility is that the frequency shifts are the result of static stress bias, which is the result of nonlinear effects acting through the third-order elastic stiffness coefficients. Another possibility is that static stress bias induces nonlinear effects that arise through the third-order elastic stiffness coefficients, and that these nonlinear effects induce the observed frequency shifts. In any event, one 'due to' could be eliminated and the relationship of the causal factors clarified.

5.6.16 'And/Or'

'And/or' means either together or as an alternative [38]. Its use is discouraged [40] [280], with Gowers [142, p. 188] criticizing it thus:

> It is always ugly, usually unnecessary and sometimes downright wrong. It is also discourteous, because it nearly always forces readers to read the sentence twice, which they would not have had to do if the writer had taken a little more trouble.

Example 5.136 shows how it is often employed.

✗

> *c* is a contrast parameter proportional to the relative deviations in mass and/or bonding strength.

Ambiguous

Example 5.136. *Ambiguous 'and/or', from a materials science article [698].*

The phrase 'mass and/or bonding strength' means 'mass or bonding strength or both', but the 'and/or' needs the reader to pause to decode the alternatives. In the context, a simple 'mass or bonding strength' might have sufficed. If not, the longer 'mass or bonding strength or both' would still have been easier to interpret than the 'and/or' version.

The use of 'and/or' can lead to more serious confusion as in Example 5.137.

✗

> The most common chronic inflammatory disease of the CNS is multiple sclerosis (MS), a condition that leads to reversible and/or irreversible disability depending on the type and stage of the disease (Box 1).

— Impossible?

Example 5.137. *Apparently impossible 'and/or', from a neuroscience article [673].*

At first reading, the phrase suggests that a disability can be reversible or irreversible or both, which seems an impossible combination. One interpretation is that during a relapse (an exacerbation of the disease) the disability experienced is transient if the relapse is acute but permanent if the patient does not recover from the relapse or is in the progressive phase. If so, then 'or' rather than 'and/or' would have sufficed.

5.6.17 '-Ize' and '-Ise'

Some verbs are formed by using 'ize' to modify the end of an adjective or noun: for example, 'colonize', 'crystallize', 'homogenize', 'localize', 'optimize', 'minimize', 'maximize', 'normalize', 'randomize', 'stabilize', 'summarize', 'synchronize', 'synthesize', 'visualize' [38].

The majority of verbs ending in 'ise' or 'ize', whether from French, Latin, or Greek, or an English imitation, derive their termination from the Greek, and in principle take the 'ize' form [40]. The 'ize' form is the spelling preferred by Oxford University Press. The occasional exceptions, such as 'advertise', 'advise', and 'analyse', are not formed in this way and so take the 'ise' form. In practice, though, spelling conventions vary across publishers and countries, and conventions outweigh etymology.

5.6.18 'However', 'But', and Other Contrasts

When 'however' is used to signal a contrast, it is usually placed immediately after the item that is the subject of the contrast [280] [284], as illustrated in Example 5.138.

Contrast with preceding

> In practice, however, this requires the SNR of each individual image to be large enough for accurate spatial alignment (79).

Example 5.138. *Correctly placed 'however', from a materials science article [567].*

Here 'however' emphasizes what is said to be required in practice against what is possible in theory. The commas around 'however' are mandatory [40]. Without them it can be interpreted as 'in whatever way'; that is, in Example 5.138 'in whatever way this requires . . .'.

Despite its blandness (or perhaps because of it), 'however' is popular in scientific and engineering reports. After 'but', it is the most frequently used of all adverbs, prepositions, and conjunctions expressing a contrast or concession, even more popular than 'although'.[47] Yet there are many alternatives to 'however', each with slightly different meaning and effect: 'although', 'but', 'by contrast', 'despite', 'even so', 'even then', 'even though', 'for all that', 'in contrast', 'in spite of', 'nevertheless', 'nonetheless', 'notwithstanding', 'still', 'though', and 'yet', A more complete list of such adverbs and propositions can be found in [297, §8.137]. All are more interesting than 'however'.

If 'however' does need to be used, then it should be placed carefully [284]. At the very least, Zinsser's stricture [391, p. 73] should be considered:

> Don't start a sentence with 'however'—it hangs there like a wet dishrag. And don't end with 'however'—by that time it has lost its howeverness.

Whatever the choice of contrast, the distinction between those words that can act as conjunctions and those that cannot should always be observed.

5.6.19 'As To'

Occasionally 'As to' and other variations provide a useful device for introducing a new subject [142], as in Example 5.139.

Introduction

As to the $\mu'' - f$ spectrum, two resonance peaks are observed . . .

Example 5.139. *Introductory 'As to', from a materials science article [634].*

More commonly, however, 'as to' appears as padding or as part of roundabout expressions [142], as in Example 5.140, where it can be omitted.

Redundant

✕ The question then remains as to whether particular biochemical mechanisms of halophytes are better activated . . .

Example 5.140. *Redundant 'as to', from a plant physiology article [537].*

In Example 5.141, too, it can be omitted and the sentence recast more economically, to read as in Example 5.142.

Redundant

✕ Groups were also comparable as to the number of errors of commission in NoGo-trials.

Example 5.141. *Redundant 'as to', from a neurosciences and neurology article [696].*

Groups also had comparable numbers of errors of commission in NoGo-trials.

Example 5.142. *Edited version of Example 5.141.*

The original phrase in Example 5.141 'were also comparable as to the number' has been replaced in Example 5.142 by 'also had comparable numbers', which has three fewer words. The sentence might be improved further by replacing 'comparable' by 'similar'.

5.6.20 'Carried Out'

'Carried out' is often used in a laborious way to create a verb from a noun that already has a corresponding verb [79, p. 200]. It is simpler and more immediate to use the original verb (§2.3).

In Example 5.143, the combination of 'Assessment' and 'was carried out' can be replaced by 'were assessed', as in the edited version in Example 5.144.

Noun — Verb

✗ Assessment of reach-to-grasp movements was carried out as described in our previous publication.

Example 5.143. *Unnecessary 'carried out', from a neurosciences and neurology article [523].*

Original verb

Reach-to-grasp movements were assessed as described in our previous publication.

Example 5.144. *Recast version of Example 5.143.*

In Example 5.145, the combination 'analysis' and 'must be carried out' can be replaced by 'must be analysed', as in the edited version in Example 5.146.

Noun

✗ The analysis of the stability properties of neural networks must be carried out.

Verb

Example 5.145. *Unnecessary 'carried out', from a computer science article [400].*

The stability properties of neural networks must be analysed. — Original verb

Example 5.146. *Recast version of Example 5.145.*

The conventional meaning of 'to carry out' is to perform or execute [38], but the phrase is sometimes used in a nonstandard way, as in Example 5.147.

✗ Stability conditions of $e_{p,m}$ $(A_z h)$ for $p > 1$ can be carried out in the same way as for $p = 1$. — Nonstandard

Example 5.147. *Nonstandard 'carried out', from a control systems engineering article [406].*

The conditions referred to in Example 5.147 are prerequisites and therefore cannot be performed or executed. More appropriate alternatives to 'carry out' might be 'established' or 'derived'.

In some situations, 'carry out' performs no useful function at all. In both Examples 5.148 and 5.149, it is superfluous and can be omitted.

Unnecessary

✗ Tables I–IV summarize the work carried out by several researchers.

Example 5.148. *Unnecessary 'carried out', from an instrumentation engineering article [520].*

✗ This condition appears to be crucial in the analyses carried out in Section V.

Unnecessary

Example 5.149. *Unnecessary 'carried out', from a control systems engineering article [643].*

If it is deleted in Examples 5.148 and 5.149, the following prepositions 'by' and 'in' might be replaced by 'of'.

5.6.21 'Based On'

'Based on' is one of the most common constructions in academic writing. It is used in a prepositional way [40] similar to that employed, incorrectly, with 'using' (§5.4.1), allowing one thing to be vaguely related to another thing without saying which thing or how (§2.3.3).

✗ In the closed-loop control, the term . . . in (2) is calculated based on the actual measured capacitor voltages.

Vague relation

Example 5.150. *Vague 'based on', from an electrical and electronic engineering article [472].*

In Example 5.150, the problem is identifying to what 'based on' applies. It could refer either to 'the term', in which case 'calculated' can be omitted, or indirectly to the calculation, in which case 'based on' can be replaced by 'from', as in the edited version in Example 5.151.

> In the closed-loop control, the term . . . in (2) is calculated <u>from</u> the actual measured capacitor voltages.
>
> Definite relation

Example 5.151. *Edited version of Example 5.150.*

There is a similar confusion in the Example 5.152.

✗

> Although new predictions can be generated and tested <u>based on</u> the theory . . .
>
> Vague relation

Example 5.152. *Vague 'based on', from a psychology article [726].*

Here 'based on' could refer to the generation and testing, but it seems more likely that it refers to 'predictions'. If so, it can be omitted and 'the theory' moved to its natural place earlier in the sentence, as in the edited version in Example 5.153.

> Although new predictions of the theory can be generated and tested . . .

Example 5.153. *Edited version of Example 5.152.*

In some instances 'based on', as with 'carry out', performs no useful function at all. In Example 5.154, either 'based on' should be omitted or, better, 'based on using' replaced by 'that use' or 'that exploit'.

✗

> Additionally, there are several modulation techniques <u>based on</u> using multiple reference waveforms.
>
> Redundant

Example 5.154. *Redundant 'based on', from an electrical and electronic engineering article [472].*

In a curious compounding of the problem, 'based on' may be combined with 'carried out', as in Example 5.155.

Vague and verbose

✗ The insertion/bypassing of the SMs is carried out based on capacitor voltage measurements [39], [41].

Example 5.155. *Vague and redundant phrasing, from an electrical and electronic engineering article [472].*

The whole phrase 'carried out based on' could be replaced by a more compact and explicit 'determined by' or 'governed by' or something similar.

5.6.22 'Former' and 'Latter'

For good reason, the general advice is not to use 'former' and 'latter' to refer to preceding items, unless they are very close [40] [284]. A reference to 'former' or 'latter'—especially 'former'—almost always forces the reader to work backwards to find the right match.

In Example 5.156, 'former' probably refers to the direction of motion, but that can only be determined by returning to the previous sentence and analysing it.

✗ Further investigations indicated that pseudopod formation is dominated by independent biases by the current direction of motion (Fig. 2*J*) and chemicals (chemokine) secreted by the cells (Fig. 2*K* and *SI Appendix*, Table S2). Evidence of the former was found by computing the angle . . .

Needs decoding

Example 5.156. *Remote referent, from a biophysics article [704].*

Identification is harder still when there are several plausible matches [142]. In Example 5.157, 'former' could refer to electrolysis or photovoltaic devices. The reader is left to choose.

✗ Whether water splitting is achieved by electrolysis using electricity generated by photovoltaic devices or dye-sensitized solar cells, or by artificial photosynthesis remains at this point unclear. While technology for the former is . . .

Ambiguous decoding

Example 5.157. *Uncertain remote referent, from a chemistry article [485].*

5.6.23 Plurals

Plurals of upper-case abbreviations and acronyms such as SEM are made by adding an 's' without an apostrophe, i.e. 'SEMs'; likewise with calendar years, e.g. '1900s' [49, §7.14] [263]. With lower-case abbreviations, an apostrophe is usually added, e.g. 'sem's', although the result is not always satisfactory.

Unit symbols, e.g. 'm', do not take a plural form, as explained in Section 3.6.5. Forming the plural of mathematical quantities such as a variable '*x*' is a little more complicated and is discussed in Section 6.2.3.

Some imported Latin words retain their distinct plural forms; for example, 'appendix' and 'appendices', 'basis' and 'bases', 'criterion' and 'criteria', 'index' and 'indices' (but 'indexes' in non-technical contexts), 'matrix' and 'matrices', 'phenomenon' and 'phenomena', and 'radius' and 'radii' [38] [284].

A few Latin plurals are in a transitional state, most notably, 'media' and 'data', which, depending on the research field and context, are treated as singular or plural. The plural interpretation of 'data' is the more common. It appears as plural in, for example, official data-management guidelines [148] [301], and it is specified as plural in some publisher and journal guides for authors [49, §5.220] [290].

Some words representing uncountable or noncount quantities [297, §5.2] appear only in singular form: for example, 'information' (the plural does not exist), 'notation' (the plural exists but would be used only to refer to different notational schemes or a specialist application), and 'noise' (the plural exists but in the sense of disagreeable sounds rather than signal processing).

Examples 5.158–5.160 all represent nonstandard usages.

Not plural

✗ These informations are exchanged using communication channels.

Example 5.158. *Nonstandard plural of 'information', from a computer science article [484].*

✗ The model included two events of photoelectric absorption and the notations were as follows.

Not plural

Example 5.159. *Nonstandard plural of 'notation', from an instrumentation engineering article [639].*

✗ For example, road irregularities are often described by white noises.

Not plural

Example 5.160. *Nonstandard plural of 'noise', from a control systems engineering article [721].*

In contrast, 'signal', which in engineering articles often appears in conjunction with 'noise', is a countable quantity, and can take the plural 'signals'.

5.6.24 Confusions

'Effect' is used routinely, perhaps excessively, in scientific writing.[48] Its main use is as a noun, when it means something accomplished, caused or produced, a result. But it can also act as a verb, when it means to bring about or accomplish. Because it can act as a verb, it is sometimes confused with 'affect', which as a verb means to influence or make an impression on something [38].

✕

> Etalons can also introduce spectral phase distortions which can effect the quality of the output pulse.

'affect'

Example 5.161. *Misuse of 'effect' for 'affect', from an optics article [488].*

In Example 5.161, the verb should be 'affect'. Confusingly 'affect' can also act as a noun, but mainly in psychology, when it means an emotion or a mood [38].

'Transform' and 'transformation' are also sometimes confused. A transform is a thing derived from another by the action of a transformation [38]. Thus the members of a Fourier transform pair are related by a Fourier transformation, as in Example 5.162.

Result

> The 2-D Fourier-transform pair is defined by
>
> $$F(k_x, k_y) = \mathrm{FT}_{2D}\{f(x, y)\} \tag{4}$$
>
> where FT_{2D} indicates 2-D Fourier transformation.

Gives result

Example 5.162. *Distinction between 'transform' and 'transformation', from an electrical and electronic engineering article [670].*

The Fourier transformation FT_{2D} (better $\mathrm{FT}_{\mathrm{2D}}$) of $f(x, y)$ gives the Fourier transform $F(k_x, k_y)$.

Somewhat analogously, there is a distinction between an estimate and an estimator. An estimator is a statistic used to estimate a value, usually a parameter of a distribution, and an estimate is the value of an estimator for a particular sample of data [360].

Gives result — From this example, it seems that the upper bound for the dynamic <u>estimator</u> $\overline{V}$ gives an <u>estimate</u> of the minimum steady-state covariance. — Result

Example 5.163. *Distinction between 'estimate' and 'estimator', from a control systems engineering article [675].*

The estimator $\overline{V}$ in Example 5.163 gives an estimate. The more general term 'estimation' refers to the use of a formula to calculate from observed data a value or values for an unknown parameter or parameters [360].

5.6.25 Padding

Gowers wrote at length on verbosity [141, ch. 6]. The following list of wordy expressions and their shorter parenthetical alternatives is drawn from [79] [142]:

- as a consequence of (because of)
- as a result of (because of)
- based on the fact that (because)
- by virtue of (by, under, because of)
- for the purpose of (for)
- for the reason that (because)
- give rise to (cause)
- in association with (with)
- in case (if)
- in connection with (concerning)
- in light of the fact that (because)
- in order to (to)
- in the absence of (without)
- in the course of (during)
- in the event that (if)
- in view of the fact that (because)
- it is clear that (clearly)
- a large number of (many)
- on the basis of (by)
- on the grounds of (because of)
- owing to the fact that (because)

- prior to (before)
- referred to as (called)
- small number of (few)
- subsequent to (after)
- utilize (use)
- with a view to (to)
- with the exception of (except)

A third of these expressions have 'because' as an alternative.

Where possible, the shorter alternatives are preferred [79, ch. 31] [142, ch. 6] [391, ch. 2, 3], but sometimes the longer expression delivers a useful distinction; for example, 'in order to' rather than 'to' [40, p. 399], as in Example 5.164.

Avoids 'to analyse to prove'

> This will be the main expression to analyse in order to prove a local well-posedness result.

Example 5.164. *Deliberate 'in order to', from a mathematics article [424].*

Here the longer 'in order to' avoids the awkward conjunction of 'to analyse' and 'to prove'.

5.7 Punctuation

Punctuation helps to define the relationship between the various elements of a sentence and set the tone [340, p. 70]. Its function is primarily logical rather than rhetorical [279, pp. 6–8]. Thus a comma is used to separate words, phrases, and clauses, not to indicate where the reader might take a breath.

Punctuation acts best in a supporting role and cannot easily introduce new relationships that are not already present in the words [284, p. 448]. Yet wrong punctuation can be disproportionately effective in distracting the reader. Worse still, it may be taken as evidence of a broader carelessness that extends to the author's research [229, p. 2].

5.7.1 Space

The punctuation marks full stop (also known as a period) '.', comma ',', semicolon ';', colon ':', question mark '?', and exclamation mark '!' are normally followed by a space but not preceded by one, as in some languages [89]. Opening brackets '(' and '[' are preceded by a space but not followed by one. Closing brackets ')' and ']' are followed by a space or another punctuation mark such as a comma or full stop but are not preceded by one. Inner nested brackets have no spaces.

✗

Missing spaces — That is to say the probabilities of the samples belong to each group.And then,the sample belongs to the group that has the highest probability[15][16].

Missing space

Example 5.165. *Missing spaces, from an unpublished electrical and electronic engineering master's report.*[49]

In Example 5.165, there should be a space after the full stop in 'group.', after the comma in 'then,', and before the square bracket in '[15]'.

5.7.2 Full Stop

Sentences begin with a capital letter and end with a full stop, and generally contain a finite verb.[36] Yet there are sentences that do not contain any verb and still end with a full stop.

In both Examples 5.166 and 5.167, the verb is implied.

No verb — First, some notation. The true residual sum of squares is . . .

Example 5.166. *Sentence without an explicit verb, from a computer science article [399].*

The first marginal condition is satisfied. Similarly for the second. — No verb

Example 5.167. *Sentence without an explicit verb, from a mathematical physics article [452].*

The use of verbless sentences has a long history but they are best employed sparingly [122].

5.7.3 Comma

Phrases and clauses that are non-defining or nonrestrictive are normally surrounded by commas [279, p. 35] [340, pp. 89–93]. Example 5.127 in Section 5.6.13 shows their use with a non-defining clause and Example 5.168 shows their use with a non-defining phrase.

Non-defining phrase

The light source, an LED, was attached to the moving probe . . .

Example 5.168. *Commas with a non-defining phrase, from a biomedical engineering article [513].*

The phrase 'an LED' is surrounded by commas. It can be omitted from the sentence without changing the identity of the light source that was attached to the moving probe.

By contrast, in Example 5.169, the phrase 'x_1 and x_2' has no commas. It cannot be omitted without introducing uncertainty into which values were calculated.

Defining phrase

The values x_1 and x_2 were calculated by applying the equation . . .

Example 5.169. *Absent commas with a defining phrase, from a chemistry article [494].*

In Example 5.170, one of the paired commas is omitted. It is unclear whether the clause 'leading to atrophy of the central nervous system (CNS)' is defining (if it is, the comma before 'leading' should be omitted) or non-defining (if it is, a comma should be inserted after '(CNS)').

✗ Loss of neurons due to cell death and axon degeneration, leading to atrophy of the central nervous system (CNS) — Ambiguous phrase
is primarily discussed . . .

Example 5.170. *Incorrect or missing comma, from a neurosciences and neurology article [673].*

The first comma of a parenthetical pair is, however, sometimes omitted when it comes immediately after a conjunction that has a preceding comma [279, p. 22], as in Example 5.171.

Interestingly H protein did not seem to be significantly degraded during stresses (Spots 2 and 3), but as shown — Non-defining phrase
earlier, the lipoic acid group . . .

Example 5.171. *Conventional omitted comma, from a biochemistry and molecular biology article [691].*

The absence of the comma preceding 'as shown earlier' does not change the nonrestrictive status of the clause. A comma should, though, be added after 'Interestingly' [40].

For all its utility in modifying the status of phrases and clauses, the comma cannot be used to join two finite clauses that would otherwise form two sentences. Its use in Example 5.172 is therefore incorrect. The error is called a comma splice [263]. The comma needs to be replaced by a conjunction such as 'and', by a semicolon, or by a full stop (and a new sentence started).

✗

> This machine learning algorithm has gained wide attention recently, it has contributed to the renaissance of interest . . .

Inadequate comma

Example 5.172. *Inadequate comma between two sentences, abridged from an unpublished electrical and electronic engineering doctoral transfer report.*[50]

Commas may be included unnecessarily before and after some adverbs. The following guide to inclusion was extracted from [40] [279]:

- also (rarely)
- even so (optional)
- namely (unnecessary after)
- nevertheless (none)
- therefore (none)
- thus (unnecessary after)
- too (rarely)
- yet (none, except before when acting as a conjunction)

Even so, stylistic reasons may take precedence over syntactic ones [279, p. 22].

When two phrases or clauses are linked by 'and' or 'but' the preceding comma can usually be omitted, especially if the subject in a clause is the same, but it should be included if there is a risk of confusion or a need for emphasis [279, pp. 29–31] [340, pp. 78–81].

In Example 5.173, there is no comma before the first 'and' since the subject is the same, whereas there is a comma before the last 'and' to add emphasis.

Conventions vary on whether a comma should precede a final 'and' in a sequence of three or more phrases or clauses. Partridge is against [279, p. 16];

> The odorant hexanal [(6)CHO] is near the center of this mitral cell's receptive field <u>and</u> evokes the strongest activation . . . whereas butanal [(4)CHO] and octanal [(8)CHO] are within its inhibitory surround, <u>and</u> hence evoke a net inhibition.

Same subject — (first 'and')

Same subject, but emphasis — (second 'and')

Example 5.173. *Omitted and added commas with 'and', from a neuroscience article [449].*

Burchfield is for [40, p. 162]. The advantage of using what is known as the Oxford comma [263] is that it consistently removes ambiguity.

In Example 5.174, if there were no comma before the 'and' in the last of the three-element sequence it would take a careful reading to deduce where the reported holes were drilled.

Three countersunk holes of 2.5 cm in diameter are drilled in the areas of the first, second and third, and fourth and fifth metatarsal heads.

Oxford comma

Example 5.174. *Added comma with 'and', from an orthopaedic surgery article [613].*

The ampersand '&' is occasionally suggested as a substitute for 'and' in sequences like those in Example 5.174. In reality, however, it is no longer used as 'and' in printed text except when it occurs in names of companies, journal titles, and journal author lists [284, pp. 35–6].

5.7.4 Semicolon

The semicolon is used to join related or coordinated clauses, particularly when commas and conjunctions would be confusing [279] [340].

In Example 5.175, commas might have been used to separate the main clauses, 'The signal w contains . . .', 'the output z is . . .', 'y is . . .', and 'u is . . .', since the last one is linked to the rest by 'and', but the commas in the parenthetical clause 'including disturbances . . .' would make parsing difficult. Semicolons are more effective in demarcating the main clauses, while maintaining their coordination [340].

The semicolon is used similarly to demarcate phrases containing commas [340, pp. 103–4]. In Example 5.176, it would be difficult without the semicolons to separate the statistical analyses from the factors in the analyses.

The signal w contains all external inputs, including disturbances, sensor noise, and commands; the output z is an error signal; y is the measured variables; and u is the control input.

Example 5.175. *Semicolons demarcating clauses, from a control systems engineering article [482].*

At a time when sentences were longer, the semicolon was deployed in many other ways [279], but its use declined as average sentence length decreased [284].

> One-way ANOVAs were used for the analysis of glucose, leptin, and insulin content; liver and white fat weight; and locomotor activity.

Example 5.176. *Semicolons demarcating phrases, from a neurosciences and neurology article [456].*

Although it could replace the comma in Example 5.172, a conjunction or new sentence might be better.

The semicolon cannot introduce a list [142, p. 155], as in Example 5.177. It should be replaced by a colon, as explained in Section 5.7.5.

Incorrect punctuation

✕

> The primers for various deletion mutants were as <u>follows;</u> 5′-AGAAGAACAATGGCCACTATGGTCGCTCCT-3′ and 5′-ACTCGAGGGAATCGGTAAGGTCAG-3′ for D1 . . .

Example 5.177. *Incorrect semicolon, from a biochemistry and molecular biology article [581].*

5.7.5 Colon

The colon has two principal uses: to announce a list or summary and to introduce an explanation or definition [279] [284]. In Example 5.178, the colon announces the list that follows.

Annunciatory

> Each patient had scans in three <u>conditions:</u> baseline, premonitory phase and migraine headache.

Example *5.178. Annunciatory colon, from a neurosciences and neurology article [598].*

In Example 5.179, the colon introduces the explanation that follows. Fowler picturesquely summarized this function as 'delivering the goods that have been invoiced in the preceding words' [122, p. 589].

> The large discrepancy in concordance rates between MZ twins and siblings can be explained <u>thus:</u> inherited or *de novo* mutation in the parental germ line affects MZ twins alike, whereas sibling and DZ concordance rates represent a mixture of modalities.

— Explanatory

Example 5.179. *Explanatory colon, from a genetics article [735].*

Opinions differ on whether the text that precedes the colon should be a finite clause or not. Partridge uses both [279]. Stilman and others argue for a finite clause only [340]. Thus Lester and Beeson [224, p. 262] suggest this test. If the colon is replaced by a full stop, does the text then form a complete sentence? Only if it does should a colon be used. Accordingly, Examples 5.178 and 5.179 represent correct usages; Example 5.180 does not.

Nonstandard

✗ The variables of interest were: constant temporal error, peak EMG, EMG duration, time to peak EMG, root mean squared (RMS) EMG and amplitude of the peak-to-peak motor-evoked potentials.

Example 5.180. *Nonstandard colon, from a neurosciences and neurology article [600].*

The colon interrupts the sentence, leaving it dangling. Either it should be omitted or if the list needs an announcement, 'these' or 'as follows' should be inserted between 'were' and the colon.

5.7.6 Apostrophe

The apostrophe is used mainly to indicate possession, and there are comprehensive guidelines on how it should be applied [40] [311] [340].

With units of measure that are not symbols but spelled out, the consensus is to include an apostrophe, as in Example 5.181.

Therefore, emergent fibrinolytic therapy after CRAO of more than 2 hours' duration is inappropriate.

Possessive

Example 5.181. *Apostrophe with a measure of time, from an ophthalmology article [602].*

When in doubt, Lester and Beeson [224, p. 248] suggest this test. If the order is reversed and 'of' or 'for' introduced to paraphrase the original, then an apostrophe should be used. In Example 5.181, the reversed order is 'duration of 2 hours': hence the apostrophe. Whether an apostrophe should be used in forming the plural of abbreviations and numbers is considered in Section 5.6.23. Its use with symbols is considered in Section 6.2.3.

More generally, the decision to indicate possession with an apostrophe or a preposition such as 'of' or 'for' depends on context and meaning. Thus 'the sensor's response' emphasizes the sensor and 'the response of the sensor' emphasizes the response.

5.7.7 Emphasis and Quotation Marks

The general advice is to avoid quotation marks, italics, or boldface to signal emphasis or special meaning [122] [279]. The result can be distracting. Rather, the words themselves should be organized to achieve the required effect [311, p. 155]. The only exception should be in mathematical and philosophical texts, where usage may be technical.

If a special use of a word or phrase does need to be indicated, say by quotation marks, they should be used in the first instance, but not subsequently [311] [340], as in Example 5.182.

Quotes on first use — No quotes on second use

> Each agent's heading is updated using a local rule based on the average of its own heading plus the headings of its "neighbors." . . . each agent's set of nearest neighbors change with time as the system evolves.

Example 5.182. *Quotation marks on first use only, from a control systems engineering article [551].*

5.7.8 Dashes, Hyphens, and Minus Signs

Dashes, hyphens, and minus signs have different lengths and functions. In increasing order of length, they appear as follows, with Unicode values in parentheses [359]:

- hyphen '-' (2010)
- minus sign '−' (2212)
- en-dash or en-rule '–' (2013)
- em-dash or em-rule '—' (2014)

The function of hyphens is discussed in Section 5.7.9. Minus signs are self-explanatory. The en-dash is used mainly to denote a span, as in '1–9' (§7.3.5) [311]. The em-dash, which usually occurs in pairs, signals a parenthetical remark, as in Example 5.183.

Parenthetical remark

> Together, they form a highly unusual magnetic state — a "Coulomb phase" — whose excitations are point-like defects.

Example 5.183. *Em-dashes demarcating a parenthetical remark, from a physics article [620].*

Dashes set off parenthetical material more firmly than commas [279]. They should be used sparingly and never as a universal punctuation symbol [142].

The combination of a dash and colon to introduce a list ':—' is obsolete [40] [284].

Many other kinds of dashes are available in the Unicode character set [359].

5.7.9 Hyphenation

Hyphenation is complicated and practice varies with country and publisher, but there is general agreement that it should be used consistently and only where necessary [40] [279] [284].

Hyphens appear in the formation of compounds, often linking an adjective with an adjective or participle that modifies a subsequent noun, as in Example 5.184.

Necessary hyphen

Fluorescence-based sensors reported prior to the work of Ruddock and Han [6]–[8] employ single-photon excitation.

Necessary hyphen

Example 5.184. *Hyphens removing ambiguity, from an instrumentation engineering article [461].*

Without the hyphen, 'Fluorescence' would modify 'based sensors' rather than forming a compound with 'based' [280]. Likewise, 'single' would modify 'photon excitation' rather than modifying 'photon'.

✗ All the major cholesterol guidelines recommend that, except for people with diabetes and very high low density — Missing hyphen lipoprotein (LDL) cholesterol concentrations . . .

Example 5.185. *Missing hyphen, from a general and internal medicine article [678].*

In Example 5.185, there is a missing hyphen between 'low' and 'density'. The incongruous effect is that both 'high' and 'low' modify 'density'. It could be argued that readers familiar with the topic would group 'low' with 'density' and 'high' with the compound 'low density', but they would also not need the definition of LDL.

The hyphen can extend over multiple elements to save repetition in the definition of compound quantities, as in Example 5.186.

Hanging hyphens

> The authors used the salt-, drought-, cold-, and ABA-responsive *rd29A* promoter fused to the firefly luciferase gene to report plant responses to salt treatments.

Example 5.186. *Hanging hyphens, from a plant physiology article [537].*

The hyphen used in this way is sometimes called a hanging hyphen [284]. It is criticized for being clumsy [142, pp. 168–9], although it reads logically. For it to be avoided in Example 5.186, the phrase would need recasting, possibly as in Example 5.187. But the parenthetical 'responsive to salt, drought, cold, and ABA' in this version loses the continuity of the original.

> The authors used the *rd29A* promoter, which is responsive to salt, drought, cold, and ABA, fused to the firefly luciferase gene to report plant responses to salt treatments.

Example 5.187. *Recast version of Example 5.186.*

The reverse hanging hyphen, unlike the hanging hyphen, is neither efficient nor logical. In Example 5.188, the appearance of '-dependent' forces the reader to work backwards to find the matching adjective.

✗

> Two pathways of transcription termination, factor-independent and -dependent, exist in bacteria.

Reverse hanging hyphen

Example 5.188. *Reverse hanging hyphen, from a genetics article [582].*

Some authors insert an unnecessary hyphen between an adverb ending in 'ly' and an adjective modifying a subsequent noun [40, p. 23] [142].

> The role of $P(T, M)$ in local and integrated transmission is demonstrated for deeply-localized waves.

Unnecessary hyphen

Example 5.189. *Unnecessary hyphen, from a physics article [672].*

In Example 5.189, the hyphen is logically unnecessary as the adverb 'deeply' can modify only 'localized' ('deeply waves' is nonsense), yet this usage persists.

Hyphenation is discussed in [49] [311].

5.7.10 Solidus

The solidus '/', also known as the forward slash or oblique, is used to link words that are alternatives [49] [284] [311]. It can be more effective than a dash in separating alternative phrases [279]. Example 5.190 illustrates a careful application.

This finding has led to an alternative explanation that contrasts a global/holistic/configural processing orientation . . . with a local/analytic/featural processing orientation.

(Annotation: Parallel solidi — bracketing "global/holistic/configural" and "local/analytic/featural")

Example 5.190. *Effective solidi, from a psychology article [427].*

The descriptors 'global', 'holistic', and 'configural' in the first underlined phrase have related meanings but come from alternative conceptual frameworks. Likewise with the descriptors 'local', 'analytic', and 'featural' in the second underlined phrase, which are in corresponding order. In both phrases, the solidus is clearer than 'or' in indicating the relationship between the conceptual alternatives and the contrast between them, global v. local, or holistic v. analytic, or configural v. featural.

The solidus can, however, cause confusion if it is misused for 'and' [311] or as a kind of all-purpose connective (§2.3.3), as in Example 5.191.

✗

UV/Vis and luminescence spectroscopies exploit the interaction of light. (= 'and', hyphen better?)

Selectivity/sensitivity, and stability typically have the most influence. (= 'and' or 'or'?)

The polymer film has to be conductive or composite (combined conductor/insulator). (= 'and', hyphen better?)

The Spreeta sensor requires connection to a computer controller/power supply. (= 'and'?)

Example 5.191. *Ambiguous solidus, from an instrumentation engineering article [714].*

The solidus functions here more as 'and' than as 'or', and a hyphen would be better with some pairs.

The particular problems with 'and/or' are considered in Section 5.6.16.

5.8 Checklist for Use of English

Person

1. Emphasis on what was done and not on who did it?
2. Use of 'I' and 'we' restricted to opinion pieces and special cases?
3. Sexist language avoided?

Verb agreement and tense

4. Subject and verb agree?
5. Past tense for methods and results? ('Samples were measured')
6. Present tense for analysis, facts, and established understanding? ('Tunnelling arises from the wave nature of matter')
7. Present perfect for findings that are still true? ('Recent work has shown . . .')

Active and passive voice

8. With active voice, agent has priority? ('Intercalation weakened bonding')
9. With passive, object of action has priority? ('Bonding was weakened')
10. Any sentence top-heavy?
11. 'It was' or 'There were' overused as dummy subjects?

Uncertain referents

12. Any relative pronouns ('that' or 'which') refer to the wrong thing?
13. Any participles ('-ing' words) refer to the wrong thing?
14. Any pronouns 'this' and 'these' refer to the wrong thing, or nothing at all?

Coordination

15. Any 'not only . . . but also' clauses incorrectly paired?
16. With repeated phrase omitted (ellipsis), text correct when the phrase is reinserted?

Word choice

17. Same word used consistently for the same thing?
18. Simple Anglo-Saxon phrases preferred ('worse than', not 'relatively poor in comparison with')?

19. Adverbs 'however', 'therefore', or 'thus' wrongly used as conjunctions?
20. Overuse of 'however' (instead 'but', 'yet', 'nevertheless', etc.)?
21. Vague 'based on' and 'carried out' eliminated?

Punctuation

22. Spacing before and after parentheses and other punctuation marks correct?
23. Sentences joined wrongly with a comma?
24. Semicolons ';' confused with colons ':'?
25. Ambiguous solidus '/' eliminated (especially 'and/or')?

6

Use of Mathematics

In principle, writing mathematics is no different from any other kind of scientific writing [150] [163] [238]. In practice, there are several presentational issues peculiar to the use of mathematics that need to be addressed separately or with a different emphasis. They are to do with purpose, organization, notation, grammar, and punctuation. This chapter is concerned with these presentational issues. It is not concerned with mathematical content per se.

It is useful to keep in mind two precepts offered by Gross [140]:

- Have mercy on the reader.
- Have mercy on the editor and publisher.

In other words, write for the reader's benefit, not your own; and accept normal typesetting limits, not challenge them.

Some of the examples in this chapter show that violating these precepts is surprisingly easy in mathematics. Section 6.1 deals with motivation and organization. Section 6.2 deals with notation and Section 6.3 with grammar and punctuation. Section 6.4 deals with document preparation. Section 6.5 contains a checklist for the items in this chapter.

6.1 Motivation and Organization

Whether the mathematics is pure or applied, you need to have something to say, and to say it in a way that can be understood, as clearly as possible [150, §2, §4].

6.1.1 Motivation

Mathematical research problems are best expressed in terms of decisive examples and counterexamples rather than generalities [150, p. 25]. With tangible motivation, the reader gains insight into the significance of the resulting analysis or proof of a theorem. If you have a particular revealing case, develop your argument for that case. After you have finished with it, the generalization can then be stated as a passing remark [134, §3.2].

An illustration of this approach is given in Example 3.41, reproduced here for convenience as Example 6.1.

> Consider the problem of distributing one pound of gold dust at random among a countably infinite set of beggars. Let the beggars be enumerated and let the procedure for distribution be as follows: the first beggar is given a random portion of the gold . . .

Example 6.1. *Opening with a research problem, from a mathematical statistics article [532].*

Apart from the declared infinite set of beggars, the research question is concrete and eminently understandable.

6.1.2 Insight

Unhappily, mathematics may be deployed in applications when there is little to say. Instead it is used for decoration or obfuscation, a chance to show off [140]. Reeves calls this usage[51] the anti-Ockham error:

> Someone has a perfectly reasonable data set and hypothesis, but embeds it all in a sea of irrelevant mathematics. . . . I cannot recall seeing this error in any of the really good papers, but it obscures many of the second-rate papers.

Mathematics should be organized to maximize the insight of the reader, not to serve as a distraction, elegant or otherwise [150, p. 23].

6.1.3 Development

It might be assumed that the logical development of mathematical material is stricter than in non-mathematical disciplines. That is, you might first establish some preparatory definitions, lemmas, and propositions; then derive the main theorems or results; and finally present some examples or applications.[52] The effect is a clean and ordered exposition but one that is hard for a reader unfamiliar with the work to follow. Although an axiomatic abstract style, which proceeds from the general to the specific [422, p. v], has its own attractions, it may not be the most suitable for communication [149].

The better practice is to aim to be conceptual rather than exhaustively computational [238]. You might first outline the structure of the argument; comment on its key components; and then guide the reader through each of the steps [163, §3.2]. To aid understanding, you could interleave questions, examples, counterexamples, and even figures [607]. At every step, the reader is kept informed about progress [134, §3.5].

6.1.4 Interruptions

There are many ways to interrupt a logical development. One of the most common is what Gillman calls a roadblock [134, §3.5]; that is, using a term without definition or explanation, and then repairing the omission only later. Example 6.2 shows a typical instance.

> Consider a stochastically perturbed system of the nominal system (1) represented using an Itô stochastic differential equation
>
> Undefined symbols —
>
> $$dx = f(x,t)dt + B(x,t)dW, \qquad x(0) = x_0 \qquad (6)$$
>
> and the conditions for existence and uniqueness of a solution to (6)
>
> . . . — Four lines of mathematics omitted
>
> Late definition — where $B: \mathbb{R}^n \times \mathbb{R} \to \mathbb{R}^{n \times d}$ is a matrix-valued function, $W(t)$ is a d-dimensional Wiener process, and x_0 is a random variable independent of W [41].

Example 6.2. *Late definition, from a control systems engineering article [464].*

The symbols B, W, and x_0 are not introduced before equation (6). On meeting them, therefore, the reader does not know whether to work backwards, in the belief that they were missed, or to work forwards, in the hope that they will appear later. Fortunately they do, albeit six lines later. It might be argued that the form of the equation is already familiar to the reader, but in that case, an explanation is unnecessary.

Halmos refers to the late 'where' statement as a sign of a lazy afterthought [150, p. 38].

It is, though, a balance. When the statement confirms existing notation or conventional interpretation (e.g. $\bar{x}$ for the mean of several x values), it seems unobjectionable.

Roadblocks have their parallel in non-mathematical writing (§2.1.5).

6.2 Notation

Good notation is half the battle. But good notation needs thought. Symbols need to be chosen carefully, combined logically, and not used unnecessarily [134, §2.4, §5.1] [150, §5].

One useful guide is that symbols should be used only for precise mathematical statements or when words are unwieldy. Conversely, words should be used when they occupy little more space than symbols [163, §3.6]. Abbreviations such as 'w.r.t.' (with respect to) and 'iff' (if and only if) do not aid comprehension [140].

6.2.1 Alphabets

The two main alphabets used in mathematics are the Latin alphabet (A, B, C, a, b, c) and the Greek alphabet (A, B, Γ, α, β, γ). Each can appear in bold face (**A, B, C**); in roman or upright, as here; and in italic or sloping type (*A, B, C*). The Latin alphabet can also appear in script or calligraphic form ($\mathcal{A}, \mathcal{B}, \mathcal{C}$); in Fraktur ($\mathfrak{A}, \mathfrak{B}, \mathfrak{C}$), where some upper-case characters are more easily confused than lower-case ones; in sans serif ($\mathsf{A}, \mathsf{B}, \mathsf{C}$); and others. Letters in both alphabets can be further annotated with superscripts, subscripts, and other marks [370].

Nevertheless, the best notation is no notation or the very least [150, §15], and the temptation of a potentially infinite alphabetic resource should be resisted. With care, it is often possible to work with just a portion of the Latin alphabet, together with occasional Greek letters for special cases. For example, a 15-page article by Milnor on isometries of inner product spaces [608] contains around 28 Latin, mostly lower-case, italic letters and only five Greek letters.

Notation is not universally fixed, although some letters do have default identities that aid recognition: for instance, π as the mathematical constant 3.141... and as a variable signifying probability; and e as the mathematical constant 2.718... and as the symbol for the identity in group theory. Other letters are used more freely, but continue to be preferred for certain roles, such as i, j, m, n for integer variables; u, v for members of a vector space; a, b for scalars; and f for a function.

In applications, some letters have a conventional interpretation as physical variables, such as t for time; x, y for spatial coordinates; u, v for frequency coordinates; λ for wavelength; and ω for angular frequency. Still, they should not be regarded as being frozen in those roles [150, §5].

The preferred way of introducing symbols is gradually, during the argument [134, §2.4], as in [608]. In longer, more complicated mathematical papers, however, a glossary of symbols placed at the end of the paper can be helpful (at the beginning it may be a little daunting).

6.2.2 Letter Fonts

Conventionally the interpretation of symbols depends on their font, roman, italic, or bold. The following recommendations are adapted from the National Institute of Standards and Technology (NIST) [358], and others [68] [190]:

- Scalar variables and quantities are set in italic, e.g. x, y.
- Descriptive terms are set in roman, e.g. in the quantity E_k representing energy, the roman subscript k stands for 'kinetic'.
- Standard mathematical functions and operators, and those with more than one letter, are set in roman, e.g. $\log x$, $\log_{10} x$, $\exp x$, $\sin x$, $\min x$, $\mathrm{Var}\, x$, $\mathrm{d}x/\mathrm{d}y$, $\mathrm{SD}(x)$, SEM.
- Numbers and units are set in roman, e.g. 4.2×10^{-3} N m rad^{-1} s (§3.6.5).
- Vectors and matrices are set in bold, e.g. $\boldsymbol{x} = (x_1, x_2, \ldots, x_n)$.

Although NIST recommends bold italic for vectors, in practice bold roman is also used, e.g. $\mathbf{x} = (x_1, x_2, \ldots, x_n)$, as in [99], and normal italic too, if the context allows, e.g. $x = (x_1, x_2, \ldots, x_n)$, as in [612]. Notice that the differential symbol d in '$\mathrm{d}x/\mathrm{d}y$' is set in roman, not italic, as it is not a variable.

In the mark-up language LaTeX (§6.4.2), the standard mathematical functions and operators automatically appear in roman when preceded by a backslash, e.g. \log.

Where there is no risk of ambiguity, single-letter descriptive subscripts or superscripts are sometimes set in italic. Pairs of symbols that differ only in whether a subscript or superscript is set in roman or italic, e.g. E_k and E_k, are therefore best avoided in the same paper.

A common notational error is to incorporate unit symbols within a mathematical expression without adjusting the font (§3.6.5). The unit symbols may then appear by default in italic, as with *mA* and *ms* in Example 6.3, where technically they then function as products (§6.2.5).

Wrong italic — Wrong italic — Wrong italic — Wrong italic

✗ $$E_{saving} = 38mA \times 36.875ms = 1401.3mA{\cdot}ms$$

Example 6.3. *Incorrect rendering of unit symbols, from an electrical and electronic engineering article [587].*

The edited version in Example 6.4 shows both the unit symbols and the subscript label 'saving' correctly rendered in roman (§3.6.5).

Roman — Roman — Roman — Roman

$$E_{\mathrm{saving}} = 38\ \mathrm{mA} \times 36.875\ \mathrm{ms} = 1401.3\ \mathrm{mA{\cdot}ms}$$

Example 6.4. *Edited version of Example 6.3 with correctly rendered symbols.*

6.2.3 Plurals of Symbols

Some guides [49, §7.14] [370] advise that to form the plural of symbols you should add an apostrophe and the letter 's', e.g. x_i's. Without the apostrophe, the composite x_is might be seen as a product. Even with the apostrophe, misinterpretation is possible and the extra punctuation mark in the composite symbol looks awkward.

There is an alternative. In normal use, the range of a variable x or index i is known or understood, so that the symbol x_i can be read as either singular or plural, depending on the context. Example 6.5 shows where the variable a_n in the phrase 'all the a_n' is read unambiguously as a plural, without an added apostrophe 's'.

Implied plural

This neighbourhood contains all the a_n except for a finite number of values of n.

Example 6.5. *Implied plural, from a mathematics textbook [422, p. 13].*

6.2.4 Subscripts and Superscripts

For clarity, no more than two levels of subscripts and superscripts should be attached to a symbol [370]. Example 6.6 shows a typical instance. The second level of subscripts can just be resolved.

$$\rho_s = Z_s^{-1} e^{-H_s/k_B T_s}$$

Second level subscript

Example 6.6. *Second level of subscripts, from a physics article [720].*

The second level can be avoided with use of the exp function, as in Example 6.7.

$$\rho_s = Z_s^{-1} \exp(-H_s/k_B T_s)$$

Single level subscript

Example 6.7. *Edited version of Example 6.6 with just one level of subscripts and superscripts.*

If the subscripts or superscripts are multi-letter indices or labels, then they should be separated by a comma or thin space and all should be on the same line [370]. In Example 6.8 the subscript l is an index and 'red' is a label, and the two are set on the same line.

$$S_{l,\mathrm{red}}^{C} = \frac{1}{|[G,G]|} \sum_{n \in [G,G]} C_{l,\mathrm{red}}(n)$$

Comma

Example 6.8. *Multi-letter indices, from a physics article [426].*

Ideally, second levels of subscripts and superscripts should be avoided altogether [238].

6.2.5 Multi-Letter Variables

Multi-letter variables can lead to confusion if they are set in italic, since juxtaposed italic letters are normally interpreted as products, for example, $AX^2 + BXY + CY^2$. Except where their use is common practice, multi-letter variables are discouraged in mathematics.

If multi-letter variables are essential, then, like multi-letter functions and operators, they can be set in roman [55, ch. 11], as in Example 6.9, where MSE stands for mean squared error.

Roman

$$\mathrm{MSE} = E[(Y - c)^2]$$

Example 6.9. *Multi-letter variable set in roman, from a statistics textbook [309, §4.4.2].*

Some disciplines routinely use multi-letter variables and labels set in italic rather than in roman, but occasionally the result can be confusing, as illustrated in Example 6.10.

Wrong italic — Wrong italic

✗

$$CH_j = \left\{ i \,\middle|\, \max_{i \in H \cap N_j^R} E^t_{res,i} \frac{P_{rx_{i,j}}}{P_{tx,i}} \right\}$$

Wrong italic — Wrong italic

Example 6.10. *Confusing notation, from a computer science article [525].*

The conventional interpretation of the main constituents of Example 6.10 would be as follows. The expression CH_j is the product of variables C and H_j; the expression N_j^R is a variable with indices R and j; the expression $E^t_{res,i}$ is a variable with indices t and i, and possibly composite index *res*; the expression $P_{rx_{i,j}}$ is a variable with indices r and $x_{i,j}$; and the expression $P_{tx,i}$ is a variable with indices t, x, and i.

Only one of these interpretations is correct, that for N_j^R. As for the rest, the expression CH is actually a single variable indexed by j; the expression $E^t_{res,i}$ is a variable E with label *res* and indexed by t and i; the expression $P_{rx_{i,j}}$ is a variable P with label *rx* and indexed by i and j; and the expression $P_{tx,i}$ is a variable P with label *tx* and indexed by i.

Example 6.10 can be edited according to the conventions in Sections 6.2.2 and 6.2.4. One version is shown in Example 6.11, where, apart from N_j^R, the indices are all set as subscripts and the labels as roman superscripts. The sizing of the curly braces and vertical midline have been adjusted to better match the height of the upper and lower elements in the expression.

Roman Roman Roman

$$\mathrm{CH}_j = \left\{ i \;\middle|\; \max_{i \in H \cap N_j^{\mathrm{R}}} E_{t,i}^{\mathrm{res}} \frac{P_{i,j}^{\mathrm{rx}}}{P_i^{\mathrm{tx}}} \right\}$$

Italic Roman

Example 6.11. *Edited version of Example 6.10 with multi-letter variables and labels set in roman.*

Another version of Example 6.10 is shown in Example 6.12, where the indices i, j, and t are reset as arguments and the labels are set as subscripts.

Indices as arguments Indices as arguments

$$\mathrm{CH}(j) = \left\{ i \;\middle|\; \max_{i \in H \cap N_{\mathrm{R}}(j)} E_{\mathrm{res}}(t, i) \frac{P_{\mathrm{rx}}(i, j)}{P_{\mathrm{tx}}(i)} \right\}$$

Example 6.12. *Edited version of Example 6.10 with important indices set as arguments.*

With the layout of Example 6.12, the distinction between variables and labels is more obvious.

6.2.6 In-Line and Displayed Equations

Equations, inequalities, and similar expressions that are set off from the rest of the text on a separate line are said to be displayed, as is the equation in Example 6.2. They are displayed because they are important, referenced later, or difficult to set in-line with the text [49, §12.21] [163, §3.7].

Displayed expressions need to be numbered only if they are referenced later [49, §12.24] [150, §15]. Unnecessary numbering is a distraction.

When expressions are set in-line with the text, it is best [179] to avoid stacked fractions (where one expression appears above another) and two or more levels of subscripts or superscripts. Example 6.13 shows how the exp function and solidus '/' can be used to avoid distorting the line spacing.

Two levels of superscripts avoided Stacked fraction avoided

The numerator $Z_j^{\infty} + (Z_j^{0} - Z_j^{\infty}) \exp(-\mathbf{r}_j^2/\sigma_j^2)$ is the position-dependent screened effective charge.

Example 6.13. *In-line expression, from a physics article [736].*

In both in-line and displayed expressions, there is no need to include the horizontal bar over the square-root sign; that is, $\surd 2$ suffices for $\sqrt{2}$ and $\surd\,(2\pi\sigma^2)$ suffices for $\sqrt{2\pi\sigma^2}$ [311]. In this way, no extra vertical space is required between lines.

6.2.7 Free Variables

Even with a restricted alphabet, more symbols may be used than necessary. Example 6.14 illustrates the introduction of what are sometimes called free variables.

✕

The short-time Fourier transform (STFT), also called the Gabor transform in its sampled variant, is a well-known, valuable tool for displaying the energy distribution of a signal *f* over the time–frequency plane.

Free variable

Example 6.14. *Free variable in an opening statement, from an electrical and electronic engineering article [410].*

The text is from the opening statement of the article. The symbol *f* fulfils no role and carries no information. Its presence can mislead the reader into believing that it does have significance, yet it remains absent from the remainder of the introduction of the article [410]. Free variables can be avoided by observing the simple rule 'use no letter only once' [150, p. 41].

6.2.8 Awkward Combinations

When symbols from one family are combined with those from another, it is sensible to use the same convention with each. Example 6.15 illustrates a combination of symbols where different conventions are used.

Awkward combination

✕

$$f(x_1, x_2) = ax_1 + bx_2 + c, \quad (x_1, x_2) \in W,$$

where

$$a = 6\mu_{10} \,/\, L(L+1)\,(2L+1),$$

$$b = 6\mu_{01} \,/\, L(L+1)\,(2L+1).$$

Example 6.15. *Awkward combination of symbols, from an optics article [642].*

The alphabetically ordered coefficients a and b would normally be associated with alphabetically ordered variables x and y, but instead they are combined with numerically indexed variables x_1 and x_2. If numerical indexing has priority, then the coefficients should also be numerically indexed [150, §5], as in the edited version in Example 6.16.

$$f(x_1, x_2) = a_1x_1 + a_2x_2 + a_0, \quad (x_1, x_2) \in W,$$

where

$$a_1 = 6\mu_{10} / L(L+1)(2L+1),$$

$$a_2 = 6\mu_{01} / L(L+1)(2L+1).$$

Example 6.16. *Edited version of Example 6.15 with harmonious combination of symbols.*

With further coordination still, the coefficients could be given double indices, that is, a_{10} for a_1, and so on, to better reflect the definitions of the coefficients μ_{10} and μ_{01}, shown in the bottom two lines.

Admittedly, awkward combinations of symbols are not wrong, but because they suggest more novelty than is truly present, they divert the reader's attention away from what matters.

6.2.9 Consistency

An accidental change in font, for instance, from roman to italic or to bold, can mislead the reader into believing that there has been a change in function. For clarity, therefore, the advice is to use one symbol (one letter and one font) for one and only one thing [49, §12.4].

6.2.10 Two Kinds of Epsilons

There is a conventional distinction between the round-backed epsilon $\in$ meaning 'is a member of' or 'belongs to' or simply 'is in', as in $(x_1, x_2) \in W$, and the non-round-backed epsilon ε used, for instance, to represent a physical constant, the permittivity of the vacuum ε_0. The two should not be confused.

6.2.11 Decoding Logic Symbols

Example 6.17 contains a statement with two logic symbols, the symbol $\exists$ for 'there exists' and the symbol $\forall$ for 'for all'.

Logic symbols

✕

$$\exists L_1 > 0, \forall t, \forall x_1, x_2 \in \mathbb{R}^n:$$
$$\|f(x_1,t) - f(x_2,t)\| + \|B(x_1,t) - B(x_2,t)\|_F \le L_1\|x_1 - x_2\|, \ldots$$

Example 6.17. *Unnecessary logic symbols, from a control systems engineering article [464].*

The symbols are easy to write, but the reader is left to decode them in order to understand the content. As Halmos points out [150, §13], no one thinks in terms of ∃ and ∀. The statement in Example 6.17 is not from an article in formal logic, where the use of these symbols might be expected, but from control systems engineering.

A decoded version of Example 6.17 is shown in Example 6.18. Although longer, it is easier to read.

There exists $L_1 > 0$ such that for all t and for all x_1, x_2 in $\mathbb{R}^n$,

$$\|f(x_1,t) - f(x_2,t)\| + \|B(x_1,t) - B(x_2,t)\|_F \le L_1\|x_1 - x_2\|, \ldots$$

Example 6.18. *Decoded version of the text in Example 6.17.*

Notice that the round-backed epsilon ∈ in the first line has been replaced by 'in'. It is a matter of personal choice, context, and practice whether to use one or the other. Halmos [150, §16] suggests that the symbol ∈ should be reserved for the verb phrase 'is in' not the preposition 'in', as here; hence the substitution.

An analogous problem occurs with the equals symbol. Gillman [134, §4.2] is more permissive, allowing '=' to mean either 'is equal to' or 'equal to'.

6.3 Grammar and Punctuation

The American Mathematical Society offers a general rule that when symbols are translated into words, they conform to the standards of good literary composition [237]. Grammar and punctuation are both important, as is choice of words [150], but punctuation can present particular problems.

6.3.1 Necessary Punctuation

For mathematical expressions to be treated in the same way as non-mathematical text, they need to be punctuated correctly [49, §12.3] [150, §16]. Displayed expressions such as equations and inequalities need commas and full stops, as illustrated in Example 6.19.

Comma

$$
\begin{aligned}
&\text{If } A_k = I_j, \text{ then}\\
&\mathcal{K}_j \le (A_{k+1}, A_{k+2}, \ldots) \le \mathcal{K}_{j+1} \text{ whenever } \sigma_j = +1,\\
&\mathcal{K}_j \ge (A_{k+1}, A_{k+2}, \ldots) \ge \mathcal{K}_{j+1} \text{ whenever } \sigma_j = -1.
\end{aligned}
$$

Full stop

Example 6.19. *Punctuating symbolic clauses, from a mathematical physics article [611].*

The three lines constitute a single sentence. The second line is equivalent to a clause, since '≤' means successively 'is less than or equal to' and '(which) is less than or equal to', and the line ends with a comma. The last line is also equivalent to a clause for the same reason, and it ends with a full stop, exactly as a sentence would in non-mathematical writing.

The omission of a conjunction 'and' or 'but' between the second and third lines in Example 6.19 is an example of unlinked coordination [297, §III.6]. Some authors would prefer a conjunction to avoid overworking the punctuation, as in the edited version in Example 6.20.

$$
\begin{aligned}
&\text{If } A_k = I_j, \text{ then}\\
&\mathcal{K}_j \le (A_{k+1}, A_{k+2}, \ldots) \le \mathcal{K}_{j+1} \text{ whenever } \sigma_j = +1,\\
&\underline{\text{and}}\\
&\mathcal{K}_j \ge (A_{k+1}, A_{k+2}, \ldots) \ge \mathcal{K}_{j+1} \text{ whenever } \sigma_j = -1.
\end{aligned}
$$

Conjunction — and

Example 6.20. *Edited version of Example 6.19 avoiding overworked punctuation.*

Overworked punctuation is discussed in Section 6.3.3.

6.3.2 Unnecessary Punctuation

Where one or more symbols are included in a line of text, they need not be surrounded with commas [179]. In Example 6.21 there is no comma before or after the string 'T, P, H_Q, σ, E, and H' and no comma before 'G'.

> Pressure-temperature (P-T) phase diagrams and the stability limits of materials can be derived for any of the parameters T, P, H_Q, σ, E, and H simply from the relaxed function G.

No comma — No comma — No comma

Example 6.21. *No unnecessary punctuation, from a materials science article [658].*

Commas are essential within the string 'T, P, H_Q, σ, E, and H'.

6.3.3 Overworked Punctuation

In non-mathematical writing, punctuation cannot readily introduce new relationships that are not already present in words (§5.7). The same holds for mathematical writing. It should not fall to the smallest of marks, a full stop or comma, to do the work of separating one sentence or clause from another. Example 6.22 shows two overworked full stops.

Overworked full stop

✗

> a is normalized as the frame area to be 1.0. $\boldsymbol{x} = (\xi, \eta)$ is normalized as the length of the diagonal segment of the frame to be 1.0. λ is given as follows . . .

Overworked full stop

Example 6.22. *Overworked punctuation, from a computer science article [655].*

Both full stops may be missed by the reader, especially as they look like part of a compound symbol, first '1.0. $\boldsymbol{x}$', second '1.0. λ'. The same error could occur with a comma or semicolon.

The general advice is never to start a sentence with symbol [49, §12.7] [150, §16] [237], something that is done three times in Example 6.22. Instead, first insert the name of the quantity symbolized, as illustrated in the edited version shown in Example 6.23.

> The area a is normalized as the frame area to be 1.0. The location $\boldsymbol{x} = (\xi, \eta)$ is normalized as the length of the diagonal segment of the frame to be 1.0. The distortion λ is given as follows . . .

Example 6.23. *Edited version of Example 6.22.*

The numbers and symbols are now separated and the sentences are much easier to distinguish, and more informative. Even when they do not form distinct sentences, adjacent mathematical expressions are best separated by words [49, §12.8] [150, §16], as in Example 6.24.

Word separator Word separator

$y_{i+1} = \psi(y_i, x_{i+1}, \lambda_i)$ for $i = 1, \ldots, n - 1$ with $y_1 = x_1$

Example 6.24. *Expressions separated by words, from a control systems engineering article [601].*

Where several short expressions need to follow one after the other, word separators may be omitted, but a little extra spacing inserted for clarity.

6.3.4 Conditional Statements

Example 6.25 shows a common form of conditional statement beginning 'If' followed later by 'then'. In non-mathematical writing, 'then' is optional [297, §15.32].

'if-then' pair

If Γ is abelian, then techniques already used in [27] should give a proof.

Example 6.25. *Conditional statement, from a mathematics article [630].*

By contrast, in mathematical writing, it is safer, always, to include it [134, §4.3] [140] [150, §16]. Thus in Example 6.26, if the 'then' is omitted, the second and third equalities, $N_1 = m_{n-1}(p-1)$ and $N_2 = 0$, could be seen as part of the 'If' statement. Only when the reader reached the end of the sentence would it be clear which part was the consequence of the 'If' statement.

'if-then' pair

If $m_n = pm_{n-1}$, then $N_1 = m_{n-1}(p - 1)$ and $N_2 = 0$.

Example 6.26. *Conditional statement, from a mathematics article [630].*

A similar argument can be made for statements beginning 'Assume that' and 'Suppose that'. A subsequent 'Then' removes the ambiguity [150, §16].

As for the more specific 'if and only if' in place of 'if', it is unnecessary in definitions [134, §4.4], but must be retained in theorems, propositions, corollaries, and lemmas [422].

6.3.5 I and We

Mathematical exposition involves an active engagement with the reader rather than a passive reporting of past events or findings. The reader expects the present tense, not the past [140], and the imperative mood; for example, ‘Let’, ‘Set’, and ‘Consider’ [150, §12]. In principle, there should be less need for ‘I’ and ‘we’ (§5.1.1). Nevertheless, the inclusive authorial ‘we’ [297, §6.18] does have a special place in the development of formal arguments [134, §4.1].

Examples 6.27 and 6.28 illustrate this use of ‘we’.

> Choosing a point x in U which maps to a periodic point in this neighborhood of the origin, we see that $\omega(x)$ is a periodic orbit disjoint from A.

Inclusive authorial ‘we’

Example 6.27. *Referring jointly to the author and reader, from a mathematics article [610].*

Inclusive authorial ‘we’

> To prove the result for this case, we substitute $\beta = q$ and $\gamma = \alpha' = q \times \alpha$ into Eq. (9).

Example 6.28. *Referring jointly to the authors and reader, from a theoretical physics article [710].*

Even so, both examples can be recast without ‘we’. Examples 6.29 and 6.30 show two of several possible alternative versions of Example 6.27.

> If a point x in U is chosen that maps to a periodic point in this neighborhood of the origin, then $\omega(x)$ is seen to be a periodic orbit disjoint from A.

Example 6.29. *Edited version of Example 6.27 without personal pronoun.*

> Choose a point x in U which maps to a periodic point in this neighborhood of the origin. Then $\omega(x)$ is a periodic orbit disjoint from A.

Example 6.30. *Edited version of Example 6.27 without personal pronoun.*

The imperative version in Example 6.30 is slightly shorter.

Example 6.31 is an edited version of Example 6.28 in which 'we' has simply been omitted to form the imperative.

> To prove the result for this case, substitute $\beta = q$ and $\gamma = \alpha' = q \times \alpha$ into Eq. (9).

Example 6.31. *Edited version of Example 6.28.*

Ultimately the use of 'we' is a matter of personal choice. The case for completely avoiding first-person pronouns in mathematical exposition is clearly set out by Halmos (§5.1.1, quotation).

6.4 Document Preparation Systems

The choice of which document preparation system you use to create, edit, and print your manuscript depends on how much mathematics needs to be set, the requirements of the journal, repository, or agency to which you plan to submit your manuscript, and the conventions of your discipline. The main choice is between word-processing software applications and mark-up languages.

6.4.1 Word-Processing Applications

A word-processing application such as Microsoft Word [253] provides an immediate formatted view of your text. There are modules and add-ins with separate menus that allow you to enter mathematical material, albeit with limited control.

The expression in Example 6.32 (reproduced for convenience from Example 6.11) and indeed all the mathematical expressions in this guide were produced by such a system but then set by the publisher's typesetter.

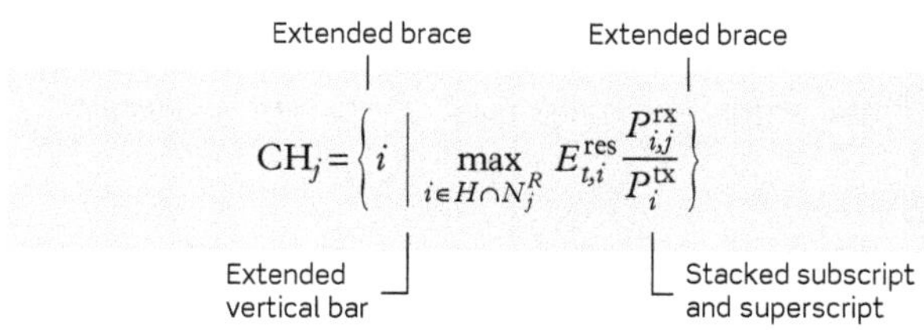

Example 6.32. *Reproduction of Example 6.11.*

Notice that the vertical separator extends to the height of the contents, the curly braces extend just beyond, and the subscripts and superscripts on the symbols N, E, and P are pulled in so that they are stacked instead of being staggered [49, §12.38].

6.4.2 Mark-Up Languages

A mark-up language such as Latex [213] [217], conventionally written LaTeX or as a more complicated logo, provides high-quality typesetting of mathematical material.[53] It is used routinely in mathematics and computer science, as well as in astrophysics, condensed matter physics, high-energy physics, and mathematical physics, but less so elsewhere. It is not a word processor. Rather, it is a suite of macros that controls a lower-level typesetting system, called TeX, created by Knuth [206].

A LaTeX document prepared with a text editor needs to be compiled into something that can be viewed or printed. The extent to which this is done automatically depends on the implementation. There are software packages that display formatted LaTeX as you type.

By way of illustration, the LaTeX code for the expression in Example 6.32 would look something like that in Example 6.33.

Expanding brace command

Extended vertical bar command

```
\[ {\rm CH}_j = \left\{ i \,\biggm| \, \mathop{\max}_
{i \in H \cap N_j^R} E_{t,i}^{\rm res}
\frac{P_{i,j}^{\rm rx}}{P_i^{\rm tx}} \right\} \]
```

Automatically stacked subscript and superscript

Expanding brace command

Example 6.33. *Possible LaTeX code for the expression in Example 6.32.*

The \left and \right commands in the LaTeX code ensure that the curly braces expand to height of the contents, and the \biggm command is used in place of the usual \mid to produce a vertical separator of the right size. The subscripts and superscripts are pulled in automatically.

The arXiv preprint server [10] expects LaTeX for document submission. But if the journal or repository receiving your paper has no preference for format, then the decision about which to use depends on how much mathematics you have to set and the control you want on the quality of the typesetting.

Advanced use of LaTeX or TeX formatting is probably best left to the publisher, since any nonstandard adjustments you make may affect the portability of the document [140].

6.5 Checklist for Use of Mathematics

Organization

1. Development logical, with equal steps in difficulty?
2. Variables and other quantities defined before first use?

Notation

3. Simplest set of Latin and Greek symbols?
4. Scalar variables in italic, labels in roman?
5. Multi-letter functions and variables in roman?
6. Symbol units in roman?
7. No more than two levels of subscripts and superscripts in displayed equations, no more than one level for in-line equations?
8. No equation numbered that is not used elsewhere?
9. No free variables in definitions?
10. Harmonious combinations of variables?
11. Round-backed epsilon for 'belongs to' or the equivalent?
12. Symbols used consistently?

Grammar and punctuation

13. Mathematical statements grammatical when translated into English?
14. Equations and other expressions correctly punctuated?
15. No commas around strings of variables?
16. No symbol at beginning of a sentence or clause?
17. 'If' statement followed by 'then'?
18. Limited use of 'we', only for author and reader jointly?

7

Data Description and Statistical Inference

What have data description and statistical inference to do with communication in science and engineering? For experimental reports, almost everything. Raw numerical data are rarely reported in research articles. Instead, some kind of summary is offered. It might take the form of a measurement on the data, such as an average value, or a claim about the properties of the population from which the data were drawn, for example, whether an effect is 'real' [99, p. 4]. Yet it is easy to mislead. A mean can be inflated by extreme data values, a pattern in a plot can be distorted by a fitted curve, and a statistically significant effect can prove irrelevant in practice. Thought needs to be given to the best way of representing data and drawing inferences safely.

This chapter introduces some basic statistical tools to help achieve that goal. Although different statistical approaches are outlined, the emphasis is not on statistical principles per se, but on their application in reporting. Chapter 8 deals specifically with graphical description.

The computations described in this chapter can be implemented in several numerical and statistical computing environments, including those provided by the commercial software package MATLAB [245] and the open source statistics package R [298] and its precursor S [364]. With several of the examples, there are links to illustrative MATLAB code in the notes at the end of this guide or to downloadable MATLAB and R packages.

Section 7.1 outlines the main approaches to statistical analysis. Section 7.2 deals with averaging data and Section 7.3 with measuring variation or scatter in data. Sections 7.4 and 7.5 deal with generalizing results beyond what is available. Section 7.6 deals with fitting curves to data and Section 7.7 with measuring their goodness of fit. Section 7.8 deals with identifying trends and describing the shape of data. Section 7.9 reviews the notions of statistical significance and the power of statistical tests. Section 7.10 contains a checklist for the items in this chapter.

7.1 Fundamentals

Data analyses can be classified as descriptive, exploratory, inferential, predictive, causal, or mechanistic [221]. Only the two most common kinds are considered here: descriptive and inferential analyses.

7.1.1 Description and Inference

Numerical data can be summarized in many ways. Typical measures are the average, the spread, the trend in values over time or with respect to some other explanatory variable, and the correlation between data values. These are all descriptive statistics, which do not depend on making assumptions about the data [8] [19] [175]. Data can also be summarized by graphical methods. Thus curve fitting is usually classified as descriptive in character [45]. A careful application of descriptive statistics enables data reduction that is principled and informative rather than ad hoc and potentially misleading [26] [101].

For inferences to be made about the population from which the data were drawn—that is, the parent population—a different approach is required, which does depend on making assumptions [201] [283]. Inferences may be to do with the value of a certain parameter of the population distribution [332] or whether the effect of a treatment on a sample is due to chance [381]. Alternatively, inferences may be about the real size of an effect, expressed in terms of confidence intervals [127] [168] or about the goodness of fit of a model with a particular set of parameter values [156], with or without alternative hypotheses [234].

Nonetheless, the distinction between descriptive and inferential is not always sharp [45, §3.4]. Some descriptive statistics may be used to make inferences about populations, either directly or indirectly, and some inferential statistics may be used descriptively.

It does not help that description and inference may be confounded, albeit unintentionally, owing to what Fisher called a purely verbal confusion [119, p. 311]. This is the common use of the same terms—mean, standard deviation, correlation coefficient—for both the particular value from a sample and the true but unknown value which is to be estimated. In this guide, sample statistics and population estimates are generally indicated as such.

7.1.2 Transparency

Even within disciplines where specific methods for the description of uncertainty and the estimation of error are well established, there is rarely one ideal method of analysis [293, §2.07] [303, §IV.A]. Different circumstances may lead to different assumptions about data and different descriptions or predictions. There are advantages, therefore, in performing and reporting more than one analysis [355], though making multiple analyses does not mean that you then choose the one that delivers the statistically most significant result [137, ch. 5]. Improper practices of this kind are discussed in Section 7.9.

Whether one or more analyses are reported, transparency is essential. Sufficient information should be presented that the reader can make an informed assessment of the trustworthiness of the findings, whether they are due to chance or represent something more systematic [69, §1.7, §8.5]. Transparency can be supported by making raw data available through data repositories (§12.7).

7.2 Averages

The average is often seen as a trivial operation, with little thought given to the ways it can represent and misrepresent a set of numerical data. Yet as Gosset observed [343, p. 1], the answer to a research question may turn on the value of a mean. But which mean?

The averages reviewed here are the arithmetic and geometric mean, the median, mode, trimmed mean, and weighted mean [86]. They are generically called measures of location [360]. As the examples show, they differ in their robustness to extreme values and changes in scale. A brief comparison is given in Section 7.2.6. A comprehensive description of measures of location may be found in [309, §10.4].

As a general caution, for an average to provide a useful summary, data values should not be too sparse or scattered [86]. If they are, then a fuller description may be needed [101].

7.2.1 Arithmetic Mean

The arithmetic mean (or simply the mean) is the most commonly used average. For a set of n data values $x_1, x_2, \ldots, x_n$, the arithmetic mean $\overline{x}$ is defined by

$$\overline{x} = \frac{1}{n}\sum_{i=1}^{n} x_i\,. \tag{7.1}$$

Take the set of ten numbers shown in Example 7.1. They are from measurements of visual acuity with Snellen letter charts [648, Table 1, left eye]. A value of 6 on this scale constitutes normal acuity and a value of 24 poor acuity.[54]

6	12	24	4.5	7.5	7.5	7.5	6	7.5	4.5

Example 7.1. *Acuity data for ten patients, from an ophthalmology article [648, Table 1].*

The mean of the data values is 8.7, but as a representative of this particular set, it has two defects. The first is that it is sensitive to extreme values [309, §10.4.1]. The value of 24 is much larger than the rest, and if it is excluded, then the mean falls to 7.0. Naturally, extreme values, large and small, cannot be arbitrarily excluded, and there are principled ways of dealing with outliers, that is, values that seem inconsistent with the rest of the set [21].

The second defect of the mean is that it depends on the measurement scale. For the data in Example 7.1, there is an alternative, more natural scale obtained by a logarithmic transformation [171] [713]. Each value is divided by 6 (the normal value for acuity) and the logarithm taken to the base 10. The transformed values are shown in Example 7.2.

0	0.30	0.60	−0.12	0.10	0.10	0.10	0	0.10	−0.12

Example 7.2. *Normalized logarithmic transform of the data in Example 7.1.*

The mean of the transformed values is 0.10, which, when transformed back to the original scale, becomes 7.6. This value is rather less than the original mean of 8.7 and closer to the most frequent value of 7.5. The logarithmic transformation reduces the asymmetry or skewness of the original data distribution [8, §11.3]. If the largest value of 24 is again excluded from the calculation, then the mean falls to 6.7. Although the effect is smaller than with the untransformed data, the mean is still sensitive to extreme values.

The mean obtained by logarithmic transformation, averaging, and then transforming back is simply the geometric mean. For a set of positive data values $x_1, x_2, \ldots, x_n$, the geometric mean, denoted here by $\tilde{x}$, is defined by

$$\tilde{x} = \left(x_1 x_2 \cdots x_n\right)^{1/n}.$$

The advantages and disadvantages of data transformations are discussed in [8] [230] [332].

7.2.2 Median and Mode

The median and mode are two other kinds of averages [86]. The median of a set of numbers is the middle value of the ordered set of values. More precisely, for a set with an odd number of values, say $2m + 1$, it is the mth value. For a set with an even number of values, say $2m$, it is the average of the mth and $(m + 1)$th values [309]. For the values in Example 7.1, the median is 7.5.

The median is robust against extreme values. The median of the data in Example 7.1 remains 7.5 whether or not the value of 24 is included (and whether or not the measurement scale is logarithmically transformed).

The mode is the most frequent value in the set, and, depending on the distribution of values, there may be more than one mode or no mode. For the values in Example 7.1, the mode is 7.5, the same as the median.

The mode or modes in a set of values can be important as indicators of clustering or the mixing of different components. Several methods are available for detecting and distinguishing modes, a task referred to as 'bump hunting' [335].

7.2.3 Trimmed Mean

Another average that is robust against extreme values is the trimmed mean [356]. For a particular fraction, say 10%, the 10% trimmed mean of a set of numbers is obtained by discarding the lowest 10% of values and the highest 10% of values and taking the arithmetic mean of the remaining values.[55] More generally, if data values $x_1, x_2, \ldots, x_n$ are arranged in ascending order, so that $x_1 \leq x_2 \leq \cdots \leq x_n$, then a $100\alpha\%$ trimmed mean, $\overline{x}_\alpha$ say, where α may be as small as 0.05 but is typically between 0.1 and 0.2, is given [309] by

$$\overline{x}_\alpha = \frac{1}{n - 2n\alpha}\left(x_{n\alpha+1} + x_{n\alpha+2} + \cdots + x_{n-n\alpha}\right).$$

It is approximate if $n\alpha$ needs to be rounded to a whole number. Alternatively, intermediate values may be interpolated. Notice that as α approaches 0.5, the $100\alpha\%$ trimmed mean approaches the median. The value $x_{n\alpha}$ is called the 100αth percentile (or centile), sometimes denoted by x_α, for α between 0 and 1.

With Example 7.1 a 10% trimmed mean is well defined, and its value is 7.3, very close to the median and mode of 7.5 and the geometric mean of 7.6.

Instead of discarding the lowest 10% of values and the highest 10% of values in a 10% trimmed mean, they can be replaced by the nearest retained value. More generally, a$100\alpha\%$ Winsorized mean is given [356] by

$$\overline{x}_\alpha = \frac{1}{n}\left(n\alpha x_{n\alpha+1} + x_{n\alpha+1} + x_{n\alpha+2} + \cdots + x_{n-n\alpha} + n\alpha x_{n-n\alpha}\right).$$

Whether a trimmed mean or a Winsorized mean is better depends on the distribution of data values. See [21, §3.2.1] [379] for more discussion about robust measures of location.

7.2.4 Weighted Mean

A weighted mean or weighted average is an average that assigns more weight to some data values than to others. That is, if w_i is the weight associated with each observation x_i, where $i = 1, 2, \ldots, n$, then the weighted mean $\overline{x}_w$ is defined [360] by

$$\overline{x}_w = \frac{\sum_{i=1}^{n} w_i x_i}{\sum_{i=1}^{n} w_i}. \tag{7.2}$$

If all the weights in Eqn (7.2) are equal, then this definition coincides with the arithmetic mean in Eqn (7.1).

The values of the weights might come from separate estimates of the precision of each measurement. For example, if each observation x_i is in reality the mean of several independent measurements, say n_i in all, then $w_i = n_i$.

7.2.5 Circular Data

For data that are distributed along a line, bounded or not, the mean and median are well defined. For data that are distributed on the circle, for instance, orientation or phase angle, the definitions need to be modified since they depend on how angles are represented. To see why [240, §2.1], suppose that the angles θ_1 and θ_2 are represented by values in the interval $0° \le \theta < 360°$, and suppose that θ_1 = 1° and θ_2 = 359°. Then their mean is 180°, which is patently wrong. Suppose, instead, that the angles are represented by values in the interval $-180° < \theta \le 180°$. Then $\theta_1 = 1°$ and $\theta_2 = -1°$. The angles are the same as before, but now their mean is 0°, which is correct. How, therefore, to choose the right representation for several angles?

One way to proceed is to convert the angles to a Cartesian coordinate representation, as follows [240, §2.2]. Suppose that $\theta_1, \theta_2, \ldots, \theta_n$ are angles. Set

$$\bar{c} = \frac{1}{n}\sum_{i=1}^{n} \cos\theta_i, \quad \bar{s} = \frac{1}{n}\sum_{i=1}^{n} \sin\theta_i.$$

Then, providing that $\left(\bar{c}^2 + \bar{s}^2\right)^{1/2}$ is greater than zero (the mean is not defined otherwise), the mean angle $\bar{\theta}$ is given by

$$\bar{\theta} = \begin{cases} \tan^{-1}(\bar{s}/\bar{c}) & \text{if } \bar{c} \ge 0, \\ \tan^{-1}(\bar{s}/\bar{c}) + 180° & \text{if } \bar{c} < 0, \end{cases}$$

where $\tan^{-1}$ takes values in the interval $-90° < \theta \le 90°$.

Circular data are usually distinguished from axial data, for which each direction is equivalent to the opposite one, that is, angle θ is equivalent to angle θ + 180° [240, §1.1].

Summary measures, data display, and inferential analyses with circular and axial data are considered in depth in [118] [240].

7.2.6 Which Average to Report?

When the distribution of data has little or no asymmetry or skewness, the arithmetic mean, median, and trimmed mean all have similar values [101, §1.4] [309, §10.4.5]. When the distribution is asymmetric or has more than one mode, the choice of which average to report may depend on additional considerations.

Thus with the skewed data in Example 7.1, the arithmetic mean gives the poorest descriptor, and the geometric mean, the median, mode, and trimmed mean give better or the best. The geometric mean, representing the arithmetic mean on a logarithmic scale, would actually be the most natural to report. Even then, a trimmed mean would be more robust against outliers.

If there are other costs associated with the data, however, they may outweigh the benefits of robustness. For example, with medical treatment data, purchasers and providers need to know the total cost of treatment. Because the distribution of costs is skewed (with most treatments being moderate-to-low cost and a few being very expensive), the arithmetic mean is larger than the median. Yet when multiplied by the total number of patients, only the arithmetic mean—not the median, mode, or geometric mean—delivers the total cost [414].

In practice, the arithmetic mean remains the most frequently reported summary of a set of data, much more so than the median.[56] Why is it so popular? There are several reasons [101]. First, it is well understood and easy to calculate. Second, the mean can be used in other analyses, for instance, in forming the mean of several data sets given their individual means and numbers of data values. Third, it is used as a parameter in many common probability distributions [17], most notably the normal or Gaussian distribution, and, as explained in Section 7.4, it is needed in inferential analyses that depend on those distributions.

Nevertheless, if the arithmetic mean is used, it should be only after an examination of any asymmetries in the data and possible outliers. If an average other than arithmetic mean is used, it should be unambiguously identified.

7.3 Measures of Variation

The degree of scatter or spread of a set of numerical data can be measured with or without reference to an average [101]. The measures reviewed here are the standard deviation and the mean absolute deviation, which refer to an average, and the range and interquartile range, which do not. They are generically called measures of variation [168], spread [360], or dispersion [309]. They may not be defined for all distributions [17].

A brief comparison of measures of variation is given in Section 7.3.6.

7.3.1 Standard Deviation

The standard deviation is not the most obvious measure of scatter, but it has some theoretical and computational advantages. It is also by far the most frequently reported such measure. For a set of n data values $x_1, x_2, \ldots, x_n$, the standard deviation SD is defined [360] by

$$\mathrm{SD} = \left[\frac{1}{n} \sum_{i=1}^{n} (x_i - \bar{x})^2 \right]^{1/2}, \tag{7.3}$$

where $\bar{x}$ is the mean of the x_i, as in Eqn (7.1). Notice that the divisor in Eqn (7.3) is n. With this definition, the standard deviation represents the spread of a particular set of data, and is not intended as an estimate of the spread of the population from which the data might have been drawn. That estimate replaces the divisor n in Eqn (7.3) by $n-1$, as explained in Section 7.4.1. Confusingly, the term 'sample standard deviation' (or 'sample variance') is used by some authors [360] for the standard deviation of a set as in Eqn (7.3) and by others [86] for the population estimate derived from a set. An example of where the standard deviation of a set is reported is given in Section 8.4.1.

The standard deviation of the set in Example 7.1 is 5.5, which may seem large on inspection of the values in the set, most of which fall on or close to the mean of 7.5. Even more than the arithmetic mean, the standard deviation as a measure of scatter is susceptible to extreme values [309, §10.5], because they appear in a squared term. If the largest value of 24 is excluded from the set in Example 7.1, the standard deviation falls from 5.5 to 2.1.

Although the abbreviation SD is used in this guide, other sources use Sd, sd, and s.d. For some analyses, the square of the standard deviation, the variance, denoted by Var or var, is a more appropriate quantity. Illustrations of how the standard deviation is reported are given in Example 7.3 and Example 7.6. The use of the plus-minus sign '±' in reporting the standard deviation is discussed in Section 7.4.2.

7.3.2 Mean Absolute Deviation

The mean absolute deviation about the mean (sometimes called the mean deviation) is a more intuitive measure of scatter than the standard deviation. For a set of n data values $x_1, x_2, \ldots, x_n$, the mean absolute deviation about the mean MAD is defined [360] by

$$\mathrm{MAD} = \frac{1}{n}\sum_{i=1}^{n}\left|x_i - \bar{x}\right|, \tag{7.4}$$

where, as before, $\bar{x}$ is the mean of the x_i, as in Eqn (7.1).

The mean absolute deviation about the mean for the values in Example 7.1 is 3.7. The interpretation of this value is simple [101]: it is the average distance of the values from their mean. The mean absolute deviation is less sensitive to extreme values than the standard deviation, and the value of 3.7 from Example 7.1 is smaller than the value of 5.5 for the standard deviation, albeit still influenced by the largest value of 24 in the set. If the value of 24 is excluded, the mean absolute deviation falls from 3.7 to 1.6.

7.3.3 Interquartile Range

For a set of n data values $x_1, x_2, \ldots, x_n$, the 100αth percentile, with α between 0 and 1, is the value x_α such that there are $n\alpha$ values less than or equal to x_α. It is often approximate, as $n\alpha$ may need to be rounded to a whole number. The interquartile range is the difference between the 75th and 25th percentiles [309]. That is, for $x_1, x_2, \ldots, x_n$, the interquartile range IQR is given by

$$\text{IQR} = x_{0.75} - x_{0.25}. \tag{7.5}$$

For sets with even numbers of data values, the interquartile range is exactly equal to the difference between the median of the upper half range and the median of the lower half range. Like the median, the interquartile range is robust against extreme values.

The number of values in Example 7.1 is rather too small for the interquartile range to be useful. A routine calculation delivers a value of 7.5 – 6.0 = 1.5.

The median and upper and lower quartiles are the main components of the graphical device known as the boxplot (§8.4.3).

7.3.4 Median Absolute Deviation from the Median

Another measure of variation that is robust against extreme values is the median absolute deviation about the median, ambiguously abbreviated MAD, the same as for the mean absolute deviation about the mean. For a set of n data values $x_1, x_2, \ldots, x_n$, with median $\tilde{x}$, the median absolute deviation about the median is defined [309] as the median of values $|x_1 - \tilde{x}|, |x_2 - \tilde{x}|, \ldots, |x_n - \tilde{x}|$.

As with the interquartile range, the number of values in Example 7.1 is rather too small for the median absolute deviation about the median to be useful. Its value is also 1.5.

7.3.5 Range

The range is simply the difference between the largest and smallest data values [360]. The range of the values in Example 7.1 is 19.5.

The range may be reported as a span, that is, as minimum and maximum values, typically separated by an en-dash, as in Example 7.3.

> Based on these observations, the average delay for insulin stationary effects was 2.5 min (SD 1.1 min, range 1–5 min).

Example 7.3. *Reporting standard deviation and range, from a physiology article [570].*

Giving the minimum and maximum values for the delay in effects as well as the standard deviation gives the reader a more complete understanding of the variation and its practical relevance.

Notation can, however, sometimes lead to confusion. In Example 7.4, the formulation of the span is incorrect, since the en-dash does not function in the same way as the word 'to'.

✗

Incorrect — Most of these instruments operate in the frequency range from 1–20 MHz.

Example 7.4. *Ambiguous range, from an instrumentation engineering article [624].*

As it stands, the statement means that the frequency range of the instruments is some number between 1 MHz and 20 MHz. To represent a span, it should have been written 'range from 1 MHz to 20 MHz', or 'range 1–20 MHz' (§3.6.7).

Despite its intuitive character, the range is not commonly used to report scatter, for it has several defects [101]. First, by definition, it depends only on the extreme values in the set and captures nothing of the other values. Second, its value tends to increase as the size of the data set increases, with the result that sets that differ in size cannot be readily compared. Third, it is unstable and patently susceptible to outliers. Still, in some situations, the range can be informative [168, pp. 300–4], as shown in Example 7.3.

7.3.6 Which Measure of Variation to Report?

The advantages and disadvantages of the standard deviation as a measure of variation parallel those of the mean as a measure of location [168], for example, in their place in statistical theory and in their sensitivity to extreme values.

The standard deviation is especially susceptible to outliers, which can dominate the sum in Eqn (7.3). Both the mean absolute deviation of Eqn (7.4) and the interquartile range of Eqn (7.5) are more robust. The performance of the standard deviation can be improved by discarding a proportion of extreme values, as with a trimmed mean (§7.2.3). In practice, the trimmed standard deviation is rarely recorded.

The standard deviation remains the most frequently reported measure of variation in a set of data, very much more so than the mean absolute deviation and other similar measures.[57]

7.3.7 Precision and Significant Figures

The number of significant figures[58] reported with numerical data should be commensurate with experimental precision [26]. In Example 7.3, the mean and standard deviation are shown with two significant figures. Given the size of the standard deviation, little would be gained by introducing a third significant figure, unless the data were to be used in a further calculation.

Different values of a measurement should therefore normally be expressed with the same precision [8, p. 13]. The data in Example 7.1 appear to violate this rule, but the fractional values represent nominal interpolations on a non-linear scale. Thus 4.5 represents a value between adjacent standard values of 4 and 5 on the scale and 7.5 a value between adjacent standard values of 6 and 9 on the scale.

7.4 Generalizing Results

One of the most basic questions a reader can ask about a set of results is whether they would hold for other samples of data. Paradoxically, most reported data are not obtained by random sampling from a well-defined larger population [101, §7.6]. Rather, they are more likely to represent what was available at the time or acquired by testing a particular material, device, or procedure.

Even so, it is possible to report estimates of the uncertainty or error associated with each measurement under the assumption that there is a population from which the data could have been drawn, and that the sample is representative of that population [244, §1.2]. These estimates may be derived from general theory. Alternatively, they may be contingent on specific assumptions about the statistical distribution of the data, or involve empirical computationally intensive resampling procedures, including the so-called bootstrap method.

7.4.1 Population Estimates

If a particular set of data consisting of n values $x_1, x_2, \ldots, x_n$ is a random sample drawn from a parent population, then the arithmetic mean $\overline{x}$ in Eqn (7.1) gives an unbiased estimate of the population mean μ [309, §7.3]. In other words, if samples were drawn repeatedly from the parent population, then the average of the $\overline{x}$ would be expected to coincide with μ.

Unlike the mean $\overline{x}$, the standard deviation SD of Eqn (7.3) provides a biased estimate of the population value. But if the definition in Eqn (7.3) is amended [309, §7.3] so that the standard deviation s is defined by

$$s = \left[\frac{1}{n-1} \sum_{i=1}^{n} (x_i - \overline{x})^2 \right]^{1/2}, \tag{7.6}$$

where the divisor $n - 1$ takes the place of n, then s^2 gives an unbiased estimate of the population variance, σ^2 say.[59] For moderate-to-large values of n there is little numerical difference between the standard deviation SD of Eqn (7.3) and the estimate s of Eqn (7.6) [309, §10.5].

The standard deviation of a sample mean is usually called its standard error [360], denoted by SE or by SEM to emphasize that it is the standard error of the mean. In fact, the standard error can refer to the standard deviation of any quantity calculated from a sample [8]. The standard error of the mean is estimated from a sample of size n by

$$\mathrm{SE} = \frac{s}{\sqrt{n}}, \tag{7.7}$$

where s is defined by Eqn (7.6). The standard error for the set of ten values in Example 7.1 is 1.83.

By contrast with the sample standard deviation, the standard error generally gets smaller as the sample size n increases [5]. The decrease in SE makes sense: as more of the population contributes to a sample, the estimate of the population mean becomes less uncertain, and therefore has a smaller standard deviation.

How often does a reported value $\bar{x}$ fall near to the population mean μ? The question cannot be answered exactly unless the population distribution is known, but if the sample size n is large, then the mean $\bar{x}$ is approximately normally distributed [309, §7.3]. Recall that the probability density function $f(x)$ of a normal distribution with mean μ and standard deviation σ is bell-shaped and governed by the formula

$$f(x) = \frac{1}{\sigma\,\sqrt{(2\pi)}} \exp\left[-(x-\mu)^2 / 2\sigma^2\right], \tag{7.8}$$

where $-\infty < x < \infty$. If the variable x is replaced by the sample mean $\bar{x}$ and the standard deviation σ is replaced by the estimated SE from Eqn (7.7), then Eqn (7.8) provides an estimate of how often $\bar{x}$ comes close to μ. Thus an interval ± 1 SE about $\bar{x}$ should contain μ on 68.3% of occasions; an interval ± 2 SE on 95.5% of occasions; and an interval ± 3 SE on 99.7% of occasions.

If the number of values n is less than 10 or the parent population is non-normal, then other methods for estimating the accuracy of the reported mean may need to be considered [101] [309]. Bootstrap measures, which can be used with both small n and an unknown parent population, are described in Section 7.4.4.

7.4.2 Standard Error or Standard Deviation?

Except in trivial cases, the standard error is always smaller than the standard deviation. Which you report, however, is not a matter of which looks better but of what you want to say about your data.

To recapitulate, the standard deviation describes the variation in individual data values and the standard error the variation in the means of data values [5]. It is essential to tell the reader which measure is being used, as in Example 7.5. Here standard errors are provided, which describe the variation in average slope values.

> In particular, the average slope for low-passed pictures was 1.69 (SEM = 0.24) and for high-passed pictures it was 1.70 (SEM = 0.30).

Example 7.5. *Uncertainty defined by standard error SEM, from an experimental psychology article [469].*

Where several means are reported each with its SD or SE, the shorthand plus-minus sign '±' is sometimes used, though the choice of error measure still needs to be declared, as in Example 7.6. Here standard deviations are provided, which describe the variation in individual wavelength values.

> Means across species (±1 SD) of wavelengths in the green-yellow and far-red spectra at which r^2 maxima occurred were 550 ± 12 and 712 ± 4 nm for reflectance, 580 ± 14 and 706 ± 7 nm for transmittance, and 561 ± 8 nm and 714 ± 6 nm for absorptance.

Example 7.6. *Uncertainty defined by standard deviation SD with ± notation, from a plant sciences article [437].*

The use of the ± notation is deprecated [5], since it is sometimes used without explanation. In Example 7.7 it is not immediately obvious whether the percentage error refers to the SD or SE or something else.

✗

> Errors in particle volume fraction and bubble volume fraction are ±3% and ±5%, respectively.

Example 7.7. *Unexplained ± notation, from a materials science article [701].*

In practice, once the error measure has been defined, as in Example 7.6, the ± notation does fulfil a useful function, offering an intuitive and compact indication of reliability and one which is common in the physical sciences.

7.4.3 Confidence Intervals

Measures of variation such as the standard deviation can refer to individual samples and to a parent population, but some inferential measures are defined only for population parameters.

A confidence interval for a population parameter such as the mean μ is an interval which is calculated from a sample and which contains the parameter value with some specified probability [360]. For example, a 95% confidence interval contains the parameter value with probability 0.95. In other words, if samples were repeatedly drawn from the parent population, then 95% of the calculated intervals would be expected to contain the true value. See [19, §7.2] [175, §7.3] [309, §7.3] for further discussion, and [283, §5.10] for some interpretational issues. Notice that a confidence interval is a random variable: different samples yield different values.

From Section 7.4.1, it follows that if the parent population is approximately normal and the sample size n is at least about 30, then an approximate 95% confidence interval CI for the population mean μ is given [309, §7.3] by

$$\mathrm{CI} = [\bar{x} - 1.96\ \mathrm{SE},\ \ \bar{x} + 1.96\ \mathrm{SE}], \tag{7.9}$$

where SE is defined by Eqn (7.7). This is an example of a standard normal confidence interval. The value of -1.96 is called the 2.5% point of the distribution and 1.96 the 97.5% point of the distribution [360]. In general, for $0 \leq \alpha \leq 1$, the corresponding 100α% percentage point is denoted by $z(\alpha)$. Thus $z(0.975) = 1.96$.

By way of illustration, since the mean of the data set of Snellen acuities in Example 7.1 is 8.70 and the standard error is 1.83, it follows from Eqn (7.9) that a 95% standard normal confidence interval is given by

$$\mathrm{CI} = \left[5.1,\ \ 12.3\right]. \tag{7.10}$$

The data in Example 7.1 are not normally distributed (§7.2.1) and so this estimate is likely to be inaccurate. But even if the parent population were exactly normal, the formula in Eqn (7.9) would still be approximate. The reason is that when the sample size n is small, the distribution of the sample mean $\bar{x}$ is described more accurately by the t-distribution than by the normal distribution [101, §8.5].

The t-distribution looks similar to the normal distribution but is somewhat broader with longer tails. It depends on sample size n through a parameter, its degrees of freedom (§7.7.1). The percentage point $z(0.975) = 1.96$ in Eqn (7.9) needs therefore to be replaced by the corresponding percentage point $t_{n-1}(0.975)$ of the t-distribution with $n - 1$ degrees of freedom (to allow for the constraint that deviations sum to zero). As n increases, the t-distribution converges to the normal distribution [309, §6.2]. For $n = 10$, the required percentage point is given[60] by $t_{n-1}(0.975) = 2.26$.

Accordingly, from Eqn (7.9) with 1.96 replaced by 2.26, a 95% confidence interval that takes into account the size of the set of Snellen acuities is given[61] by

$$\mathrm{CI} = \left[4.6,\ \ 12.8\right], \tag{7.11}$$

which is a little larger than the estimate in Eqn (7.10). Although the problem of sample size has been addressed, the asymmetry in the distribution has not. It is considered in Section 7.4.5 dealing with bootstrap estimates.

As a matter of interpretation, saying that there is a 0.95 probability of the interval CI containing the mean μ is not the same as saying that there is a 0.95 probability of mean μ falling in the interval CI, a situation that requires a Bayesian approach to analysis.[62]

7.4.4 Bootstrap Standard Errors

Traditional approaches to estimating the accuracy of an estimate are restricted to measures such as the sample mean. For other measures, there are no simple formulas like the one for the standard error in Eqn (7.7). A possible way forward is to simplify the problem being analysed or derive an approximate expression, which may depend on the size of the sample. Impediments may remain: for instance, the simplification may be invalid or misleading, the approximation may be hard to derive, or the required sample size may be impracticably large [331, §1.2].

The bootstrap offers an alternative solution. It is a computationally intensive resampling technique that requires little or no knowledge of the distribution itself [360].

The bootstrap can be illustrated [99, ch. 2] by making an estimate of the standard error from a sample of n data values $x_1, x_2, \ldots, x_n$, and then comparing the result with the known value from Eqn (7.7). Bootstrap samples $\mathbf{x}^* = (x_1^*, x_2^*, \ldots, x_n^*)$ are generated by taking random samples each of size n, drawn with replacement, from the original data set $\mathbf{x} = (x_1, x_2, \ldots, x_n)$. The starring indicates that $x_1^*, x_2^*, \ldots, x_n^*$ is not the original data set but a randomized or resampled version of it. To take a concrete example, consider the data set $\mathbf{x}$ in Example 7.1, which consists of these values, arranged in ascending order:

4.5 4.5 6 6 7.5 7.5 7.5 7.5 12 24.

The first bootstrap sample $\overline{\mathbf{x}}^{*1}$ might consist of the following values, again arranged in ascending order:

4.5 6 6 7.5 7.5 7.5 7.5 12 12 24.

Observe that this bootstrap sample contains 4.5 once and 12 twice, whereas the original data set contains 4.5 twice and 12 once. The corresponding bootstrap replicate $\overline{\mathbf{x}}^{*1}$ of the mean, i.e. mean$(\mathbf{x}^{*1})$, has the value 9.45. Suppose that this process is repeated for B bootstrap samples in all, $\mathbf{x}^{*1}, \mathbf{x}^{*2}, \ldots, \mathbf{x}^{*B}$, producing B replicates $\overline{\mathbf{x}}^{*1}, \overline{\mathbf{x}}^{*2}, \ldots, \overline{\mathbf{x}}^{*B}$, where B is usually between 50 and 200 [99, §6.4]. The bootstrap estimate of the standard error is then given by

$$\mathrm{SE}_{\mathrm{boot}} = \left[\frac{1}{B-1} \sum_{b=1}^{B} (\overline{\mathbf{x}}^{*b} - \overline{\mathbf{x}}^{*\cdot})^2 \right]^{1/2},$$

where

$$\overline{\mathbf{x}}^{*\cdot} = \sum_{b=1}^{B} \overline{\mathbf{x}}^{*b} / B.$$

For the data set in Example 7.1 and $B = 200$, the bootstrap estimate SE_{boot} is 1.73, which is close to but a little lower than the value of 1.83 given by the formula in Eqn (7.7). If a theoretical correction of $[n/(n-1)]^{1/2}$ is applied to this bootstrap estimate [99, ch. 2], where $n = 10$, then SE_{boot} increases to 1.82, which is much closer to the correct value. The code for this example is straightforward.[63]

Because the bootstrap is a resampling procedure, bootstrap estimates vary somewhat from calculation to calculation. For sufficiently large B, however, the variation is small.

7.4.5 Bootstrap Confidence Intervals

The bootstrap method makes it possible to calculate confidence intervals for a population mean with an unknown distribution. Bootstrap replicates are generated in the same way as in Section 7.4.4, though the number of bootstrap replications B needs to be larger, perhaps 1000 or 2000 [99, §13.3]. An approximate 95% confidence interval for the population mean μ is given by the 2.5% and 97.5% percentiles of these bootstrap replicates. That is, suppose that $B = 1000$ and the bootstrap replicates of the mean are $\bar{\mathbf{x}}^{*1}, \bar{\mathbf{x}}^{*2}, \ldots, \bar{\mathbf{x}}^{*1000}$, arranged in ascending order. Then an approximate 95% confidence interval is given by $[\bar{\mathbf{x}}^{*25}, \bar{\mathbf{x}}^{*975}]$. More generally, for a $100\alpha\%$ confidence interval, where $0 < \alpha < 1$, if $\bar{\mathbf{x}}^{*(1-\alpha)/2}$ and $\bar{\mathbf{x}}^{*(1+\alpha)/2}$ are the $[100(1-\alpha)/2]$th and $[100\,(1+\alpha)/2]$th percentiles, respectively, then

$$CI_{boot} = [\bar{\mathbf{x}}^{*(1-\alpha)/2}, \bar{\mathbf{x}}^{*(1+\alpha)/2}].$$

This is an example of the bootstrap percentile approach [99, §13.3]. For the data set of Snellen acuities in Example 7.1, a 95% bootstrap percentile interval is given[64] by

$$CI_{boot} = [6.2,\ 12.5].$$

The estimate can be improved by compensating for a slight bias in the bootstrap estimate [99, §14.3]. A 95% bias-corrected bootstrap interval is given[65] by

$$CI_{boot} = [6.5,\ 14.2]. \tag{7.12}$$

The bootstrap interval has the advantage that it automatically takes into account any asymmetry or skewness of the parent population [99, §13.3]. Both the lower and upper limits in Eqn (7.12) are larger than in Eqn (7.11). If the computation is repeated with logarithmically transformed data, as in Example 7.2, the bootstrap retains its advantage [99, §13.3], since the data are still not normally distributed.

Examples 7.8 and 7.9 show ways in which bootstrap confidence intervals are reported.

> Lianas reduced median forest-level net biomass accumulation by 76% (95% bootstrap CI, 55.1–93.8%).

Example 7.8. *Reporting a bootstrap confidence interval, from an ecology article [702].*

> This [effect size] was relatively small with a wide CI (0.512; 95% CI 0.011 to 0.997).

Example 7.9. *Reporting a bootstrap confidence interval, from a neurosciences and neurology article [542].*

Both articles give details of the bootstrap calculations in the methods sections.

7.4.6 Parametric or Bootstrap Estimates?

A parametric estimate is one where the form of the distribution is known, for instance, an estimate of the standard error of the mean of a sample from a normal distribution. The bootstrap does not depend on knowing the parametric form of the distribution and it is therefore referred to as a nonparametric method [384]. As noted elsewhere [737], a correctly specified parametric estimate will always do better than a nonparametric one simply because the parametric estimate enables more to be assumed about the data. If the distribution is unknown or permits only very approximate estimates to be calculated, then the bootstrap offers an impartial way of addressing this uncertainty.

When does the bootstrap not work? It will generally fail if the sample data are not independent or if there are outliers in the data [77, §2.6] or if the sample size n is too small [47, §9.1]. But in these situations traditional parametric methods will also generally fail.

Comparisons of the bootstrap with other resampling methods and with traditional parametric methods are available in [77] [236] [331].

7.5 Inferences about Means

In some studies it may be enough to report summary data measures or estimates of the properties of a parent population. In other circumstances a decision may be needed, for example, whether the mean of a sample differs reliably from some reference value or from the mean of another sample. Decisions of this kind are usually settled by formulating a hypothesis about the data, constructing a statistical test, and then accepting or rejecting the hypothesis according to the test result. A systematic treatment of such procedures—in particular, hypothesis testing—goes beyond the scope of this guide [8] [19] [175] [309]. Instead the aim here is to illustrate some simple applications.

7.5.1 Comparing a Mean with a Reference Value

If the distribution of the parent population is approximately normal and the sample size n is large, then the formula for the standard error in Eqn (7.7) can be used to test whether the sample mean $\bar{x}$ differs reliably from the population mean, μ say. For such a distribution, on only 5% of occasions should an absolute difference $|\bar{x} - \mu|$ as large as 1.96 SE or larger be observed (§7.4.3). If such a difference were observed, then it would be reported as being statistically significant at the 5% level and the default or null hypothesis—that the mean of the distribution from which the sample was drawn is μ—would be rejected with a significance probability or *p*-value of 0.05 [8, ch. 4] [309, ch. 9] (see also Section 7.9).

When the sample size n is small, then, as noted elsewhere (§7.4.3), the distribution of the sample mean $\bar{x}$ is more accurately described by the *t*-distribution than by the normal distribution [101, §8.5] [309, §6.2]. Thus, if $n = 10$, the 97.5% point of the distribution increases from 1.96 SE to $t_{n-1}(0.975)$ SE, that is, 2.26 SE.

Apart from rejecting the hypothesis that the population mean is μ, the *p*-value indicates little about the alternative values. A confidence interval for the population mean (§7.4.3) provides the missing information [127]. Both *p*-values and confidence intervals are reported in Example 7.10.

> Reduction in dietary sugars intake was associated with significantly reduced weight (−0.80 kg (95% confidence interval −1.21 to −0.39); P<0.001).

Example 7.10. *Reporting a confidence interval and p-value, from a general and internal medicine article [692].*

Standard errors can indirectly fulfil a similar function in allowing estimates of alternative values [69, §2.6], as in Example 7.11.

> Each serving/week of sweet foods at baseline was associated with [a] weight increase of 0.28 pounds (0.13 kg, SE 0.04 kg) and 0.19 pounds (0.09 kg, 0.04 kg) in women and men, respectively, over the 2 years (both P≤0.02).

Example 7.11. *Reporting standard errors and p-values, from a general and internal medicine article [692].*

The article [692] describes a meta-analysis in which estimates of standard errors were sometimes derived from reported *p*-values by assuming the data were distributed normally [162].

7.5.2 Comparisons between Means

Testing differences between sample means from multiple populations is traditionally addressed within the framework of an analysis of variance ANOVA [8] [244] [381]. In the absence of any prior hypothesis about a particular mean or combination of means, ANOVA tests whether any of them are reliably different from the rest. It does not, however, give information about which ones, and other kinds of testing are necessary to identify them [244] [254].

Where there is a prior hypothesis about particular sample means, test statistics can be formed from linear combinations of the means. Such analyses are generally referred to as planned comparisons [230, ch. 3]. A planned comparison could test, for example, whether one mean $\overline{x}_1$ is reliably larger than the average of several other means $\overline{x}_2, \overline{x}_3, \ldots, \overline{x}_m$. Planned comparisons are to ANOVA much as a rifle is to a shotgun: one requires a target to aim at, the other does not. Linear combinations of means in which the coefficients sum to zero are called contrasts [230]. Contrasts are often used in trend analyses, as illustrated in Section 7.8.1.

Care should be taken not to use planned comparisons in fishing expeditions, where sample means or their combinations are tested one after another for statistical significance (§7.9). Such comparisons can yield apparently significant effects when there is no underlying difference [230, p. 57]. Comparisons can still be made after the experiment with so-called unplanned or post hoc or a posteriori comparisons [332], but levels of significance need to be adjusted appropriately [175, §12.6] [244, §6.4].

Care should also be taken not to conclude that a significant difference exists between two experimental effects simply because one effect is significant and the other is not. A difference between a significant effect and a non-significant effect need not itself be significant [264].

7.6 Data Fitting

Given a set of n data values or observations or responses $y_1, y_2, \ldots, y_n$ of a variable y at corresponding levels $x_1, x_2, \ldots, x_n$ of an independent or explanatory variable x, how should their relationship be described or summarized? The usual way is by regression. The term has long outgrown its historic use in the phrase 'regression towards the mean' and refers to any model of the relationship between observations y_i and predictor values x_i in which the y_i are subject to random error but the x_i are not [93, §1.8]. The varieties of regression considered here include linear, orthogonal, generalized, and local linear regression.

7.6.1 Linear Regression

Linear regression is one of the most widely reported fitting methods. Typically, the observations $y_1, y_2, \ldots, y_n$ at levels $x_1, x_2, \ldots, x_n$ are assumed to be modelled by a straight line. That is, the estimated value $\hat{y}$ ('y-hat') of y at each value of x is given by a function f of x with parameters a_0 and a_1 that has the form

$$f(x) = a_0 + a_1 x. \tag{7.13}$$

At each level x_i, the fitted value $\hat{y}_i = \hat{f}(x_i)$. The coefficients a_0 and a_1 are the intercept and slope of the fitted line. The residuals, the differences $y_i - \hat{y}_i$ between observed and fitted values, represent random error [93, §1.2]. This is simple linear regression [309].

With the same notation, a quadratic regression has the form

$$f(x) = a_0 + a_1 x + a_2 x^2, \tag{7.14}$$

and a higher-order polynomial regression has the form

$$f(x) = a_0 + a_1 x + a_2 x^2 + \cdots + a_k x^k, \tag{7.15}$$

where k is the order of the polynomial.[66]

For any particular regression model—for example, the straight line fit of Eqn (7.13)—the task is to estimate the values $\hat{a}_0$ and $\hat{a}_1$ of the coefficients a_0 and a_1 for which the resulting fit $\hat{f}(x_i)$ is in some sense the best. The usual method [93] is to minimize the sum of the squared residuals $y_i - \hat{y}_i$, also known as the residual sum of squares RSS [309], that is,

$$\text{RSS} = \sum_{i=1}^{n} \left(y_i - \hat{y}_i \right)^2.$$

The fitted line (or curve) is therefore a least-squares fit.

With least-squares fitting, the coefficients a_j, where $0 \le j \le k$, can be estimated without assuming that the y_i at each x_i come from a normal distribution. By contrast, for confidence intervals to be calculated by the same methods, normality does need to be assumed [93].

Despite its ambiguity, the term linear regression is used in the literature both for regression of a polynomial of order 1, as in Eqn (7.13), and for polynomial regression of any order k, including $k = 2$, the quadratic regression of Eqn (7.14). The reason for allowing any order is that the regression is linear in the parameters of the model, that is, the coefficients $a_0, a_1, \ldots, a_k$ which multiply the successive powers of x that act as explanatory variables [93, §1.2]. Where the distinction needs to be made, linear regression may be described as simple, quadratic, cubic, or polynomial regression of a particular order.

The fitted curve $f(x)$ need not be defined as a polynomial in x but as a combination of trigonometric or other functions of x, which may be better suited to some phenomena. A trigonometric regression generally produces a poorer fit at the ends of the range because of the periodicity of the terms, although the fit can be improved by adding low-order polynomials [106].

Example 7.12 shows a simple linear regression in which an evolutionary index for amino-acid changes is plotted against a universal index [594, Fig. 3b]. If, in the context of the work, the changes are neutral, then the evolutionary index values should be proportional to the universal index values [594, p. 128]. By eye, how-

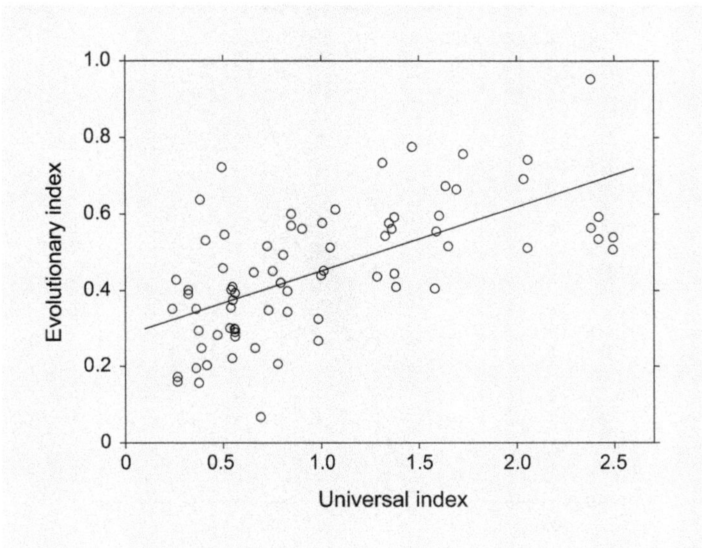

Example 7.12. *Linear regression fit. Replotted from a genetics article [594, Fig. 3b].*

ever, it seems that the straight line fit shown by the solid line has both an obvious positive intercept and a positive slope, implying a failure in strict proportionality.

Procedures for estimating the intercept a_0 and slope a_1 of the linear regression in Eqn (7.13), along with standard errors and confidence intervals, are readily available in several numerical computing environments [245] [298]. For the data in Example 7.12, the estimates[67] are $\hat{a}_0 = 0.28$ with 95% CI = $[0.23, 0.34]$ and $\hat{a}_1 = 0.17$ with 95% CI = $[0.12, 0.22]$. As both intervals exclude zero, there is a significant positive intercept and a significant positive slope, confirming the informal inspection.

7.6.2 Orthogonal Regression

The ease with which linear regression may be applied can sometimes lead to misinterpretation. The principle of minimizing the sum of the squared differences between data points y_i and fitted values $\hat{f}(x_i)$ makes sense only if the y_i and not the x_i are subject to random error. If, instead, both are subject to random error, so that x_i and y_i represent observations of equal status, then neither has the role of an explanatory variable [93]. A different approach to line fitting is then required. The aim is not to predict one data value from the other but simply to summarize the relationship between the two, although there is no single correct method [373].

One way to proceed is to minimize the perpendicular distances from each of the points (x_i, y_i) to the line. The fit is often called orthogonal regression, which is the term used here. Some authors prefer the term major-axis estimation [373] to avoid the connotations of regression. Example 7.13 illustrates such a plot.

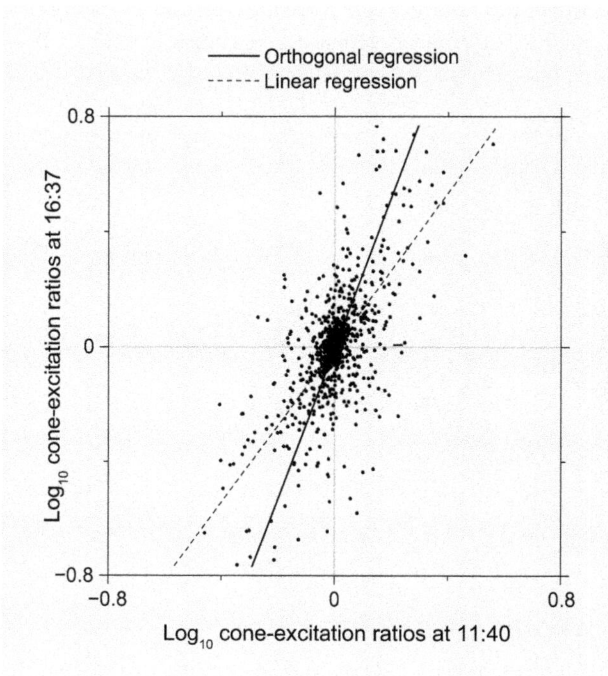

Example 7.13. *Orthogonal and linear regression fits. Replotted from a neuroscience article [507, Fig. 3, left].*

The data are from an analysis of photoreceptor excitations from daylight at two times of day [507, Fig. 3, left]. The solid line is the orthogonal regression fit and the dashed line is a linear regression fit under the assumption that at each point (x_i, y_i) it is only the y_i that are subject to random error.

The linear regression fit provides a poor representation of the data, with points distributed asymmetrically about the dashed line. In contrast, the orthogonal regression fit provides a much better representation, with points distributed roughly symmetrically about the solid line, which also defines the direction of greatest variation in the data.

7.6.3 Generalized Linear Regression

Can a regression approach to fitting be applied to other kinds of data such as event counts, relative frequencies, and dichotomous or binary data? The experimental variables differ from normal variables in that they are discrete and bounded: event counts and relative frequencies cannot take negative values, and binary variables, by definition, have only two values, conventionally represented by 0 and 1. Even so, a regression approach can still be deployed but within the larger framework of what are called generalized linear models [248].

There are three components in this approach [248, §2.2]. With the notation of Section 7.6.1, they are as follows: (1) a random component that represents the variation in the observations y_i, assumed to be independently distributed about their means μ_i, for example, a normal distribution or a binomial distribution; (2) a systematic component f that is linear in the parameters of the explanatory variables, for example, $f(x) = a_0 + a_1x$; and, crucially, (3) a function g that connects (1) and (2), so that $f(x_i) = g(\mu_i)$. The function g is known as the link function (the notation here differs a little from that in [248]). The link function takes into account the fact that the range of y may be bounded. In the present context, the relationship between x_i and μ_i can be written $F(x_i) = \mu_i$, where $F = g^{-1}f$. The link function g is usually fixed. The task, as with linear regression, is to estimate the coefficients of the linear transformation f, for example, a_0 and a_1 in $f(x) = a_0 + a_1x$.

Unfortunately, a simple least-squares approach cannot be used with data that are not normally distributed. Instead, a solution $\hat{f}$ for f is found that maximizes a quantity called the likelihood. The likelihood is proportional to the product of the probabilities of the individual data values y_i under the assumption that the fitted values $\hat{f}(x_i)$ are correct [93, §5.1] [283, ch. 6]. For normally distributed data, the maximum-likelihood solution $\hat{f}$ coincides with the least-squares solution.

Generalized linear models offer a powerful approach to regression fitting, facilitating both prediction and interpretation. But they do depend on choosing the correct link function.[68] Example 7.14 shows some generalized linear regressions where the observations are proportions.

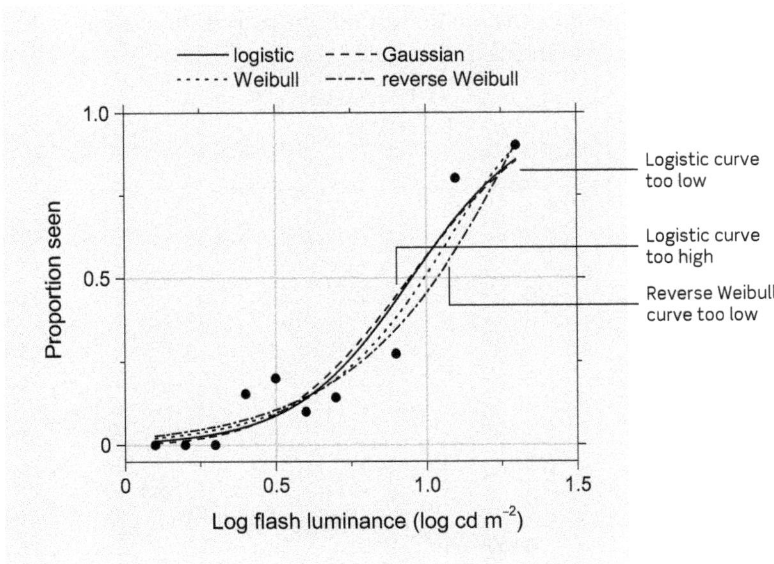

Example 7.14. *Generalized linear regression fits with four different link functions. Replotted from [737, Fig. 1]. Data from the authors of an ophthalmology article [614].*

The data are taken from a study of the light sensitivity of the human eye [614]. The plotted symbols represent the proportion of times a light flash was seen at various luminances. The solid curve shows the fit to the data by a generalized linear model in which the link function g is the inverse of a logistic function, which has the form $[1 + \exp(-z)]^{-1}$, where $-\infty < z < \infty$. This function ranges between 0 and 1. The variable z is used here to avoid confusion with the explanatory variable x. This link function is the one chosen by the authors of the study [614]. The fitted value at each x is then modelled by $\mu = F(x) = [1 + \exp(-(a_0 + a_1 x))]^{-1}$.

Software for fitting generalized linear models is available in several numerical computing environments [245] [298]. The code for Example 7.14 is included in Example 1 in [241]. It is also available in a software package associated with this guide.[69]

Other popular alternatives to the logistic function in this context are the normal (Gaussian), Weibull, and reverse Weibull cumulative distribution functions.[70] Their fits to the light-flash data are shown in Example 7.14 by the dashed, dotted, and dash-dotted lines, respectively.

All four curves have different shapes. Yet it is difficult to decide between them by their fit to the data [737]. When quantified by the methods described in Section 7.7, all the fits are statistically acceptable. Nonetheless, both the logistic curve and the reverse Weibull curve have regions where they appear to have biases, that is, regions where the curves systematically fail to follow the data closely enough [159] [336]. Section 7.6.8 describes how these biases can be avoided.

7.6.4 Disadvantages of Linear Regression

The problems of choosing the correct regression curve and minimizing bias are not restricted to data that come from discrete distributions, as in Example 7.14. Consider the data plotted in Example 7.15.

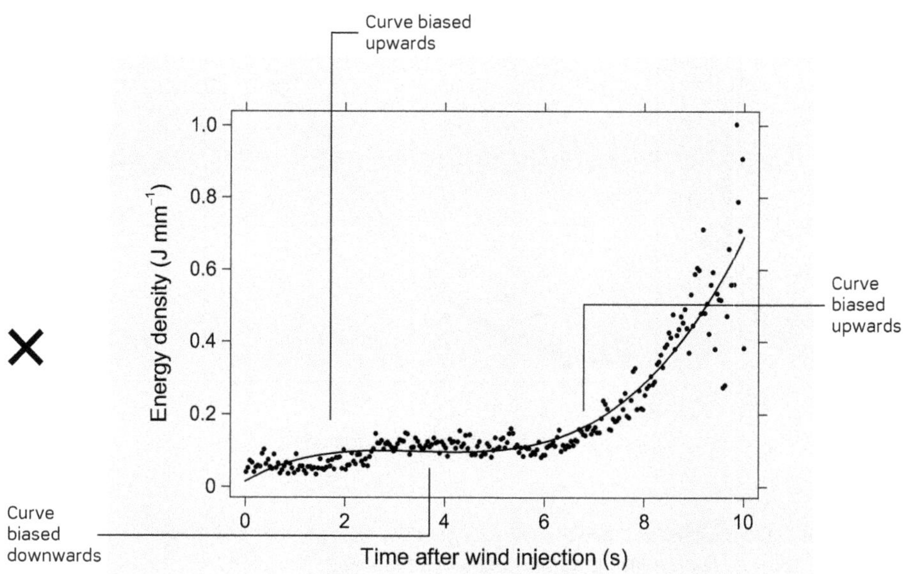

Example 7.15. *Bias in a cubic regression fit. Replotted from an electrical and electronic engineering article [733, Fig. 6b]. Data provided by S. M. Rowland.*

The data in the graph are from an investigation of the stability of low-current electrical discharges on polymeric insulating surfaces exposed to wind and rain [733, Fig. 6b]. The plotted symbols represent energy density of a discharge at various times after wind injection with a weak spray. The authors were interested in the transition between unstable and stable discharge zones, characterized by low and high energy densities, respectively.

The observed values y_i at corresponding levels x_i are fitted in Example 7.15 with a cubic regression $f(x)$ of the variable x (time), that is,

$$f(x) = a_0 + a_1 x + a_2 x^2 + a_3 x^3,$$

shown by the solid curve in the figure. It was obtained[71] in the same way as with the simple linear fit in Example 7.12.

The regression curve reveals something of the shape of the data; yet like the plots in Example 7.14, it contains biases, that is, regions where the curve falls mainly above or below the data points. As a representation of the data, the curve suggests a regularity that is not actually present.

Some of the biases arise from the variance in the observed values not being constant. There is manifestly more scatter at higher energy densities than at lower energy densities. The variance can be stabilized by applying a square-root transformation to the y_i before fitting the cubic curve [93, §13.6] [309, §14.2.2], but the failures to follow the data persist.

The problem of bias cannot be addressed by simply increasing the order k of the polynomial in Eqn (7.15) since it is unclear what the maximum value of k should be. With the higher values of k required to avoid bias (the data plotted in Example 7.15 would need $k \geq 5$), there may be unacceptable variability in the estimated coefficients $\hat{a}_0, \hat{a}_1, \ldots, \hat{a}_k$ [110, §2.3], which can take increasingly large positive and negative values. Individual observations may then unduly influence remote parts of the curve [110, §1.1].

These difficulties are not unique to linear regression, and, in one form or another, they are a feature of all parametric methods in which a strong assumption is made about a model function, a constraint sometimes called the 'parametric straight-jacket' [336, §1.2]. Thus, in polynomial regression, the model function is defined by the powers of x, that is, $x, x^2, \ldots, x^k$, and the parameters are the coefficients $a_1, a_2, \ldots, a_k$, along with a constant term a_0. Once the order k and parameters are fixed, the curve is fixed. Likewise, in generalized linear regression.

These and other problems of parametric fitting may be overcome with nonparametric methods, most notably local linear regression.

7.6.5 Local Linear Regression

There is a large literature on nonparametric approaches to fitting curves to data [110] [155] [156] [232] [336]. One of the most powerful methods is a local version of linear regression called local regression or locally weighted regression, abbreviated as loess or, in some circumstances, lowess [52] [53, ch. 3].

The idea is this [232, §2.1]. The observations $y_1, y_2, \ldots, y_n$ at corresponding levels $x_1, x_2, \ldots, x_n$ are assumed to be modelled by a smooth (differentiable) curve. The estimated value $\hat{y}$ of y at each value of x is obtained by fitting a line $f(u) = a_0 + a_1(u - x)$ (or higher-order polynomial) to the data y_i within a small neighbourhood of x of size h, say. Then $\hat{y}$ is set equal to $\hat{a}_0$, the estimated value of a_0. Notice that this procedure delivers just one estimate, that of y at x. It needs to be repeated at other points along the range of x to generate the complete fitted curve. The points may or may not include the levels $x_1, x_2, \ldots, x_n$.

The fit is local because the estimate at any point x depends only on those points x_i that are near x and not on more distant points. The fitted curve is said to be smoothed [45] [156] [336]. The solid curve in Example 7.16 is a local linear fit to the energy density data of Example 7.15.

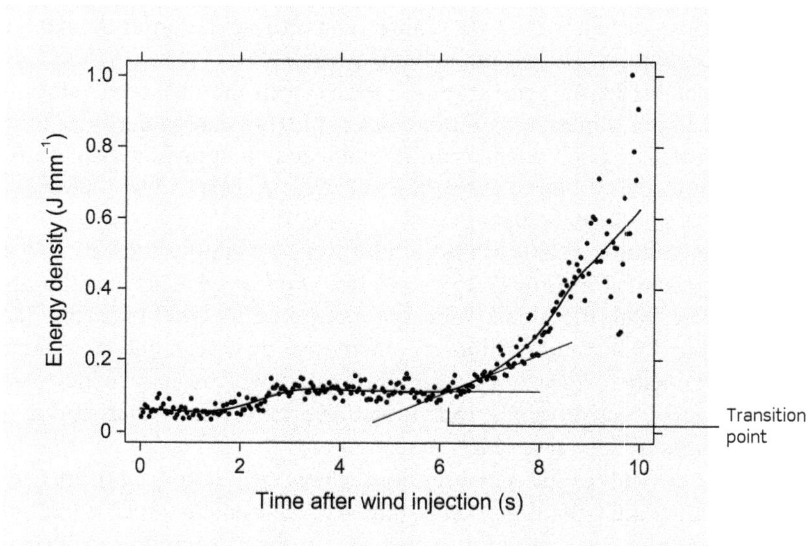

Example 7.16. *Local linear fit to the data in Example 7.15.*

This nonparametric fit is evidently free of the biases of the cubic regression fit in Example 7.15.

As with linear regression, procedures for estimating a local linear regression are available in several numerical computing environments [245] [298]. For Example 7.16, the local linear fit was obtained with a MATLAB [245] package.[72] In this fit, the polynomial is of order 1, so that $f(u) = a_0 + a_1(u - x)$, and the neighbourhood size $h = 1.9$ s.

With the loess fit, the point at which the transition occurs between low and high energy densities can be estimated by eye as about 6 s. A more precise estimate can be derived by taking successive differences along the curve to find where the rate of change in slope has a local maximum. It occurs at 6.1 s and is marked in the figure by the intersecting tangents.

7.6.6 Technical Issues in Local Fitting

The explanation of the local fit illustrated in Example 7.16 glosses over some technical issues. These are the choice of the size h of the neighbourhood ('bandwidth' in the statistical literature) controlling the local fit at each level x, the nature of the weighting of each data point according to its distance from x, and the choice of linear, quadratic, or higher-order polynomial in the local fit.

The neighbourhood size h was obtained here by a process of cross-validation [110] [388], but it could have been chosen by inspecting the fit. The effect of varying h is considered in Section 7.6.7.

The weighting of each data point is defined by a weight function. There is general agreement that its exact shape is unimportant [156, §2.3] [232, §2.2.3] [336, §3.1], although it should always decline smoothly to zero to avoid rapid changes in the fitted curve as successive data points are included or excluded, a problem that occurs with a rectangular weight function (which produces a running average). For loess, the default neighbourhood weight function is the tricube [53, p. 101]. It declines smoothly to zero and is the function used here.

As for the choice of polynomial, there is a trade-off between its order k and the neighbourhood size h. In applications, the advice is to choose k small and then adjust h for a good fit [232, §2.2.2]. The most common choices are $k = 1$ or $k = 2$. For the data in Example 7.16, the lowest order $k = 1$ suffices.

Local regression is not the only nonparametric method available for fitting curves, but it does offer advantages over other methods such as kernel smoothing, smoothing splines, Fourier series estimates, and wavelet approximations [110, ch. 2] [154, ch. 3].

7.6.7 What Neighbourhood Size?

The choice of neighbourhood size h is critical, for it controls the range over which data values influence a fit [110, §3.2.3] [154, §5.1]. For small h, the estimate of f at each x depends on only a few points near x, so that the curve is then susceptible to random variations in the data (it overfits). For large h, the estimate of f at each x approaches a polynomial fit with most points receiving nearly equal weight, so that the curve is then very smooth, and therefore potentially biased (it underfits).

This balance between variation and bias is known as the bias-variance trade-off and is a recurrent issue in fitting curves to data [232, §2.2] [321, §3.11]. Example 7.17 illustrates overfitting and underfitting with the data from Example 7.16. The data are taken from the right-hand portion of the plot.

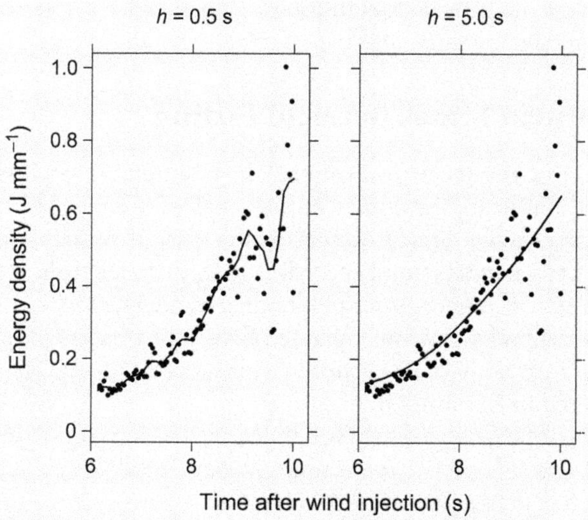

Example 7.17. *Effect of neighbourhood size (bandwidth) h on local linear fits to the data in Example 7.15, right-hand portion of plot.*

In the left-hand panel of Example 7.17 with $h = 0.5$ s, the loess curve overfits the data. In the right-hand panel with $h = 5.0$ s, the loess curve underfits the data and is biased (it falls above the data at intervals of around 6 s because of the influence of data values below 6 s). It is similar to the cubic regression curve of Example 7.15.

There are several automatic methods for choosing the right neighbourhood size. Cross-validation [388] is one. Other popular methods are the plug-in [320] and the bootstrap [115]. Even so, no one method can be guaranteed always to work, and there are good arguments in favour of each [110, ch. 4] [156, §4.2] [232, ch. 10] [336, §5.3].

Notwithstanding the availability of extensive theory, elementary graphical methods for selecting smoothing parameters should not be ignored [53, §3.6] [160, §3.4.6]. Whatever method you choose, it should be clearly identified to the reader.

7.6.8 Local Regression with Discrete Data

The local regression approach to curve fitting can be extended to discrete data such as event counts and relative frequencies. Recall from Section 7.6.3 that a generalized linear model used in curve fitting has a random component, a systematic component, and a function that connects the two. The systematic component is typically a polynomial function of x. In a local regression approach, the polynomial function is replaced by a local linear function [111]. Example 7.18 shows such a fit with the data of Example 7.14.

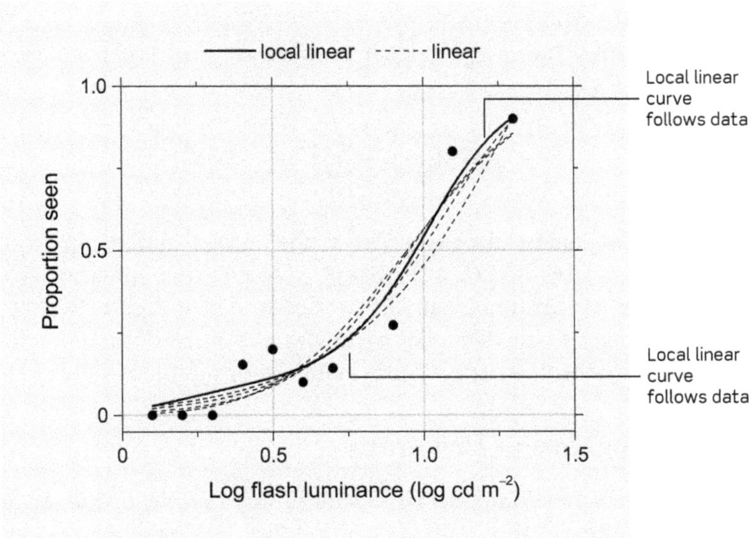

Example 7.18. *Generalized local linear fit and four generalized linear fits from Example 7.14. Replotted from [737, Figs 1 and 2].*

The local linear fit (solid curve) follows the data better than any of the other fits (dashed curves), replotted from Example 7.14. The local linear fit was obtained with a nonparametric fitting package. The code is included in Example 1 in [241] and is available in a software package associated with this guide.[69]

Some comments made elsewhere about parametric and nonparametric estimates (§7.4.6) are germane here. For a particular set of data, a correct generalized linear model will always give a better fit than a nonparametric one such as local linear regression. But in many practical situations the correct model is not known with certainty. If the wrong parametric model is adopted, then the fit may be biased and inferences may be misleading. Local linear regression offers an impartial approach to this fitting problem.

7.7 Goodness of Fit

Given a set of data and a fitted curve, the goodness of fit can be measured in several ways, allowing the reader to judge objectively how well the fit describes the data and to decide whether any data points deviate significantly from the fit or show other unusual behaviour. These measures depend in detail on the nature of the fitting procedure and on whether the observed values at each level are drawn from a normal or some other statistical distribution. In the literature, lack of fit may sometimes be emphasized more than goodness of fit [156] [336].

A popular empirical measure of lack of fit is the root mean square error RMSE [245], also known as the root mean square deviation [86]. Suppose that a set of

n observations $y_1, y_2, \ldots, y_n$ at corresponding levels $x_1, x_2, \ldots, x_n$ is fitted by a curve $\hat{f}$ with values $\hat{f}(x_i)$ at each x_i. Then the root mean square error RMSE is defined by

$$\text{RMSE} = \left[\frac{1}{n} \sum_{i=1}^{n} \left(y_i - \hat{f}(x_i) \right)^2 \right]^{1/2}.$$

The root mean square error is related to the residual sum of squares RSS defined in Section 7.6.1 by $\text{RMSE} = [\text{RSS}/n]^{1/2}$. Despite its popularity, RMSE has two drawbacks as a measure of lack of fit. It needs to be corrected for the adjustments made in fitting $\hat{f}$ to the data, that is, the degrees of freedom in the fit, and it needs to be compared with an appropriate standard or reference value.

7.7.1 Degrees of Freedom

The degrees of freedom associated with fitting a curve, or indeed any model, to a set of data refers[73] to the number of independent parameters or the equivalent that can be adjusted in making the fit [107]. For a simple linear regression, as in Eqn (7.13), the degrees of freedom is 2 since there are two coefficients a_0 and a_1, and for a polynomial regression of order k, as in Eqn (7.15), the degrees of freedom is $k + 1$, the number of coefficients [93].

The term degrees of freedom has other, related meanings. It can refer to the parameters of various families of statistical distributions [17], such as the t-distribution (§7.4.3), and in physics to the number of independent parameters required to specify the configuration of a system [218].

For local linear regression, the calculation of the degrees of freedom is more complicated than for linear regression. An equivalent degrees of freedom can be estimated from the trace of the matrix that represents the transformation from observed to fitted values, the so-called hat matrix [159, §5.4.1 [160, §3.5, App. B] [232, §2.3.2] [321, §3.13, 3.14]. The degrees of freedom in a local fit is returned automatically in some software packages.[74]

The degrees of freedom of any fit, p say (not to be confused with probability), needs to be assessed in relation to the total degrees of freedom in the data, typically the number of independent observations n. The difference between the two, that is, $n - p$, is the residual degrees of freedom. For a polynomial regression of order k, it follows that $p = k + 1$ and the residual degrees of freedom is $n - (k + 1)$. Critically, as k increases, the risk of overfitting increases. As Babyak points out [15]:

> If you put enough predictors in a model, you are very likely to get something that looks important regardless of whether there is anything important going on in the population.

As a trivial example, consider a linear regression in which the number of coefficients $k + 1$ coincides with the number of observations n. The residual degrees of freedom is then 0. The data are fitted exactly, and the model explains nothing, for it fails to distinguish between the systematic effects of the experimental variables and random effects due to noise or other uncertainties. It constitutes the most extreme kind of overfitting.

Any wider interpretation of a model's degrees of freedom is complicated. First, it is not the same thing as the model's complexity [194]. Second, its real value may need to include any preliminary elimination of explanatory variables or testing for outliers. These missing degrees of freedom are called phantom degrees of freedom [15].

7.7.2 Variance Explained

One basic way to judge the goodness of fit of a fitted curve is by comparing the residual variance in the data before and after fitting, that is, the mean square of the differences $y_i - \bar{y}$ and the mean square of the differences $y_i - \hat{f}(x_i)$, taken over the levels x_i. The proportion R^2 of the variance explained can then be defined [93] as one minus the ratio of these two quantities; that is, with a slight simplification,

$$R^2 = 1 - \frac{\sum_{i=1}^{n}(y_i - \hat{f}(x_i))^2}{\sum_{i=1}^{n}(y_i - \bar{y})^2}. \tag{7.16}$$

For the data plotted in Example 7.12, the linear regression gives $R^2 = 38.9\%$.

The value of R^2 depends on the number of degrees of freedom p in the fit, and some authors report the adjusted value of R^2 [257], that is, $1-\left(1-R^2\right)(n-1)/(n-p)$. For Example 7.12, the adjustment gives $R^2 = 36.3\%$. The reduction is small because n is much larger than p. For data with fewer observations, the adjustment may be much larger.

The proportion R^2 can mislead as a measure of goodness of fit [30] [232]. If the data have a linear trend, increasing the range of the levels x_i generally increases the variance in the data about the mean as the terms $(y_i - \bar{y})^2$ in Eqn (7.16) become larger and larger without any corresponding increase in the squared residuals $(y_i - \hat{f}(x_i))^2$. The resulting increase in the value of R^2 does not necessarily reflect any improvement in the fit [93, §5.2, §11.2].

Nevertheless, R^2 can deliver a useful initial assessment of a regression analysis, and can provide a counter to formal significance tests that may explain little of the variation or offer unrealistic predictive power [93, §1.3, §11.1].

7.7.3 Variance Compared

A more principled way of assessing goodness of fit becomes possible when there are replications of the observations available at each level x_i [156, §5.3] [232, §9.1]. The variance in the replications provides an estimate of the size of the random error. If the regression fit is good, the variance of the data about the fitted curve should be due only to random error and therefore comparable in magnitude to the variance in the replications.

Suppose that there are m replications $y_i^1, y_i^2, \ldots, y_i^m$ at each level x_i, where $i = 1, 2, \ldots, n$. Let $\bar{y}_i$ be their mean. If the random error is the same at each x_i, then an unbiased estimate of its variance, the error variance V_E, is given [156, §5.3] by

$$V_E = \frac{\sum_{i=1}^{n} \sum_{j=1}^{m} (y_i^j - \bar{y}_i)^2}{n(m-1)}. \tag{7.17}$$

This formula may be amended straightforwardly if the number of replications is not the same at each x_i.

Suppose that a polynomial regression of order k successfully models the systematic variation in the data. From Section 7.7.1, its degrees of freedom p is $k + 1$ and the residual degrees of freedom is $n - (k + 1)$. If $\hat{f}(x_i)$ is the fitted value at x_i, an unbiased model-based estimate of the variance V_M is given [156, §5.3] by

$$V_M = \frac{\sum_{i=1}^{n} (\bar{y}_i - \hat{f}(x_i))^2}{n - p}. \tag{7.18}$$

If the observations are drawn from a normal distribution, then the ratio of the model variance to the error variance, that is,

$$F = \frac{V_M}{V_E}, \tag{7.19}$$

should have a value around unity, if there are enough observations.

More precisely, the ratio of the variances should be distributed according to an F-distribution. The F-distribution is the distribution of the ratio of two independent variables each distributed as the variance in a normal sample [86]. It has two parameters termed the numerator and denominator degrees of freedom, n_1 and n_2, respectively. They are given here by $n_1 = n - p$ and $n_2 = n(m-1)$.

When the regression does not account for the systematic variation, the value of F will tend to be larger [156, §5.3]. Its value can be compared with the appropriate percentile of the F-distribution to indicate the level of statistical significance (§7.9). If n_1 and n_2 are large, then the F-distribution has its mode at 1.

If the observations are not drawn from a normal population, then other methods are available for testing goodness of fit [144, §5.2] [248, §2.3, §4.4], including the bootstrap [156, §5.4.3].

What if replications are not available at each level x_i, so that an independent estimate of the error variance V_E cannot be obtained as in Eqn (7.17)? Provided that the observations are closely spaced, adjacent values can be used to estimate V_E [156, §5.3]. At each x_i the estimated error, $\hat{e}_i$ say, is obtained [129] by taking the difference between the observation y_i and the value interpolated between the two adjacent observations y_{i-1} and y_{i+1}; that is, if a_i is the ratio $(x_{i+1} - x_i) / (x_{i+1} - x_{i-1})$, then

$$\hat{e}_i = y_i - [a_i y_{i-1} + (1 - a_i) y_{i+1}]. \tag{7.20}$$

The error variance V_E is estimated [129] by

$$V_E = \frac{1}{n-2} \sum_{i=2}^{n-1} \frac{\hat{e}_i^2}{1 + a_i^2 + (1 - a_i)^2}. \tag{7.21}$$

The ratio of the variances can then be calculated as in Eqn (7.19).

For the regression in Example 7.12, the estimated error variance from the formula in Eqn (7.21) is $V_E = 0.0177$, with 73 degrees of freedom. The square root of the variance is 0.133, which is compatible with the point-to-point variation visible in the plot in Example 7.12. The model-based estimate of the variance from the formula in Eqn (7.18) is 0.0181. The ratio F of the variances is 1.02, which is, as expected, not much greater than 1 and gives no suggestion of any lack of fit.

7.8 Trend Analysis and Smoothing

It is sometimes useful to be able to summarize the general way in which experimental observations vary across the levels of an explanatory variable without any considerations of goodness of fit. Trend analysis and smoothing can both be viewed as techniques that offer broad-brush descriptions of data, without necessarily accounting statistically for all of the variation.

7.8.1 Trend Analysis

Trend analysis is what the name suggests. Given a set of n observations $y_1, y_2, \ldots, y_n$ at corresponding levels $x_1, x_2, \ldots, x_n$ of an independent variable, where $x_1 < x_2 < \cdots < x_n$, a trend analysis reveals the underlying course of the data: for example, whether y_i generally increases with x_i, that is, has a linear trend, or whether y_i generally increases and then decreases with x_i, that is, has a quadratic trend [230].

A trend analysis can be performed by taking weighted combinations of data values to form contrasts (§7.5.2). Thus to test for a linear trend, a linear contrast $\hat{a}_1$ in the y_i is formed with coefficients c_i, that is,

$$\hat{a}_1 = \sum_{i=1}^{n} c_{i1} y_i,$$

such that $c_{i1} = (x_i - \bar{x}) / \sum_{j=1}^{n} (x_j - \bar{x})^2$ and, as usual, $\bar{x} = \sum_{i=1}^{n} x_i / n$. The null hypothesis of no linear trend is then decided by comparing the size of the estimate $\hat{a}_1$ with an estimate of its standard error, as in Section 7.5.1. An analogous expression can be obtained for a quadratic contrast $\hat{a}_2$, and so on. Calculating the standard errors of the contrasts can be approached analytically [175, §12.10] [230, ch. 3]. A trend analysis is illustrated in Example 7.19.

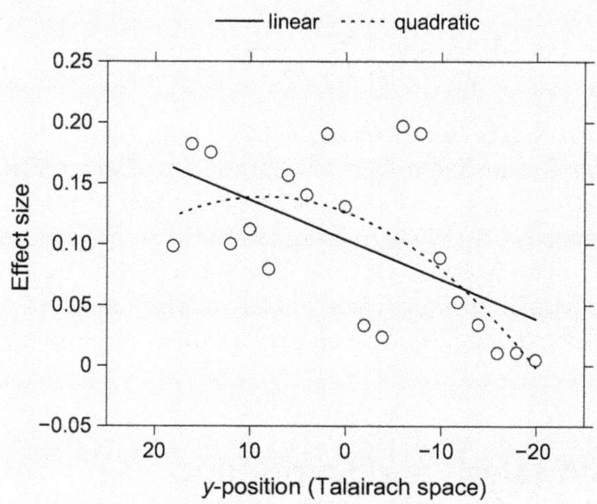

Example 7.19. *Trend analysis. The solid and dotted curves show, respectively, linear and quadratic regression fits. Replotted from a neurosciences and neurology article [467, Fig. 5, upper right].*

The plotted symbols show the recorded effects of ageing at positions along a tract in the white matter of the brain [467, Fig. 5, upper right]. Tract position has descending values on the horizontal scale. To determine whether the variation with position was gradual or abrupt, the authors tested for linear and quadratic trends, shown in the plot by the solid and dotted lines, respectively. The authors concluded that there was a significant linear trend, but no significant quadratic trend [467].

Trend analysis can be regarded as a form of linear regression. If in Example 7.19 the position along the tract is x and the effect size is y (not to be confused with the label 'y-position' for the horizontal scale), then a quadratic regression $f(x)$ has the form

$$f(x) = a_0 + a_1 x + a_2 x^2,$$

with coefficients a_0, a_1, and a_2. The presence of a significant linear or quadratic trend can be decided by examining the confidence intervals for a_1 and a_2 (§7.4.3) to see if they exclude zero.

Estimates for a_1 and a_2 are[75] as follows: $\hat{a}_1 = 0.0029$ with 95% CI = [0.0007, 0.0052] and $\hat{a}_2 = -0.00017$ with 95% CI = [−0.00039, 0.00004]. As the interval for a_1 excludes zero, there is a significant linear trend, and as the interval for a_2 includes zero, there is no significant quadratic trend, consistent with the authors' analysis.

Notice that a test for a linear or quadratic trend is not a test of whether a linear or quadratic function is a good fit to the data. Rather, it is a test merely of whether there is a tendency for *y* to increase or decrease with *x* in a linear or quadratic way [230, ch. 10]. The quadratic fit in Example 7.19 accounts for 43% of the variance in the data.

Section 7.8.2 takes a more nuanced approach to the analysis of these data.

7.8.2 Data Smoothing

In the same way that trend analysis or linear regression can be used to identify linear or quadratic (or higher-order) variations in the data without reference to goodness of fit, smoothing by local linear regression or loess (§7.6.5) can be used to reveal the underlying shape of the data.

Once the kernel and the order of the local polynomial have been chosen for the loess fit, the neighbourhood size can be chosen by eye to bring out the features of the data [53, ch. 3]. This approach can be used with the data in Example 7.19.

The result is shown in Example 7.20. The local linear fit was obtained[76] with a local polynomial of order 1 and a neighbourhood size of 22 expressed in the same units as the horizontal scale. As a fraction of the *x* range, a neighbourhood size of 22 corresponds to a fractional smoothing parameter of 0.6. In data visualization applications, the parameter varies between 0.1 and 1 [53, ch. 3] [195].

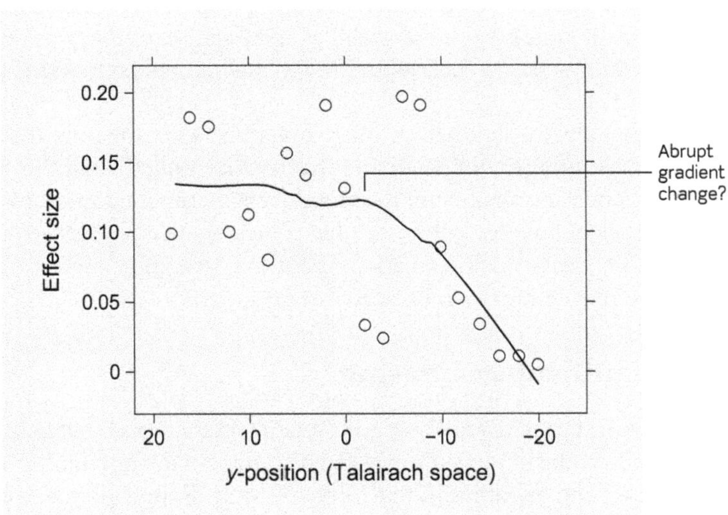

Example 7.20. *Local linear fit to the data in Example 7.19.*

This neighbourhood size gives a smooth loess fit that accounts for about the same proportion of variance R^2 in the data as the quadratic fit in Example 7.19. With the loess curve, however, it is not obvious that the change in the effect size with position is gradual; rather, it suggests an abrupt change (see also [154, Fig. 1.2]). One way to test this notion would be to construct confidence intervals for the change in gradient in the curve (§7.4.3).

7.9 Statistical Significance

As illustrated elsewhere in this chapter, statements about statistical significance usually appear in connection with hypothesis testing. The significance probability or p-value of a test statistic quantifies the difference between an observed value and the value according to some hypothesis, typically that no effect exists [244, §2.6]. The smaller the p-value in a test, the more improbable the value of the test statistic being due to chance alone. But the choice of standard p-values and their interpretation needs care [69, §8.4].

7.9.1 Typical p-values

Conventionally, p-values of 0.05 and 0.01 are assumed to represent, respectively, modest and strong evidence against the chosen null hypothesis; yet there is nothing special about these values [69, §8.4.3]. In particle physics very much smaller p-values are reported. They are usually expressed in terms of the number of standard deviations σ from the mean of a standard normal distribution, for example, 5σ or 6σ, which correspond to p-values of, respectively, 10^{-7} and 10^{-9} [405]. Although p-values as small as this seem extreme, they are necessary because p-values of 0.001 (corresponding to 3σ) have been found to be statistical anomalies [212].

More generally, evidence from other research fields suggests that p-values of 0.05 or 0.01 are unjustifiably high and that smaller values avoid the risk of a spurious significance and the failure to reproduce experimental findings [16] [197].

In the medical literature, the use of p-values alone is eschewed [303] for the reason that they give no information about the size and precision of an effect. Nonetheless, they continue to be reported [46].

7.9.2 Significance Chasing

The practice of repeatedly analysing sets of data in the search for a statistically significant result is referred to variously as p-fishing, p-hacking, significance chasing, data-dredging, and snooping [265] [309, §11.4.8]. Reporting just those analyses that produce significant p-values has little meaning [374] and may contribute to the literature on irreproducible experimental findings.

Solutions to these problems are available. One is to declare the hypothesis before undertaking measurements [83]; another is to report the number of hypotheses explored and all decisions on data collection, statistical analyses, and p-values calculated [374].

7.9.3 Power

The power of a statistical test is the probability of rejecting the null hypothesis when it is false [107] [244]. Power is important in deciding the size of a sample (§3.6.4). An underpowered test might fail to reliably detect a difference between two means simply because there are too few observations. Power can be increased by increasing the number of observations, but assumptions need also to be made about alternatives to the null hypothesis [244, §2.7]. Methods for calculating power are described in [8, §6.6] [175, ch. 8] [309, §11.2.2].

As emphasized in [69, §2.6] and illustrated in Section 7.5.1, it is often simpler and more relevant to consider the standard error of an estimate. Whether or not any formal statistical hypothesis testing is undertaken, reporting the standard error gives a useful indication of the reliability of an effect. Indeed, both *p*-values and standard errors may be reported, and there is a case for a more nuanced approach that routinely includes *p*-values along with other measures [371].

Still, as with any analysis, the broader implications should be kept in mind. The fact that an effect reaches statistical significance, even at a very high level, does not mean that it has any practical importance [69, §8.4] [374]. Significant effects were reported in a study of a drug in Alzheimer's disease [450], but, as one commentator noted, the effects were tiny and may have meant nothing for the quality of life of patients [247].

7.10 Checklist for Data Description and Statistical Inference

Averages

1. Enough data values for a mean to be usefully reported?
2. Logarithmic transformation relevant and applied before averaging?
3. Number of significant figures compatible with the precision of the data?

Variation

4. Measure of scatter identified, standard error or standard deviation?
5. Confidence interval included?
6. Estimates unbiased by large data values?

Comparing means

7. Distribution of the parent population unknown? Bootstrap tests more appropriate?

8. With p-values, power of test specified? Standard errors or confidence intervals included?
9. With multiple means, planned comparisons or an analysis of variance more appropriate?

Data fitting

10. Linear regression with a low-order polynomial justified? Evidence of bias?
11. Local linear regression (loess) for a bias-free fit?
12. With loess, neighbourhood size explained?
13. Data bounded, e.g. relative frequencies or event counts? Generalized linear model more appropriate than linear regression?
14. Number of degrees of freedom reported?
15. Proportion of variance explained?
16. Any formal tests of goodness of fit?

Trends and data smoothing

17. Underlying variation in the data captured by a linear or quadratic trend?
18. Loess curve for an accurate description of the data shape?

8
Graphing Data

In capturing the main findings in a research article, graphs can be as important as the title and abstract. As Tufte points out [353, p. 9]:

> Often the most effective way to describe, explore, and summarize a set of numbers—even a very large set—is to look at pictures of those numbers.

Well-crafted graphs allow the reader to discern a pattern or structure in quantitative data, understand the implications, and detect deviations from that pattern, all in a more efficient and compelling manner than can be achieved with a table or words alone. Yet producing good graphs is not as easy as it might seem. Like text and mathematics, graphs can sometimes confuse and mislead the reader. Occasionally they may lead to contentious debate [454] [599].

This chapter deals with some of the problems that can arise in graphing data and how they can be addressed. Because the appearance of a graph is usually determined by the author and not the publisher (in contrast with setting text in type), suggestions for formatting are also included. Chapter 7 deals with fitting lines and curves to data and goodness of fit.

Section 8.1 introduces the main kinds of graphs used in science and engineering. Section 8.2 provides examples of common graphical errors. Section 8.3 sets out some guidelines aimed at constructing effective graphs that avoid these errors. Section 8.4 is concerned with statistical indicators of spread, error bars, which are attached to plotted data values. Section 8.5 deals with graphs for categorical or labelled data, and Section 8.6 deals with the finer points of graph construction and formatting. Section 8.7 contains a checklist for the items in the chapter.

8.1 Kinds of Graphs

The most popular kind of graph is the Cartesian plot, where one quantitative variable is plotted against another in a rectangular coordinate system. The term scatter plot is used to refer to graphs where the roles of the two variables are interchangeable [45, §4.1, §4.3]. Other popular kinds of graph are those where one of the variables is not quantitative but categorical, typically in bar charts and pie charts. Examples of these and many other kinds of graphs are described in [208].

There is no standard nomenclature for the parts of a graph, and the largely self-explanatory terms introduced by Cleveland [54] are used in this guide.

8.1.1 Grammar of Graphs

The profusion of graph types does not always aid comprehension. But there is a useful distinction to be made between the physical appearance of a graph and its underlying structure. By way of illustration, a divided bar chart is a chart in which each bar is partitioned into segments whose sizes correspond to proportions and which sum to unity [208]. With a change in coordinates, it can be displayed as a pie chart [380, ch. 2]. The physical appearance differs but the underlying structure is essentially the same. The distinction between appearance and structure can be expressed more formally in a grammar of graphics [380].

Cleveland [54] and others advocate an approach to graphing that focuses on simplicity and clarity, one that is optimized for efficient pattern perception and 'communicating information' [54, p. 2]. Only the most basic kinds of graph types are therefore considered here, and in their usual two-dimensional forms. Inventive ways of representing three- and higher-dimensional data are described in [45] [53] [54] [353] [354].

8.1.2 Statistical Additions

Graphs may be augmented with lines or curves that are fitted to the data according to some statistical procedure [380, ch. 8]. Approaches can be classified as global, such as classical linear regression [93] and generalized linear modelling [248], or local, such as locally weighted regression [110] and data smoothing [336]. For more details, see Sections 7.6 and 7.8.

8.2 Common Graphical Errors

Graphing errors and lapses take a variety of forms, but can be broadly divided into omissions, occlusions, and matters of resolution. More examples can be found in [54] [208] [353].

8.2.1 Omitted Units and Hidden Data Points

Some numerical programming environments offer rudimentary plots of data with convenient default commands, but require substantial extra effort by the user to produce something more professional.

Example 8.1 shows a plot of radiant intensity against wavelength from a master's report. Default plotting parameters were used by the author.

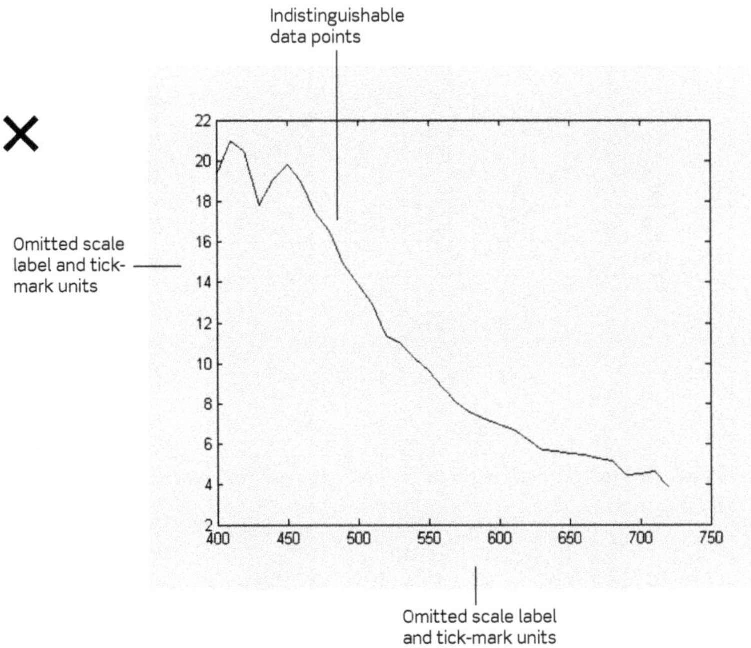

Example 8.1. *Elementary graphical omissions, from an unpublished electrical and electronic engineering master's report.*[77]

Both the horizontal and vertical scales have no units for the numbers next to the tick marks (the reference points on each scale) and no labels identifying the quantities that are plotted. There are no plotting symbols marking the data points, with the result that observations cannot be reliably distinguished from the interpolated curve. In fact, it is only where the curve has a sharp change in gradient that the coordinates of a data point can be estimated. For a reader unfamiliar with the topic addressed in the report, the plot is impossible to interpret.

8.2.2 Obscured and Occluded Data Points

When data values extend across the full range available for the horizontal and vertical scales, plotted symbols may be obscured by the lines of the rectangle enclosing the plot, the scale-line rectangle. In Example 8.2 the ratio of areas r_s from an oximetry power spectrum is plotted against an index of sleep apnoea. The plotted symbols that lie on the left, right, and bottom scale lines merge with the lines, so that the impression of the data range is distorted. The tick marks are also too short and, despite pointing outwards, may be confused with data symbols. The scale labels are hard to decipher and the single symbol r_s is uninformative.

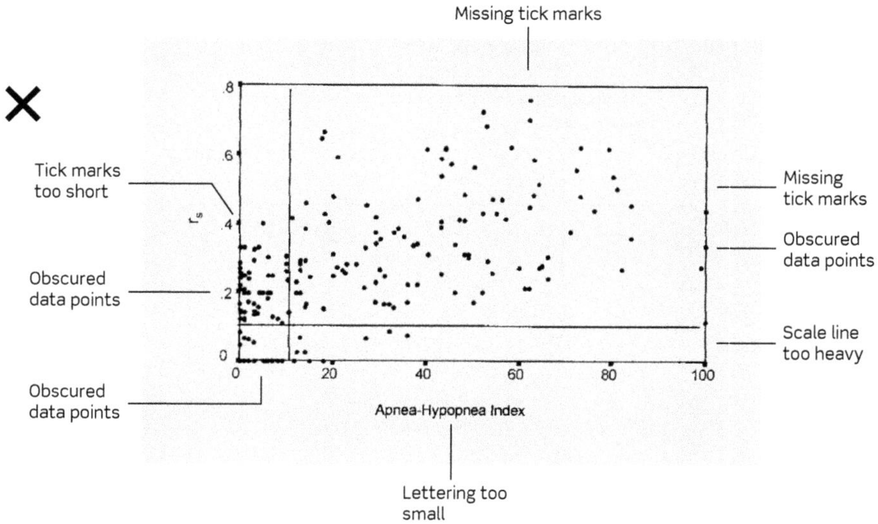

Example 8.2. *Plotted symbols obscured by the scale-line rectangle, from an experimental medicine article [728, Fig. 5].*

Lines added to a plot to aid analysis may simultaneously obscure data. Example 8.3 shows a graph of an evolutionary index of amino-acid changes EI plotted against a universal index U. Several of the plotted points fall on the fitted line, which is heavier than necessary. As these points are less easy to see, the remaining points have a disproportionate influence on the impression of the variation in the data. The absence of scale lines with tick marks at the right and top of the plotting area also makes it difficult to estimate the more extreme data values.

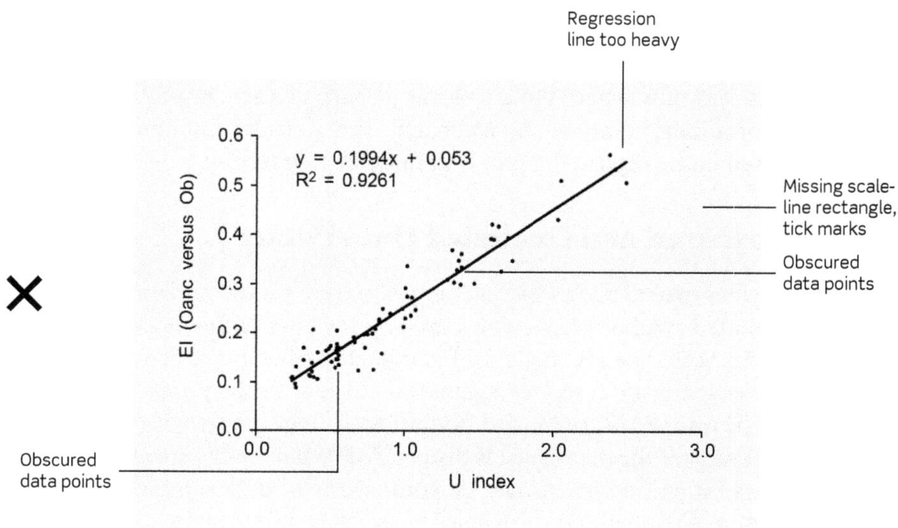

Example 8.3. *Plotted symbols obscured by the fitted line, from a genetics article [594, Fig. 3a].*

Data symbols can sometimes occlude each other. Example 8.4 shows SVC blood flow into the heart plotted against perfusion index PI. The open and filled circles represent data for infants with and without a congenital heart disorder, respectively. The solid and dashed lines show fitted lines for the two data sets. Because of their size, several of the filled symbols overlap with each other, and partially hide many of the open circles. Their visual significance is diminished and their influence on the fitted lines difficult to assess.

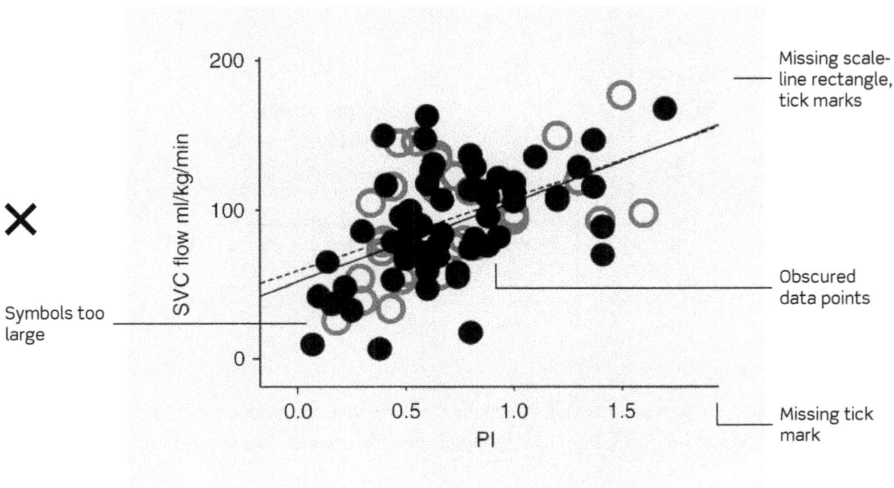

Example 8.4. *Plotted symbols obscured by other plotted symbols, from an obstetrics and gynaecology article [688, Fig. 2a]. Reprinted by permission from Macmillan Publishers Ltd, copyright 2010.*

8.2.3 Wasted Space and Resolution

Even with plotting symbols of an appropriate size, the relationships they reveal in the data depend on the choice of scale. Example 8.5 shows the probability of dark counts against bias voltage evaluated with three methods. Because the vertical scale includes zero, the resolvability of the data points is reduced to less than half of that with a vertical scale range starting at, say, 0.015. The interpretation of the differences between the symbols is further compromised by tick marks that are hard to see (and absent on the top and right scale lines). A more extreme example of a loss in resolution is shown in [54, Fig. 2.58].

There are principled arguments for and against including zero on a scale when it is not required by the data range. Thus without zero, the reader might ignore the tick marks and misinterpret the data range on the expanded scale. With zero, the reader might fail to resolve the variation in the data within the compressed range. The advice from Cleveland [54, pp. 92–5] is to include zero if it does not produce too much wasted space, and never if it leads to a loss of resolution. On balance, it is more likely that the reader will look at the tick-mark labels and interpret them correctly than ignore them altogether.

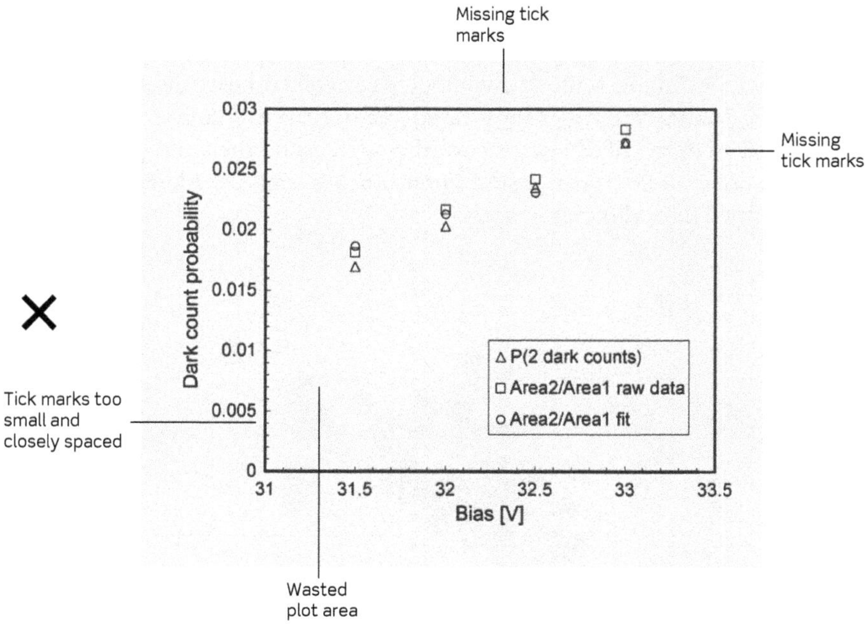

Example 8.5. *Loss in resolution with zero on the vertical scale, from an electrical and electronic engineering article [503, Fig. 13]. Symbols are colour coded in the original.*

8.3 Constructing Effective Graphs

How should you avoid plotting errors of the kind illustrated in Section 8.2? With some differences in emphasis, there is a consensus [45, §8.4] [54, ch. 2] [202] [296] [353, p. 13] on what a good graph should achieve:

- it should have a clear purpose, e.g. a description of the data or a test of a model;
- it should allow the data to stand out and not be distorted;
- it should use labelling effectively to support accurate interpretation;
- it should, where appropriate, facilitate comparisons across data;
- it should be integrated with the text and analysis in the body of the article.

Within these constraints, and without sacrificing data, a graph should be as simple as possible. All irrelevant material, non-functional perspective and shading, clutter, and pointless detail, otherwise referred to as chartjunk [353, ch. 5], should be eliminated [45, §8.4].

One way to achieve this goal is to think about Tufte's 'data-ink ratio' [353, p. 93]. If the graph is imagined as being drawn in ink, then the ratio of ink showing data to ink for the whole figure should, within reason, be a maximum.

In practice, a balance needs to be struck between simplicity and interpretability, since some redundancy may be necessary to help visual understanding. Some extreme applications of the principle of maximizing the data-ink ratio are developed in [353, ch. 6].

8.3.1 Organization of a Cartesian Plot

Example 8.6 illustrates most of the expected elements of a Cartesian plot [45, ch. 4] [54, ch. 2]. The data are taken from an electrical and electronic engineering study [606, Fig. 13]. The symbols show the breakdown voltage of a rod-plane gap plotted against the applied peak voltage for the experimental conditions indicated in the key (also called the legend). The straight line fits to subgroups of the data show the average trend for positive and negative impulses.

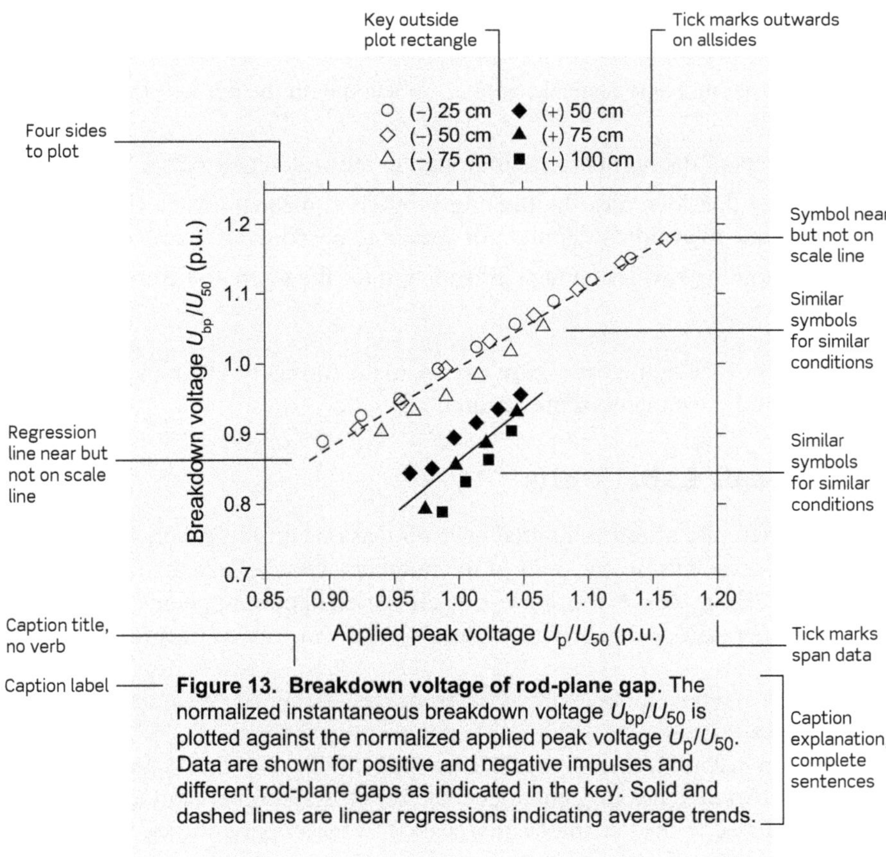

Example 8.6. *Organization of a Cartesian plot and figure caption. Redrawn from an electrical and electronic engineering article [606, Fig. 13]. Data provided by P. N. Mikropoulos.*

Most of the problems identified in the examples of Section 8.2 have been avoided:

- the plot has four sides to delimit the region within which the reader expects to see data;
- plot symbols are distinguishable from each other, with similar symbols used for similar quantities and dissimilar symbols for dissimilar quantities;
- the solid and dashed fitted lines are slightly heavier than the scale-line rectangle but do not mask the symbols;
- tick marks are attached to each of the four sides to make it easier to estimate data values; the tick marks point outwards to avoid obscuring data and are spaced well apart to avoid clutter;
- the horizontal and vertical scale ranges allow the plotted symbols to exploit the space available but not come too close to the scale-line rectangle;
- the tick marks on the left and bottom scale lines are labelled with numbers;
- the horizontal and vertical scales are labelled with the names of the variables plotted;
- the units of the numbers are indicated in the scale-line labels;
- the key that identifies the plotting symbols is placed outside the scale-line rectangle to avoid clutter and not appear to obscure data symbols;
- the caption provides enough information for the figure to be understood on its own.

Two aspects of the figure need extra attention: the format of the units in the scale-line labels and the content of the caption.

8.3.2 Scale Label Units

On axes where the numbers plotted have dimensions or depend on some scaling factor or reference value, the units of the numbers must be indicated in the scale-line label. In Example 8.6, the unit is 'p.u.'. It is used in power systems analysis and stands for 'per unit', a dimensionless quantity defined with respect to a reference value.

The unit symbol can be marked off from the label in several ways. There is a strong logical case for the solidus as separator; for example, if T is temperature measured in kelvin K, then the scale is labelled 'T/K' [190, §5.3.1]. Viewed as the division of two dimensioned quantities, the result is a number, and it is numbers that are attached to the tick marks on a scale. The use of parentheses, as in 'T(K)', and comma, as in 'T, K', are both deprecated [351, ch. 1]. Yet their use persists [572, Table 1] [591, Fig. 1] [701, Fig. 2].

None of these separators—solidus, parentheses, comma—is entirely free from ambiguity. If the differences in font are neglected, then '*T*/K' may be seen as one variable divided by another, '*T*(K)' as a function with argument K, and '*T*, K' as an ordered pair, although the last is the least plausible. The best course is probably to follow the practice in the journal where the article is to be published. At least readers will be familiar with your choice.

As for the units themselves, they should normally be SI units [190], unless, as in Example 8.6, there is a convention otherwise. Unit symbols and combinations of unit symbols are discussed in Sections 3.6.5–3.6.7. Care should be taken in using multipliers within the scale label area, for instance '$\times 10^6$', which, depending on where it is placed, could be taken to apply to the scale, so that the nominal tick-mark values need to be divided by 10^6, or to apply to the tick marks, so that the nominal tick-mark values need to be multiplied by 10^6. If parentheses are used as the unit separator and the multiplier is inserted within the parentheses, e.g. 'susceptible bacteria (10^6 cell ml^{-1})', then the meaning should be clear enough [641, Fig. 1] and a multiplication sign is unnecessary.

8.3.3 Figure Caption

The caption placed beneath a figure allows the graph, diagram, or other line drawing or image to be understood on its own and to help web and online database searches [55, ch. 15] [131]. As illustrated in Example 8.6, the caption consists of a label, 'Figure 13', so that it can be referred to elsewhere; a title, usually a short verbless sentence capturing the topic, 'Breakdown voltage of a rod-plane gap'; and then one or more sentences saying what is plotted, 'The normalized instantaneous breakdown voltage U_{bp}/U_{50} is plotted against . . .'. How much is said and any interpretation of the plot or its implications all vary according to the conventions of where the work is published [54, pp. 57–8, p. 162]. For consistency, the terms and phrases used in the caption should be the same as those used in the main body of the text [258, p. 136].

In some publications, the caption may finish with a short sentence, with or without a verb, summarizing the source of the data, for example, 'Data taken from Smith (1999)' or 'Data from 50 samples'. If a subsequent figure differs by a simple change in conditions, then its caption can be compressed, for instance, 'Figure 15. As for Fig. 14, but with different rod-plane gaps'.

8.4 Error Bars

The data in Example 8.6 show individual values, with just one value of the measured variable y_i plotted at each level x_i. Sometimes, however, independent repeated measurements are available. It is then usual to plot the mean $\overline{y}_i$ of the replicates $y_i^1, y_i^2, \ldots, y_i^m$ at each x_i, along with a vertical bar extending above and below each plot symbol to quantify the spread in the data. The spread may be defined by the standard deviation, standard error, or a confidence interval (§7.3, §7.4).

It is therefore important to say in the caption which quantity is actually plotted [54, pp. 59–60] [74], for example, 'Error bars show ±1 SD' or 'Error bars show ±1 standard error of the mean'. The inclusion of the symbol '±' is useful, for without it the full extent of the bar could be misinterpreted as one standard deviation rather than two. If error bars show confidence intervals, a statement of the kind 'Error bars indicate 95% confidence intervals' applies to the full extent of each bar.

8.4.1 Standard Deviation and Standard Error

The commonest kinds of error bars show either the standard deviation or the standard error (§7.4.2). Recall that the standard deviation SD of a set of m data values $y_i^1, y_i^2, \ldots, y_i^m$ with mean $\bar{y}_i$ is given by

$$\mathrm{SD} = \left[\frac{1}{m} \sum_{j=1}^{m} (y_i^j - \bar{y}_i)^2 \right]^{1/2} .$$

The divisor m is replaced by $m - 1$ if the standard deviation is intended as an estimate of the population standard deviation from which the sample is drawn (§7.4.1). The standard error SE of the mean $\bar{y}_i$ is given by

$$\mathrm{SE} = \left[\frac{1}{m(m-1)} \sum_{j=1}^{m} (y_i^j - \bar{y}_i)^2 \right]^{1/2} .$$

As the number of values m at each level x_i increases, the standard deviation stays roughly the same (although becoming more and more accurate), whereas the standard error generally gets smaller and smaller (§7.4.1).

The standard deviation is a statement about the uncertainty or lack of precision in individual data values y_i^j and the standard error is a statement about the uncertainty or lack of precision in the mean $\bar{y}_i$ (§7.4.2). Which of the two should be plotted depends on the focus of your analysis [74].

Example 8.7 shows a plot where the error bars represent ±1 SD, the standard deviation of the data values at each level x_i. They allow the reader to appreciate the scatter of individual values about each mean (open circles) in relation to the calculated values (solid, dotted, and dashed curves).

Error bars showing standard deviations and standard errors are used nearly equally often in the literature [210].

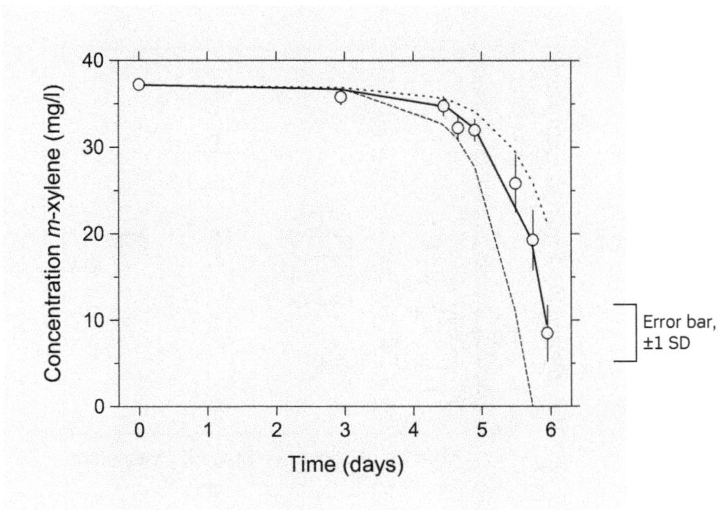

Example 8.7. *Error bars showing ±1 standard deviation of the data values. Open circles represent measured values of concentration and the solid, dotted, and dashed curves calculated values. Redrawn from an environmental sciences article [663, Fig. 1]. Data provided by M. Schirmer.*

8.4.2 Confidence Intervals

A potentially more useful kind of error bar to attach to data points is one representing a confidence interval CI (§7.4.3). A 95% CI for the mean $\overline{y}_i$ at each x_i is an interval within which the probability of finding the true value of $\overline{y}_i$ is 0.95. Error bars representing confidence intervals have the advantage over those representing standard deviations or standard errors in that any assumptions about the underlying distribution of the data (e.g. normal, binomial) are built into the construction of the error bar. It is then much easier for the reader to make inferences about the significance of the data (§8.4.4).

In the medical literature, confidence intervals are the preferred measure of uncertainty [303]. In the physical sciences, it is standard deviations and standard errors.

8.4.3 Boxplots

A still more useful representation of the distribution of individual data values $y_i^1, y_i^2, \ldots, y_i^m$ at each level x_i is provided by a boxplot, also called a box-and-whisker diagram [360]. Example 8.8 illustrates a basic boxplot in which the levels x_i are categorical, representing kinds of colon tissue. The top and bottom edges of each box represent the upper and lower quartiles of the data values (§7.3.3), the horizontal bar near the middle of each box represents the median, and the ends of the whiskers represent the minimum and maximum. Refined boxplots can incorporate even more detail [54, pp. 139–42].

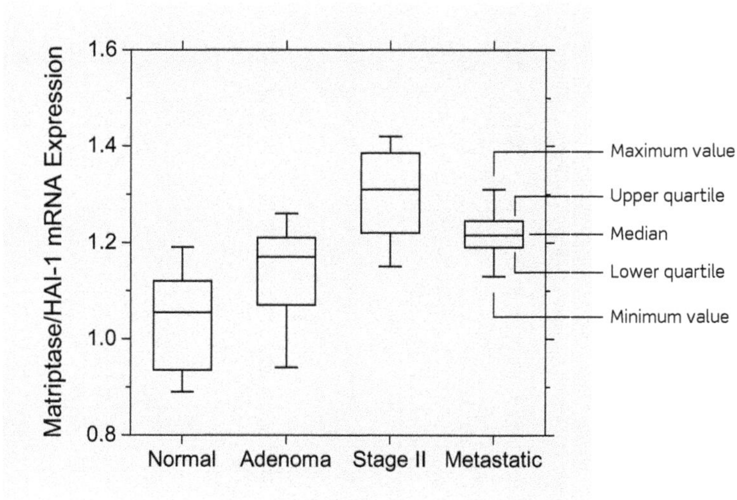

Example 8.8. *Boxplot with each box and whisker based on 12–16 data values. Redrawn from an applied biological sciences article [578, Fig. 2C, left]. Data provided by C. S. Craik.*

Example 8.8 shows data for just four levels, but boxplots can also be used with more densely sampled levels of the explanatory variable x_i [45, Fig. 3.6].

8.4.4 Interpreting Error Bars

To draw any inference from the fact that error bars do not overlap with each other (or with a reference value such as zero), the reader needs to know what kind of bar is shown and the number of data values making up each sample [74].

As a rough guide, for bars showing ±1 SE, with at least ten data values in each sample, there needs to be a gap of about 1 SE between bars for there to be a conventional statistically significant difference with a *p*-value of 0.05 (§7.9.1). For bars showing 95% CIs, there can be an overlap of 50% for the same significance level; if they just touch, the significance level is higher, with a *p*-value of 0.005 [210].

8.5 Plotting Labelled Data

After Cartesian plots, the most common graphs are for labelled data, such as bar charts. Regrettably, bar charts and other related plots such as divided bar charts and pie charts available in some plotting environments may not represent data as directly or effectively as they should. Thus with a bar chart, the strong visual cue offered by the length of the bars is misleading if the baseline is not zero [54, p. 152]. With a pie chart, differences in data values are hard to assess because they require judgements of area or of arc length across different orientations. Important features of the data may therefore be missed [54, pp. 262–9].

8.5.1 Dot Plots

Dot plots overcome several of the problems in plotting categorical or labelled data [54, §3.4]. Example 8.9 shows a plot taken from an astronomy and astrophysics article [660]. It summarizes the variations in scatter about a best-fitting model for two measures, gas density and gas temperature, and two kinds of galaxy clusters, cool core and non-cool core.

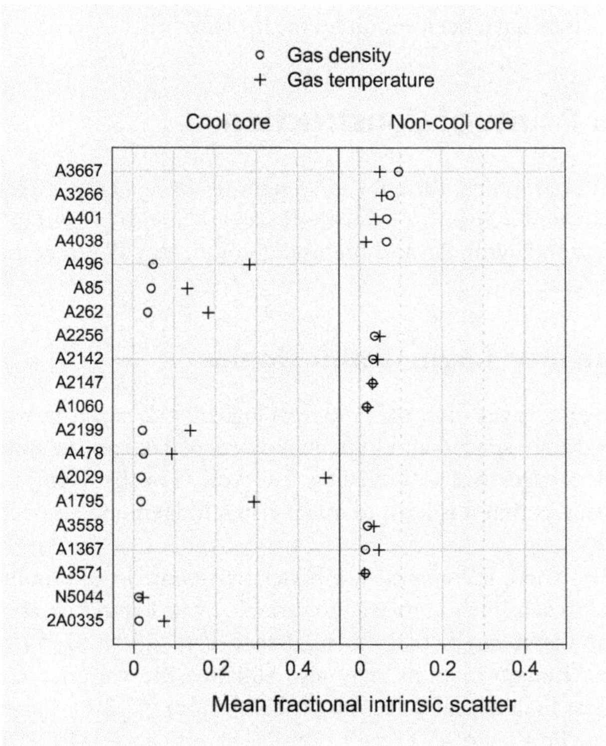

Example 8.9. *Dot plot showing labelled data. The symbols represent the scatter in gas density and gas temperature for each galaxy cluster indicated on the vertical scale. Redrawn from an astronomy and astrophysics article [660, Fig. 2, left]. Data provided by A. J. R. Sanderson.*

The galaxy clusters listed on the vertical scale are in order of decreasing scatter in density from top to bottom (not in alphabetical order). Gas density and gas temperature are shown by different symbols. Cool-core and non-cool-core clusters are plotted separately in the left and right panels with the same horizontal scales (too large for the right panel on its own). The dot plot shows that for cool-core clusters the scatter in temperature is higher than the scatter in density, whereas for non-cool-core clusters the scatter in temperature and density are similar [660].

Dot plots and Cartesian plots, in single and multiple panel formats, cover most of the kinds of plots needed in practice.[78]

8.5.2 Jittering

The data symbols in Example 8.9 are readily discriminated, but in dot plots or Cartesian plots where multiple symbols with a common coordinate overlap, the distribution of data points can be misinterpreted. To counter this effect, a small amount of random uniform noise or jitter (similar in size to the symbol width) can be added to data values before they are plotted. Jittering improves the discriminability of data points [54, pp. 158–9]. Naturally, the reader needs to be informed that the data values have been modified in this way.

8.6 Finer Points of Construction

Small changes in graphical format can sometimes have a large effect on interpretation. The choice of a linear or logarithmic scale, the use of grid lines, the choice and size of lines and symbols, and the use of colour can all affect the visual inferences readers make.

8.6.1 Linear or Logarithmic Scales

Some graphing packages offer the choice of logarithmic scales in which the major tick marks are evenly spaced and indicate powers of 10, and the minor tick marks are spaced at progressively decreasing intervals. The problem with nonlinear scales of this kind is that it is hard to make visual judgements, especially involving interpolation.

For variables where increments are better represented on a multiplicative than additive scale, it is sensible to replace the variable by its logarithm, either to the base 10 or base 2, so that it can be plotted on a linear scale [45, §3.6] [54, pp. 95–102].

Logarithmic transformations may also stabilize the variance over levels and remove skewness in data that take only positive values (§7.2.1). Other transformations can be used if they offer greater physical significance [8] [230] [332].

How should data representing dimensioned quantities then be treated? The general rule from dimensional analysis is that the arguments of functions such as the logarithm should be dimensionless (§3.6.6). Yet as long as the scale is made explicit, there should be no ambiguity. For example, if time is measured in seconds, then the corresponding scale of a plot might be labelled 'Log$_2$ time' and the units 'log$_2$ s' [53, Fig. 2.26], despite 'log$_2$ s' itself having no physical meaning.

8.6.2 Grid Lines

Six sets of data are represented within the plotting area of Example 8.6, but there are occasions when as few as two or three data sets cannot be displayed properly within a single plot rectangle. One way to resolve the problem is to plot the separate data sets in multiple panels, side by side, with each panel equipped with a grid of lines to facilitate visual comparisons [54, §3.6].

Example 8.10 shows a plot with six panels each containing a histogram from a different data set [508, Fig. 2]. With the grid lines, it is easy to judge the relative heights and widths of the histograms. The scales are necessarily identical even though for most individual panels they are not ideal in minimizing wasted space.

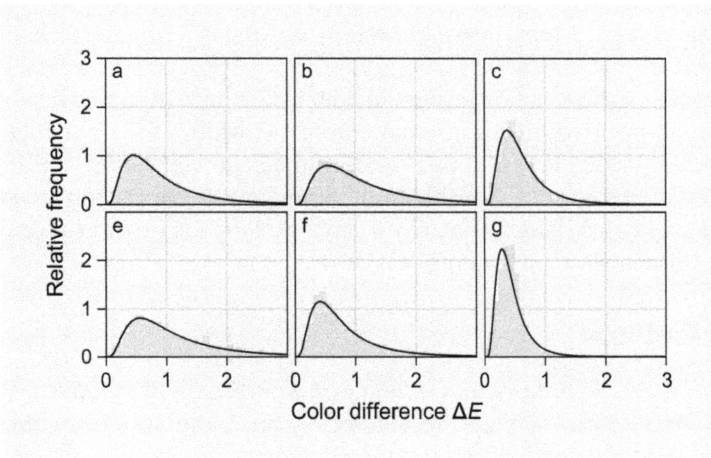

Example 8.10. *Grid lines to aid comparison of data sets. Redrawn from an optics article [508, Fig. 2]. The original contained eight panels a–h.*

With single data sets, grid lines are usually unnecessary.

8.6.3 Line, Symbol, Font, and Figure Size

Figures are normally prepared electronically at their final published size (rather than in print form at, say, twice final size to allow for photographic reproduction). The following suggestions for line, symbol, and font sizes are drawn from a range of journal guides in both the physical and life sciences. These values should be taken only as a start, after which the appearance of the figure is best adjusted iteratively. Any specific artwork instructions supplied by the journal must be observed. Font sizes are expressed in points; that is, 1 point = 1 pt = 1/72 in = 0.353 mm (approximately). Other dimensions are expressed in mm or picas, where 1 pica = 12 pt. Thus:

- line thickness for scale lines and tick marks: 0.3–0.5 pt; for fitted curves: 0.5–0.7 pt, depending on the complexity of the graph;
- grid lines: 0.7 pt grey solid or 0.4 pt dotted;
- tick-mark lengths: 1–2 mm;
- symbol diameters: 1.5–2 mm, commensurate with text;
- font size of numbers and labels: with standard fonts, e.g. Times Roman, 8 and 10 pt, respectively, or 9 and 11 pt, and with Arial, 7 and 8 pt, respectively, or 8 and 10 pt; notice that Arial appears bigger than Times Roman with the same font size, owing to Arial's larger x-height; compare a Times Roman 'x' with an Arial 'x'.

The definitive test of line, symbol, and font sizing is to print the figure, cut it to size, and then view it on the printed journal page, or electronically insert it in an electronic image of the page. Numbers and letters should be either the same size or a little smaller than the text used in the body of the article. In Example 8.6, the scale-line thickness is 0.35 pt, fitted lines 0.6 mm, tick-mark length 1.1 mm, plot symbol diameter 1.5 mm, tick-mark numbers Arial 7.5 pt, and scale-line labels Arial 8.5 pt.

If the body of the article is printed in double columns, then decide whether the figure should be sized for a single column or the width of the full page. Do not waste space by sizing it, say, to two-thirds of the page, since it may be adjusted in production to fit the space available, with unpredictable effects on visual interpretation. To avoid uncertainty, attach instructions telling the printer the desired final size (e.g. printed width in mm or picas).

8.6.4 Colour

Publishers may ask authors to use colour in figures only if it fulfils a particular function, for instance, distinguishing lines or symbols in graphs. Even then, colour should be used sparingly. First, some readers may have impaired colour vision and be unable to discriminate between some colours. Second, if readers decide to make a print copy of an electronic version of the article or to photocopy an existing colour print version, any colour differences may be lost. Finally, although the electronic version of an article may not attract a production charge for colour, the print version usually will, and colour printing can be expensive. For all these reasons, therefore, figures should be prepared so that, if necessary, they can be understood without colour.

8.6.5 Reworking

Like text, graphs need to be reworked repeatedly [54, p. 112], for example, by varying the horizontal and vertical scales, the type and size of symbols and any fitted curves, the labelling of axes and tick marks, and the overall sizing, all with the intention of maximizing clarity and insight [54] [353].

8.6.6 File Format

For the final versions of your figures, graphs, diagrams, and other line drawings, it is better to use vector graphics (e.g. Encapsulated PostScript, EPS) than bitmapped or raster graphics (e.g. Tagged Image File Format, TIFF). Vector graphics are scale invariant and therefore offer the best resolution supported by the display medium, whether electronic or print. By contrast, bitmapped graphics need to have a very high resolution, e.g. 1000 dots per inch (and consequently a very large file size) to avoid a pixelated appearance with line drawings.

Example 8.11 shows ×4 magnified images from a section of a plot redrawn from Example 7.16. The lines and circles are rendered accurately in the EPS image but in the TIFF image there are aliasing artefacts, 'jaggies', due to the discrete nature of the image, which was produced originally at 300 dots per inch (dpi).

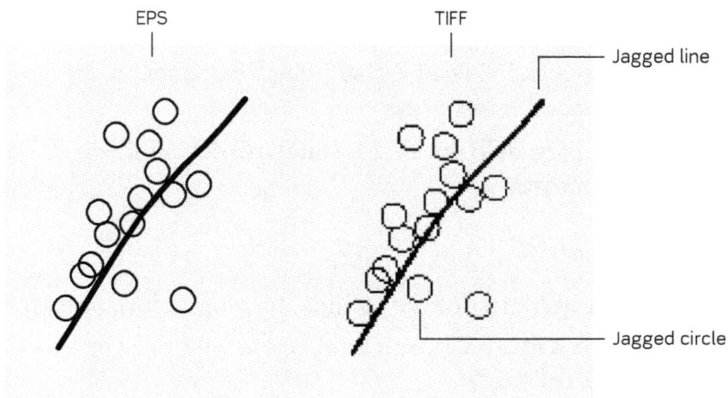

Example 8.11. *Magnified vector graphics (EPS) and bitmapped (TIFF) images from Example 7.16.*

For images acquired by a camera or other imaging device, bitmapped graphics are the normal format, but unless compression is used, file sizes can still be very large even with a minimum resolution of 300 dpi. For the version of your manuscript submitted for review, low-resolution figures are usually acceptable.

When your article enters into production, it may be prudent to verify that any line drawings in vector graphics format have not been rasterized at low resolution by default. If they have been rasterized, then offer high-resolution bitmapped originals instead.

8.7 Checklist for Graphing Data

Layout

1. Four sides to a plot, tick marks outwards?
2. Resolvable symbols, with key outside the plot area?
3. Horizontal and vertical scales labelled and units added?
4. Tick marks on horizontal and vertical scales numbered?
5. Font style and size compatible with the printed text?

6. Principled choice of fitted curve?
7. White space minimized, data-ink ratio maximized?

Caption

8. Figure number or letter matches that in the main text?
9. Figure title followed by sufficient explanation for the figure to be understood on its own?
10. Error bars defined as ±1 standard deviation or standard error or other measure?

Reproduction

11. Vector graphics format for line drawings, bitmapped for images?
12. Printed size correct, and compatible with one- or two-column layout on journal page?

9

Publication and Promotion

After you have worked long and hard on your paper, you may want to get its publication over with as quickly as possible. But where you publish and what you do after publication affect the eventual success of your paper. This chapter considers these issues in detail.

Section 9.1 compares publication in research journals and conference proceedings. Section 9.2 sets out the factors that might influence your specific choice of journal if you decide on journal publication. Section 9.3 deals with the process of submission and the subsequent stages, including peer review, revision, acceptance or rejection, article proofs, and eventual publication. Section 9.4 reviews ways of attracting attention to your article after it is published, and Section 9.5 deals with monitoring its success. Section 9.6 contains a checklist for the items in this chapter.

Conference posters, lectures, and talks are considered in Chapters 10 and 11.

9.1 Choice of Journal or Conference

Your choice of whether to publish in a journal or conference proceedings will depend on the practice in your research discipline, the nature of the research material, and the timing and requirements of the conference. It need not always be an exclusive choice. As explained elsewhere, some journals will consider submissions that represent a more substantial version of a preliminary or abbreviated conference report, or extended abstract. Many but not all journals will also consider manuscripts posted on e-print or preprint servers such as arXiv [10], which serves mainly physics, computer science, and mathematics, and bioRxiv [27], which serves mainly the life sciences.

9.1.1 Advantages and Disadvantages of Conference Publication

Conference publication has obvious advantages over journal publication:

- it reaches your intended audience;
- it may receive more attention than in a journal;

- it may generate useful criticism and suggestions for improvement;
- it may allow you to test preliminary ideas and results;
- it may be useful (and possibly necessary) for a doctoral thesis.

Yet it has disadvantages too:

- it may encourage you to submit work that is not ready for publication;
- it may compromise a later submission to a journal;
- it may represent a dissipation of your efforts.

Papers published in some conference proceedings may also appear late and be difficult to access.

9.1.2 Impact of Journal and Conference Presentations

The immediate benefits of conference presentation should be distinguished from the longer-term recognition of your published work. Most research fields attach less significance to publication in conference proceedings than in journals. Some engineering disciplines and computer science treat the two more evenly. Yet according to [125], when the impacts of the two kinds of publications are compared, it is journals that attract the most citations, at least in computer science. Additionally, conference rejection rates—often assumed to reflect research quality—show only a moderate correlation with citations [125].

9.2 Journal Publication

If you decide to submit your work to a journal, there are several issues to consider. These include whether it should be a general or specialized journal, whether it should be subscription or open access; and what weight to place on its reputation or impact.

The work to be submitted is usually assumed to be original research. If you are planning to submit a review article or an opinion piece for publication, you should consult the journal editor first (§4.1).

9.2.1 General or Specialized?

General and multidisciplinary science journals, such as *Nature, Science*, and *Proceedings of the National Academy of Sciences of the United States of America*, are aimed at a wide interdisciplinary audience, and articles need to be prepared appropriately. Subject-specific journals, for instance, those published by the Institute of Electrical and Electronics Engineers, the American Chemical Society, and the Institute of Physics, are aimed at narrower subject-specific audiences whose background knowledge can be taken for granted. Some journals fall between the two, for instance, *Proceedings of the Royal Society A* and *Proceedings of the Royal Society B*. Making a choice can be difficult. Whether articles of equal quality fare better in one kind of journal or the other is unclear.

Publishing in a multidisciplinary science journal has both advantages and disadvantages. The published article reaches readers from other subject areas and therefore has the potential for higher impact. The journal may also be perceived as more prestigious. But acceptance necessarily depends on the material being of wide interest or special significance, as well as of sufficient scientific quality. For the leading multidisciplinary science journals, rejection rates can be extremely high [132]. There is also a personal cost in submitting multiple revisions to satisfy more demanding reviewers, with consequent delays in publication and a potential loss in morale [329]. Furthermore, specialist readers who do not routinely monitor multidisciplinary science journals may fail to see the article. If they do see it, they may still consider it to be less authoritative owing to the limited technical detail available [329].

Publishing in a specialist journal has complementary advantages and disadvantages. Acceptance depends (or should depend) only on the quality of the work within the discipline. Specialist readers are likely to see the article and quickly recognize its worth and technical authority [329]. But researchers from other disciplines may never see it, and the journal itself may be perceived as less prestigious precisely because of the lack of competition between authors.

How then to choose between general and specialized? The journal's aims and scope, advice from experienced colleagues, along with bibliometric data, can inform your choice, but the style, standard, and subject coverage of a journal's published articles should be decisive. Has the journal published articles similar to yours? Have these articles been well cited by other researchers? Would you lose key readers by publishing there rather than elsewhere? Would the journal editor want changes that might compromise the direction of your research?

Whether or not you address all these points, never rely solely on the journal title. It is an unreliable guide to content, and your paper may, after a shorter or longer delay, be rejected for being out of scope. Worse still, your paper may be accepted and then never read.

9.2.2 Subscription or Open Access Journal?

Publication entails scholarly and commercial effort, which has non-trivial costs.[79] Who pays those costs depends on whether the journal is classified as subscription or open access (OA). With subscription journals, the reader—or the reader's institution or employer—pays for access to the published article, and the author does not normally pay, except perhaps for submissions where the print copy is over length or has more than the maximum number of colour figures. With most OA journals, the author (or the author's funding agency, institution, or employer) pays an OA fee or an article processing charge to cover the cost of publication, and the reader does not pay. This kind of OA is called 'gold open access'. Some subscription journals charge a publication fee but also offer a waiver [187].

Although it may not affect your choice of journal, there is another kind of OA that is available with subscription journals. It is called 'green' OA. Depending on your agreement with the publisher, a copy of the accepted author manuscript may be posted on your personal website and deposited in institutional and funding agencies' repositories (§12.8.3).

Some subscription journals operate a hybrid system in which authors can choose to either publish their work under the subscription model or pay an article processing charge for gold OA.

Which is the better option, subscription or OA? By definition, OA journals appear to offer wider dissemination than subscription journals, and ought to lead to more citations. Yet OA articles show no such increase in citations, at least in the year after publication, according to an experiment in which articles within a single journal were randomly assigned OA or subscription status [76]. Over a longer period, the average number of citations per article in OA journals seems to be lower than in subscription journals, though differences are contingent on journal age [29, Table 1, Fig. 1].

In principle, whether a paper is submitted for OA or not should not affect the quality of editorial assessment, and, in many hybrid journals, a decision about OA is made by authors only after acceptance. There do exist so-called predatory OA journals that charge authors for publication without offering any of the usual benefits of academic publication—peer review, informed editing, and professional article production and publication. Lists of predatory publishers are sometimes available online [22]. Inclusion may be contentious.

9.2.3 Journal Impact Factors

The dominant bibliometric measure of a journal's influence is the journal impact factor. The standard definition of this metric is that it is the average number of citations received by articles in the journal in the preceding two-year period [198]. Its value depends both on the journal and year and on the size of the database holding the citation data.

For example, for *IEEE Transactions on Power Systems*, the Thomson Reuters 2014 Journal Impact Factor [199] is calculated as in Example 9.1.

Number of citations in 2014 to articles published in 2012 and 2013:	2192
Number of articles published in 2012 and 2013:	779
Journal Impact Factor for 2014 (citations / articles):	2.81

Example 9.1. *Calculation of the 2014 Thomson Reuters Journal Impact Factor for* IEEE Transactions on Power Systems, *from Thomson Reuters Journal Citation Reports [199].*

According to this calculation, the average number of citations in 2014 for an article published in *IEEE Transactions on Power Systems* in 2012 or 2013 was 2.81. This impact factor ranked the journal in the top 11% of journals in electrical and electronic engineering.

By contrast, for the Scopus database [326], the equivalent metric, the SCImago 2014 Cites per Document for a two-year period, has a larger value of 4.61.

Differences of this kind may be attributable to the differing nature and sizes of the databases, with larger databases giving higher citations [333]. Some subject areas may also be better represented in some databases [216]. Reassuringly, when journals within a subject area are ranked by each of the two metrics, the correlation is high, with a correlation coefficient of around 0.9 [333].

Nevertheless, the impact factor has several defects as a measure of journal quality. As summarized in [242] [323] [363], its main problems are these:

- it has an uncertain relationship with independent measures of journal quality;
- it is open to manipulation by journal editors, legitimately and otherwise;
- what counts as an article varies from journal to journal (e.g. some may exclude editorials and brief communications, others may not);
- citation behaviour differs from field to field;
- the distribution of citations across articles is highly skewed, so that the average gives a poor indication of the most likely performance of a paper;
- a two-year window for citable articles is too short for most fields;
- it may reflect growth in some disciplines less well than in others;
- the reproducibility of some calculations is unclear.[80]

Some of these defects are addressed by other journal metrics. The differing frequencies with which authors cite each other in different fields is accommodated, for instance, in the Source Normalized Impact per Paper [200] and in the SCImago Journal Rank [200], which is an analogue of the Google PageRank algorithm. Longer citation windows are available with the 5-Year Impact Factor [198] and the three-year window of the CiteScore metric [200]. Journal citation distributions offer even more comprehensive descriptions [214].

With all its defects, the journal impact factor remains a popular measure with authors. Ninety per cent of respondents rate impact factor as 'important' or 'very important' in deciding where to submit a paper for publication, according to one author survey [11].

9.2.4 Ranking Journals or Research?

One reason for publishing in a high-impact journal is that your article gains prestige by association. But there is a difference between using journal metrics to measure the quality of a journal and using them to measure the quality of the research itself [214]. Both the European Association of Science Editors [350] and the signatories of the San Francisco Declaration on Research Assessment [323] recommend that research should be assessed on its own merits and not on the journal in which it published. Indeed, there is an argument that specialized journals carrying the most advanced (and possibly the best) research in a field have smaller impact factors as only a specialized audience can follow the work.

Unfortunately, assessing the research quality of individual articles is difficult, not least because of confounding factors, one of which is the impact factor of the journal containing the article [3]. In the long run, the accumulated number of citations an article receives gives an obvious measure of its influence, but over shorter periods the citation count is necessarily smaller and more random. Expert assessor scores then depend more strongly on journal impact factor. Tellingly, when the effect of impact factor is eliminated, assessor scores show little agreement with one other [109].

The lack of agreement may be due to experts' assessing different aspects of research [102]. If true, it suggests that research quality cannot be captured by a single metric. Therefore, in the absence of any other information, in particular data concerning citations, impact factor may be the least unreliable of the available, necessarily error-prone measures [109].

There are other metrics that include the number of website views and document downloads, mainstream news mentions, and social media discussion. Interestingly, the correlation between this kind of recognition and journal citations is moderate or weak [67] [193] [255]. Prominence in news outlets and social media may reflect cultural or social issues more than the research quality of the article [67].

9.3 Submission and Publication

Once you have decided on a journal, you need to prepare your paper for submission online. After submission, you will receive, sooner or later, the editor's decision, usually with peer reviewers' comments, on the suitability of the paper for publication. If you are invited to revise and resubmit, you then need to prepare the revision and your response to the reviewers' comments. There may be further revisions and responses until the paper is finally accepted or rejected. On acceptance, you have to deal with copyright matters, pay any article processing charges, and eventually inspect and correct the proof of the typeset version of your paper before it is published. It will appear online before it appears in print, if a print version is produced.

9.3.1 Potential Impact of Your Paper

Before submission, it is worthwhile examining your paper one last time for its potential impact on the editor, reviewers, and eventually the readers of the journal. As emphasized at various points in this guide, there are several factors to consider:

- Is there a clear research question of interest to the community?
- Is there a clear answer, a one-line summary or take-home message?
- Is there a reason for the work to be cited?
- Does it have interest beyond the research community?
- Are the core findings clear and immediate?
- Is the presentation professional?
- Are the title and opening and closing statements likely to engage readers (and possibly nonspecialists)?
- Do the figures and captions communicate a clear message?
- Is the statistical or mathematical analysis convincing?
- Have the results been validated thoroughly (on real data or real systems)?
- Is there any software or data linked to the paper that readers can exploit?

Only small edits may be needed: for example, sharpening a line or two in the abstract or conclusion; clarifying the captions to the figures; refining the title; or removing clichés. Individually the differences may be small, but cumulatively they can change the way the paper is perceived and understood.

9.3.2 Presubmission

Consult the journal's guide on preparing and submitting manuscripts. Some matters are general and need to be resolved well before submission; others are more specific and to do with the journal itself.

General issues:

- Has any material duplicated from other sources been removed or declared?
- Do any ethical requirements apply and have they been satisfied? Is there documentary support?
- Have any conflicts of interest and sources of funding or sponsorship been identified and declared?
- Has permission been obtained for the use of any material under copyright, such as previously published figures? Is there documentary support?
- Do all the references have citations and all the citations references?
- Are statistical measures, e.g. standard deviations, standard errors, and confidence intervals, appropriate and unambiguously identified wherever they occur?
- Does the text reach accepted language standards for grammatical and idiomatic use? Does a professional language editing service need to be employed?

Journal-specific issues:

- Does the submitted paper fit into one of the accepted types for the journal, e.g. original research article, research note, letter to the editor, review?
- Is the length appropriate?
- Does the submitted paper contain details of all your coauthors, including email addresses?
- Is the abstract within the required length? Is it correctly formatted, either with subheads and paragraphs or, more commonly, without? Is it self-contained with any essential references or abbreviations defined independently of the main text? Except for clinical journals, is statistical detail excluded?
- If the journal requires it, has a statement been prepared about the significance of the work in language understandable to a nonspecialist?
- Has a list of keywords been prepared? Are they compatible with any mandatory list supplied by the journal?
- Do statistical tests and reporting of statistical data comply with the requirements of the journal?
- If the journal allows or requires it, has a list of potential reviewers been prepared?

Failure to attend to these issues may lead to the submitted paper being rejected or returned for revision, or, if accepted, its publication being delayed while corrections are made. Extensive text duplication may lead to immediate rejection with no opportunity for resubmission (and perhaps other penalties; §12.3).

The manuscript does not normally need to mimic the format of the published article. For the initial submission, some journals allow you to choose your own text formatting, citation, and referencing style, although figures, tables, and captions may be expected to be embedded in the text for ease of reviewing. A correctly prepared manuscript, with separate tables and figures, may be required only on acceptance. Some other journals, however, may insist on correct formatting on submission and may supply a template for the manuscript. The journal's guide for authors should always be consulted.

9.3.3 Submission

The process of submission is itself usually straightforward, especially when the paper and supporting materials have been properly prepared. As explained in the journal's guide for authors, there are conditions attached to the submission. One of the most basic is that the work has not been previously published and is not under consideration for publication elsewhere. Another is that it complies with the principles of ethical authorship, being free from fabrication, falsification, and plagiarism (§12.1). If the paper has been previously published (or is to be published elsewhere) in preliminary or abbreviated form, for example, as a conference article (§12.2.5), then in addition to being cited in the submitted paper, the journal may want a copy of the publication [186] and possibly an explanation of how the submitted paper differs [178].

You may be expected to provide a cover letter, which includes any special factors you want the editor to take into account. It might identify ethical issues to do with the research or conflicts of interest with potential reviewers or members of the editorial board. It might mention previous related work. It might further contain a statement explaining the importance of the work. But the receiving editor is more likely to make this assessment from the title, abstract, and the paper itself, as would the reviewer and ultimately the reader. As one experienced editor noted, it's not the wrapping that counts: it's the gift inside.[81]

At a later stage, nearer to when the submitted work is published, the journal may require or encourage you to supply or make available to readers data, code, protocols, and other supporting materials [178] [186], as explained in Sections 3.11 and 3.12.

9.3.4 Review Process

What happens to your paper after it has been submitted? It is usually subjected to an initial assessment by the receiving editor, who may also be the editor-in-chief. His or her decision about its suitability for peer review depends on several conditions [303], namely that

- the work falls within the scope of the journal;
- the paper is one of the accepted types;
- the scientific quality, novelty, and presentation are adequate for peer review;
- the paper does not duplicate published work or violate any other ethical standards.

If the submitted paper fails any of these conditions, it is almost certain to be rejected without proceeding to peer review [79]. There is little point in your arguing against this rejection, known variously as immediate rejection, desk rejection, and rejection by triage. The editor's decision should be regarded as final on matters of scope, article type, quality, and ethics, all of which are fundamental to the journal's place in the research community.

If the submitted paper satisfies the initial assessment, then it goes forward for peer review. With larger journals, the receiving editor assigns the manuscript to a handling editor with the required expertise. The handling editor finds reviewers, deals with the review process, corresponds with the author, and eventually makes a decision or recommendation on the suitability of the submitted paper for publication.

With smaller journals or with journals where the receiving editor has expertise in the paper's research area, the receiving editor may act as the handling editor.

In general, the handling editor assigns at least two reviewers (also known as referees) to the paper [55, ch. 6] [178]. For multidisciplinary topics, it may be three or four reviewers, whereas for some fields, it may be only one [79].

The ethical responsibilities of reviewers have been set out by the Committee on Publication Ethics [152] and other organisations [327, §2.3] [339, ch. 10]. Reviewers are expected to have expertise in the topic or aspects of the topic of the paper, be fair in their evaluation, maintain confidentiality, and not exploit their privileged access to unpublished work. On being invited to act, reviewers are expected to declare any conflict of interest that might compromise their review, for instance, having a collaboration with one of the authors [152] [303]. Failure to act ethically can be the subject of censure [62] [308].

You may be asked to suggest reviewers, though editors are not required to accommodate your request, and if they do, they are expected not to depend on those reviewers [64]. Conversely, you may ask that certain reviewers be avoided and provide justification for your request [178]. Even then, editors may invite them in the interests of balance.

With most journals, the author does not know the identity of the reviewers. This kind of review is known as single-blind review. When the author's identity is also kept from the reviewers, it is known as double-blind review. There are arguments for and against anonymity [327, §2.3.4]. Single-blind reviewing allows the reviewer to give an honest analysis, uncompromised by collegial pressure or the seniority of the author, both of which might lead to self-censoring. Although, in principle, anonymity could encourage dishonest and abusive reviewing (and inappropriate requests for authors to cite the reviewer's own work), in practice most reviewers behave conscientiously. Occasional intemperate language is (or should be) redacted or omitted by the handling editor, with an explanation.

Double-blind reviewing is less common. It reduces the risk of ad hominem attacks, but author identity can often be inferred from the methods and references in the paper. Completely open reviewing, where both author and reviewer identities are known, is also less common, and it may suffer from reviewers' self-censoring. Interestingly, the proportions of citations to reviewers' own publications appear to be the same in open peer review articles and single-blind peer review articles [228]. For this measure at least, open peer review seems not to affect peer reviewers' capacity to suggest revisions.

Anonymous or not, reviewers constitute that special group of people who are likely to read your paper carefully and critically. Their misunderstandings and disagreements may well represent the typical reaction of readers given the same material [79, p. 129]. If reviewers misinterpret your research or do not recognize its true worth, then, almost always, that is your fault, not theirs. If they fail to understand, then very probably so will the reader [132].

9.3.5 Reviewers' Comments

What does a peer review look like? It is usually expressed in free format. It may start with a short, neutral statement about your paper; for example, its aims, results, and conclusions. It may then go on to analyse the major strengths and weaknesses of the work, perhaps finishing with a list of more detailed remarks and minor corrections [55, ch. 6]. The order of these items may vary with the journal and research area. A review would not normally contain the reviewer's recommendation about the suitability of the work for publication.

Sometimes reviewers comment on the quality of the exposition, especially if it is poor or impedes understanding, and they may offer suggestions for improving a line of argument, a technical description, or the terminology. Nonetheless, their primary concern is with the underlying science, not the presentation. An exception is with some elite journals, where the editor (subeditor or copy-editor) may offer detailed advice about the exposition once the paper is accepted [132].

What specific kinds of comments do reviewers make? Even when supportive, their observations are usually restrained. The following comments are abridged[82] from peer review reports: 'good paper', 'nice piece of work', 'important problem', 'methods solid', 'results clear', 'mathematically sound', 'well written'. Occasionally the language is less restrained: 'breakthrough in understanding', 'powerful experimental technique', 'experiments elegant, simple, and well-controlled', 'rigorous and innovative'.

More commonly, reviewers concentrate on the perceived defects in the work, for understandable reasons. An author submits a paper under the assumption that it is suitable for publication: it is the reviewers' job to test that assumption [303]. Examples 9.2–9.4 show comments abridged from the peer review reports mentioned earlier. As is conventional, defects are classified broadly as minor, major, or fatal, insofar as they are likely to guide the editor's eventual action (§9.3.6). The examples are illustrative, not exhaustive. They are each compiled from many sources, and do not represent a single reviewer's report.

Notice that the division between minor, major, and fatal defects is imprecise. It depends on context and quantity; for instance, minor errors might be so pervasive that taken together they constitute a major or fatal defect as they render the paper unready for review (§9.3.4). It also depends on the journal: what is a major defect for one may be a minor defect for another.

Some comments are more serious than they may appear on first reading. Thus in Example 9.3, the comment about the introduction not appropriately reviewing past work is classified as a major not minor defect. It is major because it suggests that previous research has been misrepresented or omitted to an extent that affects the balance of the paper.

Exposition	• Presentation is careless. • Material is repeated across paragraphs. • Sentences are too long (one has 50 words). • Descriptions are inconsistent (a factor cannot be both 'irrelevant' and 'almost irrelevant'). • Text is missing from references. • Figures are out of order. • Text fails to cite figures. • Excessive acronyms make reading hard. • Unit symbols have typographical errors ('K'confused with 'k'). • Use of English is poor. • Statements are grammatically incorrect, contain spelling errors. • Bullet points, highlights, and significance statement do not capture content of paper. • Subtitles do not represent subsequent material. • Summaries are incomplete.
Introduction	• Introduction is too long, fails to focus on objectives, includes redundant material. • Problem is not placed in context. • State-of-art summary is outdated. • Previous work is misrepresented.
Methods	• Details are missing or unhelpful. • Where are dimensions of apparatus? • What does 'very large time constants' mean? • Statement about signal (what signal?) needs to be more specific. • Number of trials is missing. • Novel unit of measurement is idiosyncratic and annoying.
Mathematics	• Technical level is too high for intended audience. • Quantity modelled is not what was defined. • Embedding theorem is wrongly attributed. • Symbols in equations are undefined. • Subscripts are too small. • Equation needs reference to explain its origin. • Variables need defining before use.
Results, analysis, comment	• Results are too detailed; not every finding needs reporting. • Results are poorly linked to model. • Analysis is difficult to follow, overgeneralizes what results allow. • Why no comment about previous discrepancies? • Linear regression fits could improve qualitative interpretations. • Multiple testing and error inflation need comment.
References	• References need updating. • Key references are missing. • More recent web link is needed. • Referencing style conflates numeric and author-date systems.

Example 9.2. *Reviewers' comments indicating minor defects in submitted papers.*[82]

Figures	• Figures are too small, contain too much detail. • Axes have no labels. • Fonts are too small. • Figure and table captions contain no explanation, force reader to consult main text. • Curves seem offset but caption does not explain. • Caption refers to coloured lines that do not exist. • Graphs could replace tables. • Are error bars 1 SD or 1 SE or ±1 SD or ±1 SE (SE standard error, SD standard deviation)?

Example 9.2. *(Continued)*

Exposition	• Paper is too long for content, poorly organised, difficult to follow. • Narrative is egocentric, referring to things known to author but not to reader. • Informal explanations should precede formal, not vice versa. • Replace dogmatic statements by nuanced ones. • Avoid long sentences, didactic asides. • Shorten to keep reader's attention. • Improve grammar and phrasing to level of native competence in English.
Significance	• How is state-of-the-art advanced? • Problem, significance, and novelty are unclear. • Problem has been previously addressed, despite authors' claim. • Work is incremental. • Experiment does not deal with research question. • Title, abstract, and summary misrepresent content. • Major claim of work is unproven.
Introduction	• Introduction does not appropriately review past work, contains textbook errors. • Recent advances need describing. • Well-known material can be omitted. • Approach needs justifying.
Methods	• Why new method instead of standard one? Is method original? Can it be implemented? Is it scalable? • Why do chance features prove important? • Why not quantify computational efficiency by conventional methods? • Device for validation is nonstandard, needs realistic examples. • Comparison of methods is unbalanced. • Need more than 'commercial program was used'. • Consider realistic testing of algorithm.

Example 9.3. *Reviewers' comments indicating major defects in submitted papers.*[82]

Human and animal studies	• Description of patients' disability seems contradictory. • Reliable statistics need more participants. • Why no quantitative thresholds? • What was matching process? • Why use same stimulus to test different functions with indistinguishable effects? • Why use drugs whose mechanism of action is unclear? • Why are inter-individual differences not larger?
Results	• Results are not credible as presented. • Standard errors are unexpectedly small. • Statements about causal mechanisms need careful wording. • Familiar techniques need real applications.
Analysis	• Statistical analysis is too weak for deductions. • To promise future work is not solution. • Theoretical diversion is unconvincing when device cannot be manufactured. • Why new performance measure? • Is sensitivity better or worse than in other studies?
Mathematics and modelling	• Theory is general, but application is trivial. • Derivation is unsystematic, with unjustified simplifications. • Mathematical understanding is naïve. • Can variables describe phenomena? • Metric is similar to earlier one, but not justified nor compared. • To say good agreement is misleading. • Is algorithm optimal? • Model parameters are missing. • How many degrees of freedom in fit?
Discussion	• Discussion is cursory, cites few studies, ignores others that explain results. • It includes irrelevant material, refers to contradictory results as being complementary, overstates conclusions.

Example 9.3. *(Continued)*

Exposition	• Quality of exposition is substandard. • Paper is illogical, poorly structured, contains previously published text. • Reasoning is hard to follow, has unsupported assumptions and conclusions. • Statements are unclear. • Work is not self-contained, forcing reader to consult other papers repeatedly. • Artwork is substandard, contains copyrighted images. • Language is imprecise and confused. • Text is obscure. • Writing lacks clarity. • Terms are used wrongly.

Example 9.4. *Reviewers' comments indicating fatal defects in submitted papers.*[82]

Significance	• Paper lacks originality, novelty, motivation, rationale, significance. • Work has no impact on existing knowledge. • It does not address question posed, is incremental, redundant, of insufficient interest for larger community. • Similar problem has already been addressed. • Claims are inadequately supported by results. • Results are too specialized. • Work is out of scope. • Theoretical questions are poorly developed. • Paper seems preliminary. • Any purely theoretical contribution must be substantial.
Introduction	• Major parts of literature are ignored. • One paper is selectively quoted. • Referencing seems random. • Claim that effect is unknown is untrue. • Descriptions of well documented phenomena are idiosyncratic. • Experimental hypothesis is unclear. • Premise is implausible.
Methods	• Methods are defective, mutually incompatible, lack technical rigour, so poorly described that data are valueless. • Experimental design contains confounding factors. • Procedure is conventionally used for different purpose. • Experimental measure is not robust, cannot quantify effect. • Experimental controls are inadequate.
Human and animal studies	• Animal experiment has unclear logic, no clinical significance. • Specific stimuli undermine generalization. • Range excludes values for best response. • Calibration is questionable. • Response bias is ignored. • Confusion between tasks compromises analysis. • Where is ethical approval for research?
Results	• Results are trivial, self-contradictory, unconvincing, purely descriptive, do not answer research question, have artefacts, multiple explanations. • Findings are not novel. • Data are incomplete. • Responses are compromised by ceiling effect. • Reported noise level is implausible. • Means and standard deviations cannot be reconstructed from significance tests.
Analysis	• Analysis is inadequate, fragmentary, based on insufficient data. • What were sample sizes? • Why no post-hoc analysis? • Connection between predictors and general case is missing. • Statistics do not make sense. • Parametric tests are applied to ordinal data. • Degrees of freedom is wrong. • Null effects dilute single effect. • Design and results do not match question posed and interpretation. • Statistical significance confused with scientific importance.

Example 9.4. *(Continued)*

Mathematics and modelling	• Mathematics has gaps, is sloppy, laboured, unclear, insufficiently detailed. • Proof of theorem is incorrect. • Hypothesis and data are obscured by irrelevant, excessive mathematics. • Models are poorly differentiated, too complicated, flawed, obsolete, do not fit data. • Too many ad hoc hypotheses are needed. • Only linear case is tested, so general case is immaterial.
Discussion	• Discussion is underdeveloped. • Explanation of results is descriptive, needs testing. • Data do not support discussion. • Results are not compared with earlier findings. • Alternative explanations need to be eliminated.

Example 9.4. *(Continued)*

9.3.6 Editor's Decision

The editor has a crucial role in bringing together the comments of peer reviewers, who may differ in their priorities and technical expectations and in the way they assess your work (§9.2.4). Editors can draw on their knowledge of each reviewer, their own reading of the manuscript, and their broader understanding of the field. In short, editors' decisions are informed by the reviews, not determined solely by them, and need not reflect the majority opinion [55, ch. 6] [132]. Even if the reviews are unanimous, the decision remains that of the editor [79, p. 128].

The editor's decision almost always amounts to one of four possibilities:

- acceptance without revision;
- minor revision;
- major revision;
- rejection.

Acceptance without revision is extremely rare for a reputable journal. Rejection is much more common, with rates varying from journal to journal. For leading journals they can be high, particularly when desk rejections are included in the total. Thus for the journal *Nature* the rejection rate was over 90% in 2013 [132], and for six journals of the IEEE Power & Energy Society, it was 85%–90% [180]. Rejection rates averaged over much larger samples of journals are lower, but still substantial: for example, 63% in 2010, a modest increase on 59% in 2005, according to an analysis of 4,200 journals [136]. Some public rejection rates may be slightly misleading with journals that reject and at the same time allow or encourage a fresh submission.

An offer to consider a revised submission is, therefore, the normal route to publication, though the amount of work required may vary markedly with the kind of revision requested. Major revision will almost certainly require a substantial re-review by the original reviewers, if they are available, or, if not, perhaps by new reviewers. The eventual acceptance of the paper is not assured. Minor revision is likely to require a lighter re-review, and the editor may decide to accept without further consultation. With some journals, the editor may offer acceptance but subject to mandatory (and rapid) revision. In that case the editor cannot step back from acceptance once the authors have complied with what was requested.

Do editors and peer reviewers and make good decisions? An analysis of papers from three leading medical journals [334] suggests that they do. Desk-rejected papers when published elsewhere gain fewer citations than papers not desk-rejected. And when papers do go on to be reviewed, those which receive recommendations from reviewers to reject ultimately gain fewer citations than those which do not.

Equally, do editors and peer reviewers offer good advice? A broader analysis of papers submitted to over 900 biological sciences journals [42] suggests that they do. Papers that have previously been rejected by higher-impact journals gain more citations than papers not previously rejected. The assumption is that the feedback authors receive with rejection leads to improvements in the paper that is eventually published.

Naturally many papers rejected from elite journals do go on to be highly cited. Occurrences of this kind are inevitable when there is intense competition for space in these journals.

9.3.7 Request for Revision

Suppose that your paper has not been rejected and you have instead been asked to revise it. The editor's letter may have a standard format but it usually contains instructions about what is required and how any revision will be handled on resubmission. It may also contain some specific editorial suggestions, which should not be taken as the only comments to be addressed.

The reviews themselves need to be read carefully and dispassionately. Although they may seem combative and unsympathetic, they should not be treated as if they are part of an exchange at a scientific conference or workshop. Rather, they should be taken for what they are—a stimulus to get the best out of your work.

This advice is from Reeves, an experienced editor (and author and reviewer):[83]

> It is important to welcome corrections and make a real effort to understand (and sometimes incorporate) the thinking behind a negative review. The author needs to realize that others, not just the reviewers, will have similar reactions, so arguments against the reviewers should be made explicit in the paper and welcomed as a chance to improve the paper. As an editor, I've seen too many people claim to do this, when in fact they have not; the effort has to be a genuine one, even if the conclusion is only that the reviewer is wrong.

How, therefore, should you deal with a request for revision? If the reviewer's comments are to do more with the exposition than with the content, then they are unlikely to be substantial and you should be able to accommodate them. Something that is obvious to you—such as a technical term, a procedure, or a deduction—may not be obvious to the reader, and explaining it will improve both the accessibility and impact of the work. Small corrections like these, especially ones that do not affect the content, are not worth disputing on grounds of principle [79, p. 129].

It may be, however, that the reviewers' comments are more substantial and to do with content. If so, you need first to decide whether they can indeed be accommodated [258, p. 71]. Some problems may be more amenable to solution than they first appear. For instance, if your reasoning is misunderstood, can it be amended? If a new measurement is requested, can it be performed or the text modified to explain why it is unnecessary? If the findings are seen as incremental, can more material be added or their significance better explained? In other words, if what is asked for can be satisfied or sensibly countered, you need then to decide whether the time and effort are worthwhile.

If, on reflection, you decide that the requested changes cannot be accommodated, you can withdraw the paper and submit it to another journal, preferably after having taken into account the comments in the reviews (§9.3.9).

9.3.8 Response to Reviews

If you do decide to submit a revised version of your manuscript, ensure that the submission is timely [55, ch. 6]. You need to prepare a separate document detailing your response to the reviews. Make it as easy as possible for the editor and reviewers to verify that you have done what was asked for [79, p. 131]. Explain any new material you have introduced to answer a reviewer's question or concern or, where you disagree, explain how the text has been improved to avoid misinterpretation. Give the page and paragraph numbers or section numbers so that the editor can readily find your changes. You can also insert a copy of the amended text into your reply or include a copy of the manuscript with the affected sections highlighted.

Reviewers' comments and author responses are not normally published. The responses in Examples 9.5 and 9.6 are consequently taken from the reply to a review of one of my own co-authored manuscripts. The anonymous reviewer's comments, which were interleaved with our responses in the reply, are not reproduced here. The response in Example 9.5 addresses a concern raised by the reviewer about a potential bias in the calculation of certain ratios of physical quantities.

> We repeated the simulations for ratios whose logarithms fell outside the interval [–0.5, 0.5], about 5% of the total. Providing that the illumination changes were symmetric, extreme ratios were almost invariant. Details are now set out in Section 3.4, Extreme ratios.

Example 9.5. *Authors' response to a reviewer's comment.*[84]

The response in Example 9.6 addresses a request by the reviewer for a fuller discussion of atmospheric effects and sampling issues.

> We have now added a comment about haze and blur and its limited explanatory role (Section 4, Discussion, para. 6). We have also added estimates of spatial sampling thresholds for each scene in Section 3.2.

Example 9.6. *Authors' response to a reviewer's comment.*[84]

Each response explains what was done and where the changes could be found in the manuscript.

When you construct your reply to reviewers' comments, do not ignore any of them or claim that you have addressed them when you have made only superficial changes or no changes at all. Ensure that whatever additional explanation or argument you put into the reply also appears in one form or another in the revised paper so that it can stand on its own [184].

Notwithstanding your best attempt at a comprehensive revision, be prepared for another request to revise, and, perhaps, eventual rejection. The fact that you have been offered the opportunity to revise does not in itself guarantee acceptance. New material that you introduce may reveal more problems with the work.

9.3.9 Rejection

Rejection is common, as noted elsewhere. But how best to deal with it? Do not immediately submit the rejected manuscript to another journal or write an angry rebuttal to the editor, whether or not you feel you have right on your side. Although comforting, neither action is for the best.

Submitting the manuscript elsewhere needs thought and preparation, as explained shortly. Complaining to the editor about the choice of reviewers or their inconsistent comments or their ignoring the evidence in your favour is unlikely to lead to a different decision. The editor chooses the reviewers, knows their expertise, and will weigh their advice in reaching a decision. Making that decision, which may well not reflect the average reviewer opinion, is the editor's job. As with desk rejection (§9.3.4), it should be viewed as final, except in the event of a breach in procedures. Even if an editor is prepared to consider an appeal, it is unlikely to be successful [55, ch. 6].

Instead, examine carefully and objectively why the paper has been rejected. Whereas the details may vary (§9.3.5), the judgement usually reduces to one of the following:

- the paper is inappropriate for the journal;
- the paper is potentially publishable but needs more work than a revision allows;
- the paper is not potentially publishable.

A paper can be judged inappropriate on several grounds. For example, it could be too specialized for a multidisciplinary science journal or too applications-oriented for a pure theory journal. These and related issues to do with scope should really be addressed before submission (§9.2.1, §9.3.2). To ease resubmission to another journal, some publishers offer an automated manuscript transfer service, whereby a paper rejected from one journal can be immediately submitted to another more appropriate journal, along with the reviewers' comments.

If the paper is judged potentially publishable, but requires substantially more work, then, as with a revision, you need to decide whether the time and effort are merited. Can the defects in the paper be repaired? It may be worthwhile trying. The feedback received may eventually make a more successful paper than one that was not rejected (§9.3.6).

Still, unless invited to resubmit by the rejecting editor, it is wiser to submit it to a different journal [79, p. 133]. Whichever you choose, the advice is never to submit an unrevised manuscript, that is, one that does not address at least some of the concerns raised by the editor and reviewers [88]. Not only is it a waste of a useful resource, you may receive the same reviewers again, who will be even more unhappy with the work.

Finally, if the paper is judged not to be potentially publishable, ask yourself these two questions:

- Will it be useful to anyone?
- Will it ever get cited?

If answer is no to either question, then it may be best to stop and reconsider. Not every piece of research needs to be published. Furthermore, publishing or attempting to publish poor research may be self-defeating and eventually damage your reputation [79, p. 133]. Know when to move on.

9.3.10 Copyright Transfer

When your paper is accepted for publication, the normal course is for you to transfer the copyright to the publisher. Copyright protects your rights and, in particular, limits who can reproduce the material in the paper, although the protection extends only to the expression of the content, not the ideas behind it [189]. By default, as creator of the work you own the copyright. The purpose of the transfer is to enable the publisher to publish the paper and to defend the intellectual property contained within it [66] [174]. As part of the transfer, you retain to a degree certain rights to share the article.

The rights you retain are contingent on the licensing agreement with the publisher. When the paper is accepted for publication in an open access (OA) journal (or an OA fee is paid to a subscription journal), you retain the right to share the article more freely. How freely depends on the nature of the licence. A Creative Commons Attribution 'CC BY' licence [70] is the most accommodating licence on offer.[85] It permits maximum dissemination of your work, letting others build on it and commercially reuse it, as long as it is properly credited. A Creative Commons Attribution-NonCommercial 'CC BY-NC' licence lets others build on your work, but not commercially, again as long as it is properly credited. The licences from learned society and institutional publishers can be less or more restrictive than from commercial publishers.

9.3.11 Proofs

The accepted version of your manuscript is definitive and it should be the version that is published [49, §2.5]. Even though it is submitted electronically, however, errors do arise in the process of copy-editing to match the house style of the journal and in typesetting for publication online and possibly in print.

To allow authors to check the accuracy of the typeset document, publishers almost always provide a proof copy, which is what the published article will look like, apart from details about the publication year, issue number, and pagination. The proof usually takes the form of a PDF or some other electronic version of the document. You should return the corrected proof to the publisher within the allotted time, typically 48 hours. Failure to do so may lead to the article being held over for publication in a later issue of the journal [368].

The errors introduced in typesetting are customarily called printer errors [68, §32.3.1]. For example, errors may be introduced with mathematical expressions and with the symbols for physical units; hypertext links within and outside the article may be misdirected; and figures may be set too large or too small or incorrectly rendered for colour. In spite of reference-checking software, errors may occur in reference details such as article numbers and page numbers. As noted elsewhere [79, p. 78], there are more mistakes in the references section of a paper than in any other.

How should you check the proof? The following is the usual advice [68] [79] [368]:

- Read the proof carefully for scientific content and logic, without referring to the manuscript.
- Compare the positions of figures and tables with where they are first referenced.
- Read the proof word-by-word against the original manuscript. Pay attention to the correctness of the author details, mathematical symbols and equations, citations and references, and numerical data, especially in tables.
- Examine end-of-line breaks in words containing hyphens, in mathematical symbols and equations, and in unit symbols containing spaces; badly placed breaks can be misleading.
- Examine the sizing and colour rendering of figures; wrongly sized figures can affect their visual interpretation.

Corrections may be marked on the PDF or other electronic version in several ways [68]. Commenting tools allow you to explain any changes unambiguously.

While reading the proof, you may discover errors of your own making. The corrections of these errors are customarily called author alterations [68]. The fact that you can make author alterations does not mean you can rewrite the paper [79], as was allegedly the view of Keynes [41]:

> I don't really start until I get my proofs back from the printers. Then I can begin my serious writing.

There are two reasons not to rewrite. First, the introduction of new content at the proof stage is seen as unethical [79, p. 140]. Neither the reviewers of the original paper nor the editor, who ultimately carries the responsibility for what is published in the journal, will routinely see the alterations. Second, author alterations can be expensive. If they are extensive, then publishers may reserve the right to transfer the cost to the author [68]. Nonetheless, if there is a scientific case for making substantial author alterations, then, with the agreement of the editor and publisher, you can introduce a 'note added in proof', which may appear as a footnote or endnote in the published article.

If there are very many printer errors, ask for a second proof. If serious errors remain or new ones are introduced, ask for another proof. But accept the fact that perfection is unlikely to be achieved, and the paper may eventually be published with minor errors, some of which you did not notice in the proof. Any serious errors detected after publication need to be dealt with by a different process (§9.3.13).

9.3.12 When Is My Article Citable?

At some point on or after acceptance of your article, the publisher makes it available online and assigns it a digital object identifier (DOI) [87]. The DOI identifies the article uniquely even though it may not have been typeset or, if it has, it exists only in proof form. Thus for a print version, it would not yet possess a volume, issue, and page numbers. The exact sequence of events depends on the publisher. Although your article is potentially citable at any stage, once it has a DOI the citation is unambiguous. Bibliographic databases can use the DOI to associate any citations an article receives during the proof stage with its final published form.

Because the date on which the article becomes available online may precede the date on which it achieves its final published form, both the online publication date and the print publication date are used as official publication dates. Some bibliographic databases start their indexing with the online uncorrected or corrected proof of the article, but offer limited supporting data; others wait for the final online or print version of the article.

9.3.13 Dealing with Errors after Publication

When your paper is published, examine whether the corrections have been implemented satisfactorily and no new, unauthorized changes have been introduced. If all is well, then no further action need be taken. Suppose, however, that you do discover an error or some other defect after publication. The action required depends on the seriousness of the problem and who was responsible.

If the problem is an error in a small, isolated part of an article that is otherwise reliable, you can ask the journal to publish a correction [327, §3.5.1]. Some journals distinguish between the sources of error. If the correction is due to author error it is entitled Corrigendum (plur. Corrigenda), and if it is due to typesetting or other production errors it is entitled Erratum (plur. Errata). The difference in titles may affect how the correction is catalogued in bibliographic databases. If new material is added, the corrigendum may instead be entitled Addendum. In all cases, corrections after publication, particularly repeated corrections, may not help the reputation of the author [55, p. 35].

If the problem is more serious—unreliable data, invalid conclusions, nonreproduceable research, scientific misconduct, duplicate publication—then you, the editor, or the publisher may arrange to issue a retraction (§12.3).

When there is significant concern about an article but insufficient evidence for a retraction, the editor may instead issue an expression of concern, which holds until the matter is decided [327, §3.5.1]. Not all expressions of concern are entitled as such.

9.4 Promoting Your Article

Ideally, the publication of your article in a good journal should be all that is required for it to be seen by your intended audience. Yet with about two million scientific articles published each year [375], some extra work may be needed to raise its visibility.

9.4.1 Dissemination

Unless you have one of the more accommodating copyright agreements with your publisher (§9.3.10), you cannot systematically distribute your published article, as if you were a publisher in your own right. But, depending on the details of the agreement, you still have considerable freedom of action in sharing your work [66] [183] [287]. You can give copies of the published article to research colleagues for their own research. You can distribute copies to your students. You can also send copies to potentially interested researchers whom you know. If you did not cite them, you may want to explain the lapse in a tactful cover note. In all cases, you should include a full acknowledgement of the source of the document and, where appropriate, a link to the final published version on the publisher's website, the version known as the article of record.

To achieve greater dissemination, you can upload your article in one of the permitted forms to your institutional or personal website, to your funder's website if the work was externally funded, and to popular non-commercial websites and some commercial websites too. What precisely you can upload (the submitted or accepted manuscript or the published article) and when you can upload it are contingent on your copyright agreement (§9.3.10).

The journal's publisher will send a table of contents containing the article details to anyone who subscribes to the journal's alerting service. If the journal is one of those covered by bibliographic databases such as Web of Science [375] and Scopus [326], they will automatically index your article and make available details to subscribers.

9.4.2 Secondary Reporting

Before or after publication, you can draw attention to your article in several ways [131]:

- Upload the manuscript to a relevant e-print or preprint server [10] [27], providing that journal submission is not compromised, or journal copyright when published.
- Refer to the research at relevant conferences, workshops, and seminars, and give the web address of an accessible version of the article.
- If you present a poster on a similar topic, attach a sign-up sheet so that visitors can indicate their interest in receiving the article and related material (§10.2.3).
- Create a video or podcast describing the research.
- Submit the article or images from it for inclusion in review magazines from your publisher.
- Submit a synopsis or announcement of the article to a professional networking website.
- Set up a website devoted to the research area.
- If the work is of general interest, consider issuing a press release, perhaps with the help of an institutional press office.
- Engage with social media, for instance, writing a blog post.

If the article is part of a much larger body of work that is ready for an authoritative overview, explore the possibility of writing a review article with the journal editor (§4.1).

9.4.3 Linked Data and Software

If the underlying research led to data sets or software that you uploaded to an external repository or to an institutional or personal website (§12.7), then ensure that you attach information about the published article. Include a hyperlink to an accessible version. Every time other researchers use your data in a publication, not only will they cite the repository record, they will (or should) cite the published article as well.

Sometimes the value of a piece of research is more in the data made available than in the analysis and conclusions reported in the article.

9.5 Measuring Your Success

The impact of an article can be measured in many ways. Download rates from the publisher's website or other repository and journal citation counts are two of the most popular numerical indicators, but as noted in Section 9.2.4, the two measures are at best only moderately correlated. Citation counts may also be several orders of magnitude smaller than the number of downloads [193] [255].

Citation counts are available from bibliographic databases such as Scopus [326], Web of Science [375], and Google Scholar [138], which also provide summary measures quantifying impact over different periods of time. Citation data from Google Scholar [138], though, tend to be less reliable owing to uncertainties in the authenticity of the citing sources.

9.5.1 Who Cites You?

To see who cites your work, search for your published article in Web of Science [375] or Scopus [326]. Each database shows the number of times your article has been cited and gives a link to the citing articles. Examine what these authors say about your work and why they cited it. The reasons can be sobering. One revealing case is described in Section 12.5.2.

When authors do cite your article for the right reason, it may not be the reason you expect. In an experimental article, it may be for the experimental methods or analysis you used rather than for the results you obtained, and in a theoretical or mathematical article, it may be for a preparatory lemma or enabling theorem you derived rather than for the principal finding. Authors are not always the best judge of what is most useful or important in their research, or indeed who is their best audience [150, p. 23].

9.5.2 Counting Your Citations

The number of citations your article receives depends both on its content and on the citation behaviour within its research area and the age of the article. Research areas that are developing quickly and have many active participants generate larger numbers of citations. According to one bibliographic database [375], articles in cell biology attract, on average, around five times as many citations as in mathematics. But articles in cell biology also tend to have shorter lives than in mathematics—periods over which they are likely to be of interest to and cited by the community.

You can judge the performance of your article in relation to its peers across different years. Some bibliographic databases make available for each research discipline percentile data, which give the minimum number of citations an article needs to receive in order to be in, for example, the top 25% or top 1% for each year after publication.

9.5.3 Personal Impact Measures

When you have been publishing for several years, you may want a measure of your overall publication activity and influence. The total citation count is unsatisfactory since the citations may be from just a few very highly cited articles or from many moderately cited ones. One popular alternative measure is the h-index [166]. It is equal to the maximum number of articles, n say, that have each been cited at least n times. It has the advantage that it takes into account both citation count and sustained productivity.

Values of an author's h-index are offered by Web of Science [375], Scopus [326], and Google Scholar [138]. All three indices may differ [18].

The disadvantage with the h-index is that it depends on research area and on age, although it can be normalized by both [299]. Average h-indices for the main research disciplines can be extracted with queries made through Google Scholar [361]. Thus in 2016, the average Google Scholar h-index for neuroscience was about 37 and for mathematics about 15.

Other indices are available offering improvements on the h-index [100] [166] [299].

9.5.4 Author Identity

Personal measures of research impact such as citation counts and h-indices are helped by a unique public identity. With it, your publications can be correctly attributed in online databases, retrieved in electronic searches, and credited in citation analyses. Without it, procedures of varying sophistication may be required for databases to identify and distinguish you from authors whose names may appear with differing numbers of initials, prefixes, suffixes, hyphens, diacritical marks or accents, apostrophes, and hyphens.

To reduce ambiguity, some authors include a middle initial or initials if they exist [79, p. 50]. Some authors with a two-syllable given name may introduce hyphens—for example, Xiao-Ling Sun [344]—to improve indexing with initialled given names. Other authors may add an initial to their given name.

Even if your own use is consistent, some databases may record a family name with two unhyphenated components (e.g. one paternal, one maternal) as if the first component is a given name. Some authors therefore concatenate a two-component family name with hyphens, for example, Iván Marín-Franch [375], or settle on one component of the family name.

Whatever you decide, be consistent. Use the same form of institutional address or other affiliation, and do not add ephemeral research group names or other associations.

One way to eliminate all ambiguity is to use a unique author identifier. The major identifiers are Author Identifier [235], which is used in Scopus; ResearcherID [306], which is used in Web of Science; and ORCID [273], which is independent of platform and to which both ResearcherID and Author Identifier can be linked (ORCID stands for Open Researcher and Contributor ID). Establishing each of the identifiers is straightforward. Errors in attribution in online databases can then be routinely corrected.

9.6 Checklist for Publication and Promotion

Submission

1. Paper complies with the journal's guide for authors? Style and content compatible with other articles?
2. Abstract self-contained, correctly formatted, and not over length?
3. Duplicated text eliminated or declared?
4. Permission obtained for copyrighted material?
5. Ethical requirements satisfied? Conflicts of interest identified?
6. Quality of the exposition professional?

Revision and response to reviews

7. Revised manuscript meets each reviewer's concerns?
8. Responses explain each correction and where it can be found?
9. Text modified to explain disagreements with reviewers?
10. Easy for reviewers and editor to verify your corrections and amendments?

Proofs

11. Proof checked for content and logic?
12. Proof compared word-for-word with the original manuscript?
13. Sizing, location, and colour rendering of the figures correct?

Promoting

14. Courtesy copies of the article sent to authors you cite and know?
15. Copy of the manuscript or published article uploaded to institutional, personal, and other websites permitted by the copyright licence?
16. Article mentioned in your conference, workshop, and seminar presentations?
17. Any open access data or software linked to the article?

10

Posters

In a poster presentation at a conference or meeting, you describe your research to small groups of participants who gather around your poster, which is usually mounted on a display board. The poster contains a mixture of text, figures, and other data that support your presentation. Like lectures and talks, poster presentations allow you to speak directly to your intended audience, but in an informal, more intimate, self-paced way. They can generate focused and productive discussion and provide an opportunity for you to meet members of your professional community at all levels of seniority [79].

A poster presentation is sometimes thought of as being less prestigious than an oral presentation [91]. In reality, it can be a more effective means of communication, albeit to a potentially smaller audience [79]. In this respect, a poster session is neither better nor worse than a lecture or a talk, merely different in style.

Although there are occasional parallels with the material in Chapter 11 on lectures and talks, this chapter is concerned solely with posters. Section 10.1 deals with the preparation of posters and Section 10.2 with their presentation. Section 10.3 suggests how you can add value to a poster presentation. Section 10.4 contains a checklist for the items in this chapter.

10.1 Preparation

A poster is not a journal paper. Though accuracy is essential, the detail you supply need not be complete—just enough to support your presentation and to make sense to a casual viewer [79].

Consider how your poster will be seen by different members of the conference or meeting [91]. Some participants may wander randomly from one poster to the next, stopping when a title or image attracts their attention. Others may be more purposeful, intent on learning, for instance, about a specific methodology. Yet others may be more involved and want the full story. The content and design of your poster need therefore to fulfil several functions.

10.1.1 Content

The main sections of a poster can be similar to those for a journal paper. For experimental work, you might have an introduction, a methods section, results, comment or conclusion, references, and acknowledgements [289], but all greatly simplified. For more theoretical work, you can have a freer structure, though still with an introduction and comment or conclusion. An abstract is unnecessary, unless required by the conference organizers [79]. Contact details should be included and a web address, if available.

As with lectures and talks, the introduction and conclusion are the most important parts of the poster for communicating your findings. The introduction explains the research question or other reason for the work, and the conclusion summarizes the outcome or implications. The space between the two should be occupied mainly by results and analysis [79], which will be the principal point of reference for your explanation. The general advice is to resist the temptation to insert as much detail as possible. Rather, you should select material sparingly and arrange it so that the logical structure is immediately evident [91]. A disorganized collection of figures, images, and text is likely to irritate the viewer.

Your poster may be photographed. With much of it consisting of results and analysis, it represents a readily accessible account of your research. You might therefore want to be cautious in presenting unpublished material, unless you are about to submit a report for publication, or have something relevant in-press, or available in the conference proceedings. Some conferences and meetings encourage electronic versions of posters to be uploaded to an associated website. If an explicit record of this kind is made accessible, it establishes your priority as the creator of the work [292].

10.1.2 Design

Before starting your poster, establish its maximum size from the conference instructions (and any limitations by the printing resources available to you). Find good examples from previous conferences to act as a guide to style, detail, and visibility, since expectations vary across research fields and conferences. Even so, there are some commonly accepted principles:

- Use images and figures more than text [79].
- Ensure that images and figures are uncomplicated and understandable from their captions alone; some guides recommend that graphs contain no more than three curves [289].
- Omit information-free decoration, unnecessary colour, patterns, and background images.
- Avoid colour combinations such as red and green that are hard to distinguish by viewers with colour vision deficiencies [353].

Use section headings in same way as you would for a journal paper, but avoid using subheadings within sections, except perhaps for methods. Sections should be brief, a few sentences each [79].

In order to guide the reader's eye from section to section, ensure that the layout of the poster provides clear visual cues. Should the reader work from top to bottom or left to right or something else? Numbered sections and physical proximity can both define the required route.

10.1.3 Text

The text in your poster should be large enough to be read by visitors at the outer edge of your audience, at say 2–3 m [288] [313]. The title should be visible from a still greater distance, at say 3–5 m [79, p. 181] [313].

Conference guides sometimes suggest a minimum font size [288] [291], though the optimum is closer to 1.5–2 times the minimum [313]. Font sizes are expressed in points (§8.6.3). Thus:

- Text in the body of the poster should be 24–36 pt for maximum visibility.[86]
- Text in the title should be 70–120 pt, with the authors and affiliations in a somewhat smaller font.
- Titles and section heads should be in a sans-serif font, e.g. Arial or Helvetica [289]. Text in the body can be in either a sans-serif font or a serif font, e.g. Times Roman.
- Avoid text set entirely in capitals as it is harder to read than lower case.
- Avoid coloured text on coloured backgrounds that affect legibility [289].

Although its main role is to support your presentation, the poster should be able to act as an extended abstract [79]. You may have visitors who for one reason or another do not hear or participate in your presentation yet who want to retrieve the gist. If it is too difficult, they will move on to the next poster.

10.1.4 Copyright and Plagiarism

Ensure that you do not include any images or line drawings protected by copyright. If you do want to include copyrighted material, then request permission from the copyright holder, usually the publisher (§12.2.7). The author of the material may also be among the visitors to your poster. Likewise, do not include any text from another source that is not properly attributed (§12.2.1–§12.2.4).

10.2 Presentation

Despite its informality, a poster presentation still needs to be prepared and rehearsed. But it is not a lecture, and your delivery—the way you communicate your findings—will vary with the visitors to your poster. How you interact with them involves a balance: you need to encourage them to learn more, though not make them feel obligated or trapped. Exercise restraint.

10.2.1 Rehearsal

You should be thoroughly familiar with the material in your poster. Poster sessions allow visitors to probe more deeply and persistently than would be possible in a lecture or a talk. Try to anticipate the questions you might be asked, especially those to do with methods and analysis. Invite your colleagues to identify the strengths and weaknesses of your presentation and to ask challenging questions about content.

10.2.2 Delivery

Let visitors pause at your poster long enough so that their interest is evident, and then ask if they have any questions or would like an explanation [91]. They will probably say yes (but do not be disheartened if they say no). Be concise, focused, and responsive:

- Ask about their research background so you can adapt the technical level.
- Ask if they want just a one-line summary, a take-home message.
- If they want more, extract the key points, for instance, 'our main aim was . . .', 'we used this method . . .', 'the important result was . . .', 'its implications are . . .'.
- Pause after each point to see if they want clarification or to move on.
- Adjust your pace continually, omitting details unless they ask.
- If they question something, answer honestly and avoid being defensive, and if appropriate, ask for their interpretation [79, p. 182].
- Accept suggestions for further work with thanks.
- Some visitors may want to monopolize your time; moderate the questioning so that others are not excluded [288].
- Be alert to signs of diminishing interest; your poster may be one of very many your visitors want to attend [91].
- At the end, thank them for their attention; if appropriate, you might exchange business cards or contact details.

You will have done your job if your visitors leave the poster remembering your take-home message and without feeling it was hard work. They may later use your findings and cite them in their own publications. They may even become collaborators or employers [79, p. 182].

10.2.3 Supporting Material

Next to your poster on the display board, place a sheet of paper or card with a heading of the form, 'Leave email address for related PDFs'. You can send PDFs of your related research articles after the session or when your visitors have returned home.

If you want to offer visitors a permanent record of your poster, leave A4 copies in a folder on the display board and include contact details and a web address, if available, for supporting material [91]. Although most of the handouts will eventually be discarded, some will be read again, and may lead to more interaction.

10.3 Adding Value

As with lectures and talks, poster presentations can provide useful feedback and inspiration for when you prepare a more comprehensive report or embark on further work. Since a poster session is much longer than an oral presentation, keep a record of questions, suggestions, and criticisms as they are made; you may forget them by the end of the session. Also consider corresponding with interested visitors when the conference or meeting is over. Remember to give due credit when you make use of their advice in a paper.

10.4 Checklist for Posters

Content

1. Research question explained in the introduction?
2. Outcome described in the conclusion?
3. Most space devoted to results?
4. Sections ordered logically?
5. Understandable without a presenter?
6. Unpublished work ready to submit or safely in-press?

Poster design

7. Maximum size used?
8. Mainly simple images and figures, understandable from the captions?
9. Clear cues to guide the reader's eye from section to section?

10. No decoration, patterns, and background images?
11. Title set in 70–120 pt, body text 24–36 pt?
12. No images protected by copyright?
13. No spelling errors?

Rehearsal

14. Presentation seen by colleagues?
15. Questions anticipated, especially about methodology?
16. One-line summary or take-home message prepared?

Delivery

17. Interests and expertise of the audience identified?
18. Main points of the research emphasized?
19. Questions invited?
20. Pace and level of detail adjusted to the audience response? Lecturing style avoided?
21. Advice and suggestions encouraged?
22. Thanks offered at the end?
23. Contact details exchanged?
24. Handouts and mailing list next to the poster?

Added value

25. Questions, suggestions, and criticisms noted during the session?
26. Correspondence pursued with interested participants after the conference?

11

Lectures and Talks

Lectures, talks, and other kinds of oral presentation have advantages for disseminating your research. Whether the occasion is a large conference or a small seminar, you can speak directly to your intended audience and draw on a range of display technologies, including animation, video, and physical models [4, p. 208]. The case you make can be more compelling than would be possible with the written word [7, p. xi]. And the response of the audience can be informed and stimulating.

But lectures and talks have disadvantages, too. The audience has only one chance to hear what you say [4, p. 209]. If they fail to follow, they cannot turn back and reread the material as they would with a journal article. Nor can they adjust your pace to suit their needs. They are committed to remaining in place for the duration of your talk.

Accordingly, an oral presentation needs to be more than a mere recitation of a written account [90]. Although it provides its own special challenges, it can prove to be an enjoyable and rewarding experience.

There are occasional parallels with the material in Chapter 10 on posters, but this chapter is concerned solely with lectures and talks. Section 11.1 deals with the preparation of lectures and talks, and Section 11.2 deals with their presentation. Section 11.3 identifies potential failure points and how they can be avoided. Section 11.4 suggests how value can be added to your presentation. Section 11.5 contains a checklist for the items in this chapter.

11.1 Preparation

Like a research article, a lecture or talk should make clear to the audience the reason for the work, the approach, the main results, and any implications [258, p. 172]. But unlike a research article, an oral presentation is not a document of record. Rather, it is an account intended to convince the audience by verbal and nonverbal means [90], albeit governed by the time available, the aids at your disposal, and the nature of the audience. The content and presentation need to accommodate these constraints.

11.1.1 Content

As with writing, you need to have something to say [7, p. 12]. The two most important parts of your presentation are the explanation of the research problem or other motivation for the work and the description of the outcome or the implications of the work [79]. You can therefore omit much of the background review (unless it is the subject of your presentation) and methodological detail [90] (unless it, too, is the subject of your presentation). Include just enough to convince the audience of your competence. If someone wants the detail, they can ask during the question period or later.

Your presentation is normally supported by a digital slide display, video, or both. A slide is a single screen of information with text and graphics, with various levels of animation.[87] The number of slides you need naturally depends on the allotted time and on how much detail you put into each slide. But avoid cramming in too much material [73, ch. 4]. As a rough guide, you might aim for one slide a minute [90], and some learned societies insist on lower rates still [286]. Nonetheless, it is sometimes easier to explain a complex idea with several simple slides than with one complicated slide [7, p. 115].

As with posters, your slides may be photographed. Give some thought to how much unpublished material you want to present, unless you are about to submit a report for publication, or you have something relevant in-press, or available in the conference proceedings. Some conferences record presentations and others encourage slides to be uploaded to an associated website. If accessible, these records establish your priority as the originator of the research [292].

11.1.2 Organization

It is customary to start with a title slide, which contains your name and affiliation and any coauthors. You would normally have received an introduction from the person chairing the conference or meeting, but a title slide helps to orient the audience [258, pp. 172–3] and allows you to acknowledge collaborators and the support of funding agencies, if you wish. Keep these preliminaries brief. The opening is precious and should not be wasted [7, pp. 156–7].

You might use the next slide to outline of the structure of your talk, the topics to be addressed, or the motivation for the work. You can follow it with a more detailed plan so that the audience can anticipate your progress [4, pp. 210–11]. Even so, audiences are unlikely to remember the detail or indeed the outline structure. Whether you decide to anticipate the outcome of the work, as with a journal article, is matter of personal choice.

Instead of outlining a somewhat formal structure, you might choose a more natural, expository style by turning first to the research question or problem, explaining it, and then describing how you are going to address it. If you take this approach and arrange your talk in sections, you can provide the motivation for each as you come to it. It imposes less on the audience's memory and encourages their continuing involvement.

Whichever approach you decide on, ensure in longer talks that you occasionally recapitulate the key points. Audience attention invariably lapses from time to time [73, ch. 9].

It is customary to end with a review or summary of what you have achieved and a conclusion, the take-home message [258, p. 173]. You may also want to identify some of the limitations or implications of the work, not least so that there is a ready starting point for questions afterwards. Some speakers prefer to give their acknowledgements after the summary slide (it is easier to say who did what), though it may test the patience of the audience and diminish the impact of your conclusion [7, p. 168]. Moreover, the summary or conclusion will not be visible on the screen during questions.

11.1.3 Text or Graphics?

Slides, or whatever other visual aids you use, need to communicate the main points clearly and effectively through images, graphs, diagrams, and of course text. But how much text and how much graphics?

A traditional text-dominated slide might contain a title and a bulleted list, somewhat like the outline of a research article, perhaps embellished with a small graphic. An alternative, graphics-dominated slide might contain a brief assertion as a heading, and consist mainly of images, graphs, and diagrams providing evidence in support of the assertion. This assertion-evidence approach may lead to better comprehension and fewer misunderstandings [128].

Keep in mind the linguistic competence of your audience, some of whom may not follow spoken English as well as the written word. Even in a graphics-dominated slide, include some explanatory text to aid interpretation, although it needs to be cogent and not duplicate what you say [7, p. 117]. In short, avoid extremes: neither all graphics nor all text [4, p. 215].

11.1.4 Memory Aids

Without experience in giving talks, you may find it hard to remember exactly what you want to say and in what order. You can deal with this challenge in several ways [7, ch. 16]:

- Prepare a short printed list of prompts on cards or a sheet of paper; they would usually be words or phrases capturing the key points or topics that you can consult at a glance.
- Use your display software or a separate monitor to view the prompts for each slide or to preview the next slide.
- Use the images and text in each slide to guide what you say; the last graphic or text on a slide can be marked discreetly as a prompt for the next slide.
- If a technical term or part of a graphic needs explanation, introduce an animated text box or graphic to act as a reminder.

Still, do not let the aids take over. Avoid focusing your attention too long on the printed prompt list, projection screen, or computer monitor, to the exclusion of the audience. Otherwise they will disengage.

11.1.5 Slide Design

Whether your slides are mainly text or graphics, they need to take into account the limitations of the projection system and viewing conditions. Display resolution is often poorer than with print media, and each slide is available for inspection by the audience only briefly. Graphs need therefore to be readily comprehensible. They should be prepared afresh and not taken directly from a journal manuscript or article [79].

More than with a journal figure, the notion of maximizing Tufte's 'data-ink ratio' is relevant (§8.3). Thus if a slide is imagined as being drawn in ink, the ratio of ink showing information to ink for the whole slide should, within reason, be a maximum.

What does all this mean for your slide design?

- Prefer images, graphs, and diagrams to text: they attract attention more than words, are more informative, and easier to remember [4, p. 213].
- Avoid clutter and overcrowding [4, p. 216] [79, p. 175]: omit all information-free decoration, unnecessary colour, patterns, and logos—they occupy useful space and confuse the eye.
- Choose text in no more than two font sizes and colours. Recommended font sizes are 18–36 pt or even larger [4, p. 216] [7, p. 121]. In practice, the smallest text in the body of the slide should be 18–22 pt for maximum visibility in a moderately large theatre.[88]
- Use the same fonts and font sizes throughout.
- Avoid text set entirely in capitals and coloured text on coloured backgrounds that affect legibility [289].
- Avoid colour combinations such as red and green that are difficult to distinguish by viewers with colour vision deficiencies [353].
- Be economical with bullet points:[89] no more than about five bullet points and preferably fewer [215]; one or at most two lines of text in each bullet point. If possible, link bullet points to images, graphs, and diagrams in the slide.
- Simplify language: no need for proper sentences or determiners such as 'the' and 'a'.
- Do not use images or line drawings protected by copyright, unless approved by the copyright holder (§12.2.7). Because an image or drawing is available

on the web does not mean it can be used freely; moreover, the author of the material may be in the audience.

- Do not include any text from another source that is not properly attributed (§12.2.1–§12.2.4).
- Eliminate spelling errors, which can capture the interest of the audience [90].

The title or heading for each slide has a special role in setting the context or the implications for what is in the body of the slide. But it should not mislead, even indirectly. Any slide entitled 'Summary' or 'Conclusion' should really be your last dealing with content. To persist with more may prove an anticlimax for the audience.

11.1.6 Slide Animation

People can read faster than you can speak. While you are still developing the first point on the slide, they may have moved to the last.

To maintain audience focus and ward off reading ahead, exploit your presentation software:

- Use controlled animation to reveal the key points, e.g. graphics, parts of graphics, or bullet points.
- Do not over-animate: making every word or every element of an image appear in sequence is irritating.
- Make the transition from one bullet point to the next or one image to the next by appearance only, nothing more elaborate: no elements that fly in, float, expand, or contract—or anything else that diverts attention.
- Remind yourself where an animated list or the slide itself ends by adding a punctuation mark or symbol to the last item.

The transition from one slide to the next should also be by appearance only. Slide gymnastics distract the audience and diminish the significance of the content [7, pp. 124–5].

11.2 Presentation

For the presentation to be effective, you need to attract and hold the attention of the audience, which, if large, can be intimidating for an inexperienced speaker. Rehearsal is essential [7, ch. 12]. It is also worth considering your delivery, the way you communicate [4, pp. 217–20], especially your use of language and gesture, your posture, timing, and intonation.

11.2.1 Rehearsal

You may know your research thoroughly, but this is not the same as knowing how to present it. Your talk needs to be thoroughly prepared and rehearsed so that it makes sense and works in practice [286]:

- Invite your colleagues, in particular those unfamiliar with your work, to comment on the strengths and weaknesses of the presentation [7, p. 149].
- Encourage them to ask hard questions about the content.
- Test your timing beforehand, and be prepared to adjust the pace and detail as you interact with the audience. Identify material you can contract or expand as needed.
- If you use a printed list of prompts, practise speaking from them until you are sure that you can deliver the material without lapses [258, p. 171] and without rote learning.
- Test whether your slides are visible and your voice is audible.

If feasible, test your slides or video on the conference projection system before your talk. If you plan to use your own computer, ensure it is compatible with the conference system. If not, take care transferring your slide presentation from your computer to the conference computer: fonts may change, images may be distorted, and macros may fail. Some projection systems may not show high-resolution images. Most projection systems render colours and fonts differently from those on your own computer monitor.

11.2.2 Delivery

The audience expects your delivery to be authentic, that is, to reflect the fact that the research is your own. You should therefore not learn your material word-by-word nor read from written notes [90]. Either could be done by a substitute who knows nothing about the material. Indeed, one learned society expressly forbids presenters from reading their presentations.[90] Audiences can also tell when you are reciting from memory [7, p. 136].

Use your natural speaking style [4, pp. 217–18], with its customary pauses, repetitions, fillers, and corrections, augmented by the hand and arm gestures you would make in everyday conversation. You gain credibility because you draw the words genuinely from yourself, rather than from some prepared script. But avoid speaking too quickly. Slower speech improves your diction and the audience's comprehension [73, ch. 17].

Ensure that you can be heard by the whole audience, including the back row. Take a stable, confident, upright stance, and try to relax [7, p. 206]. Avoid 'pushing and squeezing' your voice to gain volume [243, ch. 12]. Instead aim for clarity and resonance, with your breath coming more from the diaphragm than from the upper chest. You will reach farther and be more pleasant to listen to. Ensure you are well hydrated [7, p. 186].

With practice, public speaking becomes easier, and there are techniques for reducing anxiety in these conditions [7, ch. 15] [73, ch. 24]. Remember that the audience is still made up of individuals, most of whom are sympathetic and want to hear what you have to say [258, p. 171]. Engaging with them is just the same as engaging with an individual but on a larger scale [73, pp. 208–9]. Try visualizing yourself working professionally through the material, developing the argument, making smooth transitions from one slide to the next [4, pp. 218–19] [258, p. 172]. Thorough preparation, knowing your material and its correct order, and rehearsing answers to questions will all add to your confidence.

Do not worry if English is not your first language. A slower delivery with the occasional pause for the right word makes a more effective and compelling presentation than one delivered at high speed in idiomatic English.

11.2.3 Practicalities

There is no one correct way to give a talk [7, p. x], but there are several general points to bear in mind:

- Move position only when you need to, for example, to gesture to something on the projection screen or to address a different section of the audience. Avoid relentless pacing [169, p. 530]: although it may help you think, it distracts the audience.
- Use a remote pointer or mouse to control slide animation and slide transitions: it is unobtrusive and makes your presentation smoother.
- With graphs, explain the axes and key elements first: you know what the graph means—the audience needs orienting.
- Ensure that you are familiar with each slide (no slide should surprise you).
- Be more relaxed about personal pronouns than in a research article (e.g. 'we then tested . . .') but do not lapse into extended personal reminiscence, unless that is the object of your talk.
- Never apologize for minor errors in your slides or in your delivery, for example, a mispronunciation or misreading [73, pp. 207–8]: the audience may not have noticed.
- Do not wave a laser pointer aimlessly around the projection screen: the audience cannot follow it. Instead, hold the pointer steady when highlighting something and then turn it off.
- Talk to the audience and keep eye contact [7, p. 48]: avoid drifting into private conversations with the screen, the floor, or your notes.

You may well not remember all this advice, especially when concentrating on the content of your talk. But you can ask colleagues to give you feedback either in rehearsal or after the presentation itself. Eventually these practical matters become automatic.

11.3 Potential Failure Points

Presentations can go wrong for many reasons. You may have a bad start because of problems with the projection system, or because you receive too many interruptions, or mishandle an explanation, or find a question unanswerable. These and other challenges are all manageable, though it is sensible to be properly prepared.

11.3.1 Beginning

- Ensure your slide presentation is ready to start: asking the audience to wait while your computer boots up or you search for the right video cable is discourteous.
- Decide beforehand on your opening remarks [7, p. 156]. You could announce one or two broad goals or an illustration of the research question or problem, e.g. 'This talk is about . . .' or 'The problem I want to discuss is . . .'. Be sure to explain why the topic is important [215].
- If you do decide to outline the organization, avoid an uninformative listing: 'First I give an introduction to the topic; next describe the methods; then the results; and end with a conclusion'. It will quickly be forgotten by the audience, and you will have wasted valuable time (§11.1.2).

11.3.2 Middle

- If you are interrupted, keep a mental note of what you were saying so that you can quickly resume.
- If a questioner is persistent or your answer would be too much of a digression, suggest that you discuss the matter privately afterwards (the questioner may simply have misunderstood and needs a more detailed explanation).
- Monitor your progress against the clock and contract or omit material if necessary.
- Do not overrun [73, ch. 4]. It may limit the time for the next speaker; the audience may miss a talk elsewhere; questions may be disallowed; and

worst of all, you may be stopped by the moderator from completing your account [79, pp. 174–5]. Offering the excuse that you have too much material or insufficient time will not be greeted sympathetically by the audience.

11.3.3 End

- Your last slide may be left displayed during questions. If it is your summary or conclusion slide, it will reinforce the take-home message. If you add your contact details or web page, it may also encourage more interaction after the talk.
- Do not worry about questions. You know your own work better than anyone else, and audiences are more likely to offer supportive comment than criticism with less-experienced speakers.
- Make notes if a question has several parts: you can then address each in turn.
- Do not answer immediately or before the questioner finishes. Instead, pause, think, and then answer carefully what was asked, not what you might have wished for, and repeat the question if it was unclear to the audience [90]. Return to a specific slide that supplies the answer but only if it is immediately available, not if it entails searching back and forth through your presentation sequence.
- Avoid ungracious jokes [7, p. 56] and unhelpful clichés ('that's a good question').
- If you do not know the answer, be honest. If it is something you could address later—for example, by consulting your records—then offer to do so [90]. Alternatively, if it is something that no one can answer, explain why. Whether your ignorance is justified or not, avoid rambling or bluffing [79, p. 178].
- Accept suggestions for further work with thanks.

11.4 Adding Value

As with poster presentations, conference and seminar presentations can provide useful feedback and inspiration for future research. When your presentation is complete, make a note of the questions and criticisms you received and your answers. Both may be useful if you prepare a more comprehensive report or embark on further work. But remember to give due credit. Also consider corresponding with interested participants when the conference is over.

11.5 Checklist for Lectures and Talks

Preparation

1. Research question explained, outcome described?
2. Main elements of the argument identified?
3. Unpublished work ready to submit or safely in-press?

Slides

4. Simple images, graphs, and supporting text?
5. No decoration, no clutter?
6. Text 18–36 pt, consistent style?
7. No more than six bullet points a slide, each 1–2 lines, linked to graphics?
8. Animation by appearance only, no fade, movement, or other distractions?
9. Only the main graphics or bullet points animated?
10. No images protected by copyright?
11. No spelling errors?
12. Summary slide last?

Rehearsal

13. Prompt list or slides chosen as memory aids?
14. Number of slides suited to the allotted time?
15. Clarity and audibility satisfactory?
16. Slide order rehearsed?
17. Presentation seen by colleagues, questions anticipated?
18. Conference projection system tested?

Delivery

19. Natural presentation style supported by slide prompts?
20. Graphs explained before interpretation?
21. Laser pointer held steady?
22. Talk directed to the audience, eye contact maintained?
23. Pacing minimized?
24. Hand and arm gestures for authentic communication?

Avoiding failure

25. Slide presentation ready to start?
26. Introductory remarks prepared?
27. Progress monitored against elapsed time?
28. Persistent interruptions deferred?
29. Summary slide displayed during questions?
30. Questions answered succinctly, but with thought?
31. Ignorance acknowledged, especially if universal?

Added value

32. Questions and criticisms noted after the presentation?
33. Correspondence pursued with interested participants after the conference?

12

Ethical Issues

Research is expected to be conducted in an ethical way and the results of that research are expected to be presented honestly and transparently, with authorship reflecting each individual's contribution to the work [369]. This chapter is concerned with ethical conduct and misconduct and how it affects communication in science and engineering.

Wide-ranging advice on ethical standards in reporting is available from governmental organizations [314], learned and professional societies [59] [182] [294], publishers [81] [275] [295], and individual research institutions and scientists. Many publishers are members of the Committee on Publication Ethics (COPE), which offers advice to editors and publishers on research and publication misconduct. Discussion of individual cases of misconduct is available on the COPE website [62].

Section 12.1 identifies the main elements of research misconduct. Section 12.2 deals with plagiarism and its variations, and Section 12.3 reviews the consequences of research misconduct in published works. Section 12.4 considers the requirements for ethical research with human and animal subjects. Section 12.5 deals with citing and referencing and Section 12.6 with authorship. Section 12.7 considers data analysis and storage. Section 12.8 reviews issues to do with submission and publication. Section 12.9 contains a checklist for the items in this chapter.

12.1 Elements of Research Misconduct

There is no standard definition of research misconduct. From [117] [327, §3.1] [339, ch. 2], the main elements are the following:

- fabrication: invention of data or results and recording or reporting them;
- falsification: alteration of materials, equipment, procedures, data, or results so that the research is not accurately represented in the research record;
- plagiarism: taking of another person's ideas, processes, results, or words without giving appropriate credit;
- mistreatment of research subjects: activity that departs from the ethical standards of the appropriate institutional or national committees on research with humans and animals.

Research misconduct does not include honest errors or differences of opinion [117]. It does, however, take place on a continuum, with some forms of misconduct less obvious than others. For example, it includes replacing the data points in a noisy plot by a smoothed curve which is thought to be more representative but which in the absence of the data points gives a partial view of the original data. It includes the computer processing of images that brings out the salient aspects of the data, but which suppresses other aspects. It also includes a range of improper statistical practices such as significance chasing (§7.9.2).

Many examples of misconduct are available on the website Retraction Watch [308], which reports on retractions in the scientific press.

12.2 Plagiarism and Its Variations

Ethically, there is an implicit contract between the author and the reader in which the reader assumes that the author is responsible for all the material that he or she does not explicitly attribute to others [314]. Plagiarism is a violation of that contract.[91]

Plagiarism comes in many forms [6]. The most frequent textual varieties are word-for-word copying (direct, verbatim, or literal plagiarism), interleaved word-for-word copying from multiple sources (patch or mosaic plagiarism), and copying with word or phrase replacement while preserving the original structure (synonym or paraphrase plagiarism) [158] [203] [314]. Equally important, yet sometimes given less attention, are the nontextual forms of plagiarism, such as plagiarism of an article's structure, viewpoint, and theme, and of its content, the methods, results, and analyses. In all cases, the ethical course is to appropriately acknowledge the contributions of others, whether theoretical or practical, published or unpublished [314].

The interpretation of plagiarism varies with discipline, with journal, proceedings, or other work, with publisher, and with time [61] [172]. Occasionally contentiously, plagiarism may be taken to include what is called self-plagiarism—passing off one's own previously published work as though it were new [295] [314]. Illustrative individual cases of plagiarism are analysed by COPE [62].

The expectation that other people's ideas and work should be acknowledged does not extend to material that has achieved the status of common knowledge (§3.10). Such material can be used without citing the original studies. In cases of doubt, the consensus is to cite the source [158] [314].

12.2.1 Tests of Plagiarism

How much text can you safely copy with attribution and how much without attribution? Extreme instances of plagiarism are easy to identify. Examples 12.1 and 12.2 give an illustration of almost verbatim plagiarism of original material but in an unusual order of publication.

Alteration — Colon cancer tissues were acquired from a total of 44 male patients with an average age of 67 years (range, 49 to 77 years) with newly diagnosed, pathologically
Alteration — confirmed colon cancer. . . . We detected miRNA-302a expression levels in 44 colon tissues and 10 normal colon — Alteration
✕ tissues and found that, compared with normal colon — Alteration
Alteration — tissues, colon cancer tissues expressed lower levels of miRNA-302a (Fig. 1A).

Example 12.1. *Duplicated and altered text from the results section of a retracted multidisciplinary sciences article [682] published in 2014.*

The text in Example 12.1 is from the plagiarized article published in 2014 [682]. The underlining shows where the original text was altered.

> PCa [prostate cancer] tissues were acquired from a total of 44 male patients with an average age of 67 years (range, 49 to 77 years) with newly diagnosed, pathologically confirmed PCa. . . . We detected miRNA-302a expression levels in 44 PCa tissues and 10 normal prostate tissues and found that, compared with normal prostate tissues, PCa tissues expressed lower levels of miRNA-302a (Fig 1A).

Example 12.2. *Original text from the results section of a multidisciplinary sciences article [731] published in 2015.*

The text in Example 12.2 is from the original article, published in 2015. It describes an investigation of prostate cancer, not colon cancer as in the plagiarized version of Example 12.1. It was the earlier 2014 article—not the subsequent 2015 article—that was subjected to penalty, namely retraction (§12.3.2). The editors of the journal PLOS ONE [491], in which the articles appeared, had to investigate to determine which of the two was the plagiarized version. The case is discussed on Retraction Watch [308].

Even if large portions of text are duplicated, then as long as some changes are introduced, the result is not inevitably seen as plagiarism, at least by some potential authors, according to a survey summarized by Gibney [133] with the unhappy headline 'I'm no plagiarist, I moved a comma'.

At the other extreme, some guides insist that there is no acceptable minimum level of copying that does not require attribution [114]. Certainly, decisions about modest amounts of text duplication may be difficult. Plagiarism-detection software [191] is able to identify duplicated text and provide numerical measures of the degree of similarity overall; yet whether a particular similarity score for a paper signifies plagiarism requires human judgement. Nominal threshold scores are unreliable: a 100-word paragraph of duplicated text may amount to only 1% of a paper and still be plagiarism. A decision depends not just on the size of the score (though high scores may suffice) but also on how systematic the duplication is, what it duplicates, whether and in what form it references the original work, and whether there is some explanation for the duplication.

In a simpler approach, Knuth and his colleagues [207, p. 68] cite a test for plagiarism from Prentice-Hall's *Guide to Authors*:

> A reasonable person could not believe that you didn't have the other chap's book open in front of you as you wrote yours.

To pass this test, that is, to show that the statement is false, you need to understand the original and express that understanding in your own words. With due reference to Searle [56], an alternative test for plagiarism might therefore be expressed thus:

> Could the text have been produced from the original by a machine using a look-up table of synonyms and substitute phrases?

One or other of these tests captures most varieties of text plagiarism, notwithstanding some potential authors' uncertainty about the meaning of plagiarism [133].

Plagiarism of the literal kind is not restricted to text. Articles have been retracted for allegedly using figures from other publications without acknowledgement [440] [441] [647]. Reproducing figures and images without permission of the copyright holder can also constitute an infringement of copyright (§12.2.7).

12.2.2 Self-Plagiarism

Given that plagiarism is defined as taking someone else's work and passing it off as your own, it may be argued that self-plagiarism is impossible. You cannot steal from yourself [314]. The more neutral term 'text recycling' is accordingly sometimes preferred [61] [172]. The more important issue, though, is whether there is an intention to deceive; that is, the author reproduces his or her previously published work in a new article without informing the reader that it is available elsewhere [314].

Examples 12.3 and 12.4 illustrate text recycling.

> This paper presents a model for aircraft flap side edge noise prediction. The flap side edge noise is modeled as two components, respectively for the low and high frequency domain.

Example 12.3. *Original text from the abstract of an acoustical engineering conference article [526] published in 2011.*

The text in Example 12.3 is the original material, from a conference article [526].

> This paper presents a model for aircraft flap side edge noise prediction. The flap side edge noise is modeled as two components, respectively for the low and high frequency domain.

Text same as in Example 12.3

Example 12.4. *Duplicated text from the abstract of an acoustical engineering article [527] published in 2013.*

The identical text in Example 12.4 is from a later journal article [527]. It was then retracted on the grounds [528] that it duplicated significant parts of the earlier publication by the same author [526]. The notice of retraction (§12.3.2) is reviewed on Retraction Watch [272].

12.2.3 Repairing Verbatim Plagiarism

How can plagiarism be avoided? Example 12.5 is a verbatim extract from Montgomery's *The Chicago guide to communicating science* [258, p. 81] and is part of a discussion of the differences between the primary literature such as reports of original results and the secondary literature such as reviews. Suppose that this material is to be legitimately incorporated into an imaginary article.

> Over the past several decades, these distinctions have begun to soften. This has happened in direct response to the changing, indeed broadening, realm of scientific publication as a whole.

Example 12.5. *Original text from* The Chicago guide to communicating science *[258, p. 81].*

Examples 12.6–12.9 show progressive attempts at achieving that goal.

Simply reproducing the text in Example 12.5 would obviously constitute verbatim plagiarism. But suppose that a citation to the source is added, as in Example 12.6.

✕

> Over the past several decades, these distinctions have begun to soften. This has happened in direct response to the changing, indeed broadening, realm of scientific publication as a whole <u>(Montgomery, 2003, p. 81)</u>.

Citation added

Example 12.6. *As in Example 12.5 with an added citation.*

The text now acknowledges the author, but it still constitutes verbatim plagiarism. All the citation tells the reader is that the ideas are due to Montgomery, not the expression of those ideas. Also the fact that the original was published in 2003 makes the phrase 'past several decades' in the verbatim text ambiguous for something read at a much later date.

Suppose, instead, that some of the words and phrases in the original in Example 12.5 are replaced by synonyms and phrases with similar (though not quite identical) meaning, as in Example 12.7.

Synonym substitution   Synonym substitution

✗

> In the last third of the twentieth century, these differences began to blur. This has been a simple consequence of the developing and widening domain of scientific publication in general (Montgomery, 2003, p. 81).

Synonym substitution   Synonym substitution   Synonym substitution

Example 12.7. *As in Example 12.6 with phrase substitution.*

One of the changes in Example 12.7 is that the original 'past several decades' has been replaced by 'last third of the twentieth century', which is then correct for something written in 2003. The perfective 'have begun' (§5.2.3) has been replaced by the simple past 'began', which further improves the sense. The structure, however, stays the same as in the original, and even with the acknowledgement it still constitutes plagiarism, now synonym plagiarism. Other attempts at using paraphrasing intended to defeat plagiarism are described in [314].

Suppose, now, that quotation marks are added to the original text, as in Example 12.8.

Quotation mark

> 'Over the past several decades, these distinctions have begun to soften. This has happened in direct response to the changing, indeed broadening, realm of scientific publication as a whole' (Montgomery, 2003, p. 81).

Quotation mark

Example 12.8. *As in Example 12.6 with added quotation marks.*

With quotation marks and the acknowledgement, the text no longer constitutes plagiarism. Yet the effect is unconvincing, since it adds nothing of value to the report in which it appears. Unlike the quotations by Gibney and by Knuth in Section 12.2.1, the text is not particularly pithy or aphoristic (nor presumably was it intended to be). Instead, the reader is left with the impression of the imaginary author's perfunctory view of communication as an exercise in copy and paste.

In reality, quoted sections of text appear rarely in scientific and technical papers, and then only for specific reasons: for example, as a definition in a statement of a research question or problem, usually by an authority; as a piece of text to be critically analysed; and as an epigram or aphorism to encapsulate the work. Unsurprisingly, the introduction of quotation marks in a corrigendum to retroactively correct the publication of duplicated text [717] [718] is viewed with disapproval by many commentators [239].

So, finally, what might be an acceptable representation of the content of Example 12.5? There are two basic points in the original text: the distinction or lack of it between primary and secondary sources, and changes in the practice of scientific publication. A possible formulation, therefore, is as in Example 12.9.

> Scientific publication has been evolving, becoming more inclusive. As a result, the distinctions between primary and secondary sources are not as sharp as they used to be (Montgomery, 2003, p. 81).

Example 12.9. *Recast version of original text in Example 12.5.*

The wording and structure are no longer those of Example 12.5. The original and problematic 'Over the past several decades', which was contingent on the date of the original work, has here been replaced by a more general 'evolving' and linked to 'Scientific publication'. And the original 'these distinctions' has been replaced by a more specific 'the distinctions between primary and secondary sources'. Despite the differences in emphasis, the ideas are still manifestly those of Montgomery, as signalled by the citation. As part of a longer piece, the recast version might differ even more from the original.

12.2.4 Necessary Duplication

A case may be made that some text duplication is necessary, most commonly in the description of identical experimental methods, procedures, or analyses. Even then, this material should not be passed off as being original [61].

All that is needed is a statement that the methods, procedures, and analysis have been described previously, with a citation, and that the account in the current work is a reproduction [293, §1.10]. In Example 12.10, the authors have been transparent in identifying both the source of the experimental description and the fact of its reproduction.

Reproduction declared — Source identified

> The present work is a continuation of earlier optical investigations in Ar^+ irradiated Si reported in [19]. For the sake of completeness, the experimental details are reproduced from it. Si(1 0 0) single crystal wafers were used as the starting material.

Example 12.10. *Transparent description of experimental details, from an instrumentation engineering article [683].*

Text duplication is also necessary in some mathematical work [366]. Because the form and content of mathematical expressions are so intimately related, it does not make sense to distinguish them.

Reproduction declared — Source identified

> The solution of the equation is reproduced from the work of Boyd et al.[22] as
>
> $$F_t = 1-(6/\pi^2)\sum_{n=1}^{\infty}(1/n^2)\exp(-n^2Bt) \qquad (2)$$
>
> where
>
> $$F_t = Q_t/Q_e \qquad (3)$$
>
> and
>
> $$B = (\pi^2 D^i)\ /\ r^2 \qquad (4)$$
>
> where . . .

Example 12.11. *Transparent reproduction of a mathematical equation, from a chemistry article [636].*

In Example 12.11, the source of the equation is made evident with the phrase 'reproduced from', although the authors introduce a working variable B that is absent in the original formulation [425]. With or without the working variable, the citation is sufficient.

12.2.5 Conference Publications

Conference presentations are attractive for several reasons, but they have disadvantages as a means of publication (§9.1). Conference papers are generally short, perhaps taking the form of an extended abstract; they are available primarily to conference attendees; and they may not represent a definitive version of the work, the article of record. Later, therefore, they are often expanded and submitted for publication in a journal [61] [174] [314].

Presentation at a conference, however, usually constitutes publication. The abstract appears in the conference programme or book of abstracts and the conference paper appears in the conference proceedings.

An expanded version of the work submitted to a journal must acknowledge this prior publication, whether it is an abstract or a paper. A statement needs to be included along the lines that some of the work has been previously reported, in preliminary or partial form, in the published programme, book of abstracts, or proceedings [314].

In Example 12.12, a statement about partial previous publication is placed at the end of the introduction of the article [451]. Reference 13 in the example gives all the required information about the conference article; namely, the authors' names; abstract title and number; conference name, place, and date; and a web link.

> A partial account of this work was presented at the Society for Neuroscience Annual Meeting 2010 [13].

Example 12.12. *Statement describing partial prior publication, from a neuroscience article [451].*

In Example 12.13, the statement is placed in the acknowledgements section of the article [495] instead of in the introduction. It, too, gives enough information for the conference to be identified and the article traced.

> A partial report of the data was presented at the 12th Conference on Laboratory Phonology, Albuquerque, New Mexico, July 2010.

Example 12.13. *Statement describing partial prior publication, from an experimental psychology article [495].*

It is important that the version of the paper submitted to the journal represents a significant advance on the work reported at the conference [174]. As noted in Section 9.3.3, the journal may want a copy of the publication [186] and possibly an explanation of how the submitted paper differs [178].

12.2.6 Review Articles and Opinion Pieces

Review articles and opinion pieces are sometimes claimed to be exempt from concerns about text-recycling or self-plagiarism. One rationale is that they are necessarily derivative, so originality is less relevant. In fact, reviews and opinion pieces are expected to follow the same principles as research articles [61]. That is, the analysis and synthesis of previously published work should be fresh and original, and not a rehearsal of old material [366].

A second rationale by authors for self-plagiarism is that the original text expresses the author's thoughts best and cannot be improved by rewriting.

A counter to this view was posted by Seidman [61] under COPE's text recycling guidelines:

> You'd think that researcher/authors with MDs and PhDs would be bright enough to know how to reword.

If truly your text cannot be improved or adequately reworded, then you can either (1) omit the previously published text and direct the reader to the original or (2) preface the previously published text by an explicit statement of the kind 'The following material is reproduced from . . .', much as with the repetition of previously reported methods (§12.2.4). As a device to explain repetition it should be used sparingly; otherwise the review becomes a pastiche.

Care does need to be exercised with derivative material. Review articles have been retracted for reproducing material from other published works [579] [684].

12.2.7 Reproduction of Figures and Images

In practice, the reproduction of figures and images is more complicated than the reproduction of text. The reason is that the appearance of a figure or image (the expression of the data) is usually protected by copyright [189] unlike the content (the data). The copyright is normally held by the publisher, although the author may have a licensing agreement that permits more general use (§9.3.10).

Unauthorized reproduction is an infringement of copyright, whether or not the source is acknowledged. Infringement occurs if the figure or image is reproduced as such; for example, by electronically extracting it from the original or by scanning a print copy, or if it is redrawn so that retains the appearance of the original.

If, therefore, you want to reproduce a copyrighted figure or image, you need to obtain permission from the copyright holder. Fortunately, the process of applying for and receiving permission is largely automated. On the publisher's website there should be a permissions link from the web page of the article to a copyright licensing service [310], which asks for details about where and what is to be reproduced. If the application is standard, a decision can be made automatically. Some licences permit copyrighted material to be used freely as long as it is properly credited (§9.3.10).

When permission is granted, the publisher usually requires only that an appropriate acknowledgement be provided; for example, 'Reproduced with permission from . . .', along with the name of the copyright holder. The precise formulation is specified by the publisher. A statement of this kind is shown Example 12.14.

> This figure is replotted from the Figure 4 in ref 13 with the permission of American Chemical Society.

Example 12.14. *Acknowledgement of permission to reproduce an original figure, from a physical chemistry article [716].*

This acknowledgement is stated at the end of the caption to the authors' figure [716, Fig. 2]. The replotted version, not shown here, is closely similar in appearance to the cited original [589, Fig. 4].

If a figure is redrawn or replotted in such a way that copyright is not an issue, then an appropriate acknowledgement suffices, for example, 'Redrawn from . . .'. A statement of this kind is shown in Example 12.15.

> Fig. 1 shows the thermodynamics of the carbon assisted electrolysis process at low to intermediate temperatures [6].
>
> . . .
>
> (Figure is redrawn from Ref. [6].)

Example 12.15. *Acknowledgement of source data for a figure in the body of the text (upper extract) and in a figure caption (lower extract), from a physical chemistry article [518].*

The acknowledgement of the source for the figure is stated in the text and at the end of the caption [518, Fig. 1]. Although not shown here, if the redrawn figure and the original [498, Fig. 6] are compared, it is clear that the content is the same, but that there are differences in appearance—in the scale range, gridlines, labelling, and layout.

Judging whether replotting data from a figure infringes copyright can be difficult, and Examples 12.14 and Example 12.15 should not be taken to represent definitive usage. The safest course always is to ask the copyright holder for permission.

12.2.8 Reproduction of Text

Limited use of copyrighted material is permitted under what is called fair use, for instance for analysis or criticism, but the reproduction should not be so substantial that it diminishes the copyright holder's own publication [49, §4.79]. A restriction is therefore placed on how much text can be reproduced from a copyrighted article, even when quotation marks are used or the original text is paraphrased and the source is correctly acknowledged [314].

If, therefore, you want to reproduce more than a brief excerpt of copyrighted text, you need to obtain permission from the copyright holder. The procedure is the same as that described in Section 12.2.7 for the reproduction of a copyrighted figure or image.

12.3 Penalties for Research Misconduct

The penalties for research misconduct depend on whether the work is submitted for an academic degree or for publication in a research journal.

12.3.1 Dissertations and Theses

If a dissertation or thesis submitted for a degree or other academic qualification is found to involve misconduct, then an academic institution has several penalties available. They may range from requiring the author to submit a revised document for minor misconduct to outright rejection of the submitted work with no opportunity for resubmission. If a degree already awarded is found to have involved misconduct, the degree may be revoked.

12.3.2 Journal Articles

For a paper submitted to a research journal, the discovery of research misconduct may lead to immediate rejection by the editor, with no opportunity for resubmission. For an article that is published, the discovery of research misconduct may lead to one of several actions by the editor and publisher [367]. The most serious sanction is retraction. Retraction does not entail removal of the article from the archive. Rather, it remains available, but its retracted status is identified publicly [295]. The editor publishes a notice stating the reasons for the retraction; the notice is linked to the electronic version of the retracted article, and the electronic version has 'RETRACTED' superimposed on it. The retraction notice is published in such a way that it should be returned automatically in searches for the retracted article [367].

Authors found guilty of research misconduct may also be banned for a period from publishing in the journal or in any associated conference proceedings [174], a penalty not supported by COPE for legal reasons [65]. Apart from the loss of professional standing for the affected author, further penalties may be imposed by the host institution or funding agency; thus an author may be excluded from entering into government contracts [268].

There is some evidence that the proportion of articles being retracted is increasing over time [123] [147] and that the proportion due to fraud or error is higher with more influential journals [113].

12.3.3 High-Profile Cases

There have been several very public cases of authors losing their jobs because of research misconduct. Perhaps the most well-known case in the physical sciences is that of Jan Hendrik Schön, who was found to have duplicated, falsified, or fabricated data on at least 16 occasions [23]. The total number of affected articles is about twice that number [308]. Schön was dismissed from Lucent Technologies' Bell Labs [39] [94] and his doctorate was subsequently revoked [324]. A comprehensive account of the case is available as a book [304].

Another high-profile case, but in the life sciences, is that of Haruko Obokata, a stem-cell biologist. She was found to have committed research misconduct by fabricating data [305] and her two publications in the journal *Nature* were retracted [629]. She resigned from her post at the RIKEN research institute and a former deputy director committed suicide [260]. The head of the institute later resigned [120].

Several high-profile cases have occurred outside science and engineering. Pál Schmitt resigned as Hungarian president [347] after he was found to have included lengthy literal translations in his thesis for a sports doctorate [85]. Karl-Theodor zu Guttenberg resigned as German defence minister after admitting copying part of his law thesis [205], and Annette Schavan resigned as German education minister after having her philosophy doctorate revoked for alleged copying [36].

12.4 Research Ethics Approval

Conducting your research ethically is not in itself enough: you need also to report that you have done so. Thus if your research involves the use of human beings or animals as subjects or human tissue, you need to explain in the paper how the research adhered to the expected ethical standards, as discussed in Section 3.6.3. Requirements for compliance are normally established when seeking approval from a research ethics committee [105] or institutional review board [188] either at your own institution or place of employment, or at a national level.

Standards and operational guidance for research ethics committees are set by international bodies, such as the World Health Organization [338] and the World Medical Association [386], by national bodies such as the Office of Human Research Protections (OHRP) in the USA [267], by national associations, by professional bodies, and by learned societies.

12.4.1 Human Subjects

Ethical requirements for research involving human beings as subjects are listed in [338] [339, ch. 3]. They include the validity of the scientific methods, the qualifications of the researchers, the adequacy of monitoring and study site, the possibility of harm (physical, social, financial, or psychological), and the overall balance of risk and benefit. Other matters include the extent of any financial inducement to participate, the protection of participants' privacy, and the adequacy of procedures for securing informed consent from each participant. Informed consent means that the researcher must be able to convey to the participant in appropriate language all the important information about the research and what participation involves.

Clinical research with human subjects is expected to comply with the Code of Ethics of the World Medical Association (Declaration of Helsinki) [383] or with guidance from the equivalent national body [339, ch. 3]. If the experiment constitutes a clinical trial, including testing a device or a behavioural procedure, it needs to be registered in a public trials registry at or before the time at which the first patient is enrolled [303].

Non-clinical research with human subjects is expected to comply with guidance from national bodies such as the OHRP in the USA, from national associations, and from funding agencies [339, ch. 3]. Regulations governing the use of human tissue vary from country to country.

12.4.2 Animal Subjects

Research with animals is expected to comply with international guidelines such as the European Commission Directive on the protection of animals used for scientific purposes, along with other related legislation [222] and with national and regional regulations governing animal research, such as the Animal Welfare Act and the Health Research Extension Act in the USA [339, ch. 4]. Any institution or research facility conducting research with animals is expected to have an animal welfare and ethical review body or animal research committee or its equivalent in place.

12.5 Citing and Referencing

The practicalities of citing and referencing are discussed in Section 3.10. Although the purpose of citing is primarily to give credit to previously published research, including your own [314], there are ethical issues to do with priority, relevance, accuracy, and bias.

12.5.1 Priority and Relevance

In citing material in support of a particular theory or experimental finding, priority should normally be given to the first study [314] [390], with subsequent confirmatory or incremental studies assigned less weight. Some journals and conference proceedings using the author-date system (§3.10.2) enforce alphabetical ordering of citations, but there are good reasons to press for date order. Property rights in science are intimately related to the recognition of publication priority [252]. Failure to recognize publication priority represents a loss in author rights, and a disbenefit for society [342].

Incidentally, as noted in Section 3.10.1, this guide represents a different kind of publication. Citations are ordered numerically and both the bibliography and the list of reference works are ordered alphabetically for reader convenience, which here outweighs priority.

Whatever the system of citation, publication priority can always be recognized explicitly, as in Example 12.16.

> The earliest report on visible-light-active N-doped TiO_2 is dated back to 1986 [11]. . . . Of various methods in association with different precursors for nitrogen-doping of TiO_2, calcination or magnetron sputtering of TiO_2 target under special gas [12–15] . . .

Example 12.16. *Explicit and indirect priority of citation, from a physical chemistry article [712].*

The earliest report [11] published in 1986 is clearly identified as such and the other reports, cited as [12–15], are ordered in the list of references of the article [712], not shown here, by publication date, namely 2001, 2003, 2004, and 2005.

Recognizing priority does not of course mean ignoring the most recently reported work. Both are important, but electronic coverage of journal archives is invariably incomplete, and sometimes print copies need to be examined as well as online databases.

Only significant, published, peer-reviewed references should be cited [79, p. 78]. Unpublished material may be inaccessible to others and may not have undergone rigorous review. Too many references, as with too many results, suggests inexperience or lack of discrimination, and too few references suggests carelessness or misrepresentation [258, p. 93].

12.5.2 Persisting Errors

As part of the implicit contract between author and reader, the reader assumes that the use of a reference is honest and reflects the author's inspection of the original (§3.10.6). You should therefore never cite a reference unseen [68, §29.3.1]. To do otherwise is viewed as deception [314].

Relying on secondary sources instead of the original leads to persisting errors in the literature [68, §29.3.1]. In one case, Handoll and her co-author [153] report that a short research letter she published in 2006 had received 62 citations by 2015. All were for the wrong reason. The first citing work [711] referred to between-study heterogeneity and the *Q*-statistic, neither of which are mentioned in Handoll's letter or relevant to it. Subsequent authors continued to cite the letter for the wrong reason, presumably because they assumed that the earlier citations were correct. As Handoll and her co-author point out [153], none of the citing authors seem actually to have read the letter.

12.5.3 Citation Bias

Bias in referencing previously published work can arise in several ways. One way is to cite only those articles and sources that support your viewpoint [145]. This practice is unethical [314]. Not only does it distort the literature and mislead the reader, it may also be unproductive. As evidence of a lack of objectivity, it is likely to weaken the authority of your work.

Another way bias can arise is by excessively citing your own publications. It may be unconscious, owing to your greater familiarity with your own work, or deliberate, to acknowledge the significance of your contribution. But what constitutes excessive self-citation? Montgomery [258, p. 94] suggests that more than five or six self-citations out of 30–40 in all appears unprofessional, except perhaps for an authority in the field (who might still exercise some humility).

Excessive self-citing may be more common in opinion pieces and in larger works bringing together a body of research activity in which the author has actively participated. With textbooks and monographs, transparent motivation helps. The following is from the preface of a monograph by Brindley [428]. It explains why the author cites his own work disproportionately.

> With my own work I have been more lenient than with the work of others. This is not solely from vanity; readers have a right to expect that a book of this kind will lead them to the whole of the author's own contribution to the subject.

For routine research articles, however, excessive self-citing, sometimes called citation stuffing, is unethical [314] and may be seen by reviewers as egocentric. That said, is self-citing productive in generating citations from others? The evidence is mixed, suggesting, on the one hand, that articles with more self-citations receive more citations from other articles [124] and, on the other hand, that articles with the highest proportion of self-citations are among the least-cited [1].

Another way citation bias can occur is by over-citing the work of the anticipated reviewers of your paper, in the hope that they are more likely to recommend acceptance. This practice, too, is unethical [314]. To counter such bias, editors may deliberately seek reviewers who are not cited in the paper.

Yet another way citation bias can occur is by concentrating on work from a particular laboratory or institution or group of laboratories with which you may have a relationship or from a particular region of the world. There is evidence from some research fields that researchers tend to over-cite articles from their own country [281] [385].

Other distortions of the literature are discussed in [145]. One kind of distortion is called amplification. It entails the publication of papers, usually review articles, which repeat a hypothesis, thereby adding to the literature, but which provide no new data. Another is herding, in which citation behaviour is influenced by the behaviour of peers rather than by individual judgement [278].

12.6 Authorship

In most science and engineering disciplines, articles with a single author are uncommon. That this was not always so is revealed in a comparison of publications in 1965 and 2011 [387]. Single authorship decreased from around 46% to

13% over the intervening years and multiple authorship involving four or more coauthors increased from 7% to 50%.

With two or more authors, individual responsibilities need to be considered; for example, activities such as helping to design and perform the research, writing the paper for submission, and taking public responsibility for the published article [68, §27.7.1]. All the authors are expected to have made a significant contribution to the work [369], but complications can arise in deciding on the degree of significance, the ordering of names, and potential conflicts of interest. Decisions about each are best made before writing begins [63] [174] [303] [314].

12.6.1 Who Is an Author?

The most widely used set of rules for recognizing authorship [259] is due to the International Committee of Medical Journal Editors (ICMJE) [303]. Many non-medical journals, including those of the Royal Society [294], also adhere to these rules. They are as follows, reproduced without change from [303]:

1. Substantial contributions to the conception or design of the work; or the acquisition, analysis, or interpretation of data for the work; AND
2. Drafting the work or revising it critically for important intellectual content; AND
3. Final approval of the version to be published; AND
4. Agreement to be accountable for all aspects of the work in ensuring that questions related to the accuracy or integrity of any part of the work are appropriately investigated and resolved.

The ICMJE rules are viewed as too prescriptive, both by the Council of Science Editors [327, §2.2.1] and by COPE [63], especially where the research involves disparate disciplines or where very large numbers of individuals participated in the research.

Thus, in multidisciplinary work, one author with a particular expertise may make a substantial or significant contribution yet be unable to take responsibility for (or fully understand) the contribution of another author with different expertise.

In disciplines where there are exceptionally large numbers of contributors, authorship has a different meaning. With around one thousand authors in a genetics article [584] and five thousand in a particle physics article [392], it is difficult to know what would constitute a significant contribution for each contributor [314]. Indeed, the individuals listed as authors could fail all of the ICMJE criteria [63]. Fortunately, at least for ethical considerations, the proportion of articles with more than one thousand authors is extremely small, less than 0.01% over the period 2010–14 [316].

In short, although the ICMJE rules may be sufficient for authorship, they need not all be necessary. It may be enough for each author to be responsible for a particular section of a paper instead of the whole. Nevertheless, one (or more) of the authors must be able to take responsibility for the integrity of the work [63].

Whatever the discipline, most authorship rules require the inclusion of so-called ghost authors [327, §2.2.2], that is, individuals who in reality contributed significantly but whose names do not appear in the article's list of authors, the byline [314]. Exclusion may be to do with perceived conflicts of interest [259] or expediency or manipulation of the record [314]. Unsurprisingly, the practice is condemned [2] [59].

12.6.2 Who Is Not an Author?

Despite differences in the way that authorship rules are construed in different disciplines, there is general agreement about who should not be an author. Most rules exclude individuals who provided only routine technical and statistical support. They also exclude individuals who provided no more than the working environment or secured funding for the work [63] [314]. All would normally be recognized, subject to their agreement, in the acknowledgements section of the paper (§3.9).

Most rules forbid so-called guest, gift, and honorary authors [2] [63] [327, §2.2.2]. These are individuals who did not contribute to the work in the way expected and are included for social or other reasons unrelated to their contribution, such as their need for career progress or their administrative or managerial role [387], or the reputational gain for the paper if the guest author is internationally recognized [259], or simply as an exchange of favours ('You can be an author on my paper if I can be an author on yours').

Both guest authorship and ghost authorship (§12.6.1) are unethical and are recognized as such by editors and publishers [59] [81] [104] [174] [259] [303], with some variation in interpretation [349]. Even so, the prevalence of guest authors in one field is reported to be 21% and that of ghost authors 8% [382].

12.6.3 Order of Authors

With multiple authors, a decision needs to be made on the order of names in the byline. Unhelpfully for readers wanting to assign significance to an author's position, there is no widely accepted rule for deciding the order. Practice varies from discipline to discipline [63]. Most guides suggest that it is a matter for authors [81] [174] [293] [327, §2.2.4], although there are exceptions [68, §27.7.1].

In disciplines where papers have only a few authors, the order is generally expected to follow the significance of the author's contribution [314]: the greatest first, the next greatest second, and so on down to the least significant contribution. In a small reversal, the last position may be reserved for the administratively most senior or most experienced member of the team [63] [174] [307]. As with authorship itself (§12.6.1), departures from expected practice may be made for social or other reasons, most notably career progress. Still, descending order of significance of contribution, with or without the most senior contributor last, is overwhelmingly preferred, according to a survey of largely life-sciences researchers.[92]

Being first author attracts more attention, not least in author-date citations, but it traditionally carries significant extra responsibility, too [104] [174] [185]. The first author may be expected to take the lead in supervising data analysis, presentation of the results, writing the paper, and submitting it for publication [307]. Coauthors are expected to support the first author in each of these tasks in accordance with the requirements for authorship (§12.6.1).

In disciplines with large numbers of authors, the significance of individual contributions may be impossible to decide, and the order may be settled alphabetically either for all authors or subgroups of authors [392] [584]. The use of alphabetic ordering in other circumstances is contentious.[93]

With intermediate numbers of authors, the first and last few positions may be assigned to the most significant and most senior contributors, respectively [2], as indicated earlier. The intervening positions may then be decided at random.

Given the one-dimensional nature of the article byline, it will always give an imperfect representation of each author's contribution. Some journals address this problem by asking authors to declare, either in a footnote or near the end of the article, the individual nature of their contributions that merited authorship. This contributorship policy is advocated by both the Council of Science Editors [327, §2.2.2] and the ICMJE [303].

Example 12.17 shows an author contribution statement. The order of the authors in the byline of the article [702] is this: first, G. M. F. van der Heijden; second, J. S. Powers; third and last, S. A. Schnitzer. The contribution statement shows that although all three authors performed the research, it is the first author, G.M.F.v.d.H, who analysed the data and wrote the paper, and the second and third authors, J.S.P. and S.A.S., who designed the research. In the article, the first author was also designated as the author for correspondence.

> Author contributions: J.S.P. and S.A.S. designed research; G.M.F.v.d.H., J.S.P., and S.A.S. performed research; G.M.F.v.d.H. analyzed data; and G.M.F.v.d.H. wrote the paper.

Example 12.17. *Author contribution statement, from an ecology article [702].*

Whatever method is used to decide authorship and author order, once the paper is submitted to a journal or conference proceedings, it is unusual for the article byline to be altered. Any changes after submission of the paper need to be approved by all the authors, including any new authors and any existing authors who are to be removed [327, §2.2.5].

12.6.4 Corresponding Author

Journals expect one author to be identified as the corresponding author [2] [185] [303]. This is the author who deals with correspondence arising from the submission of the paper for publication, along with any disputes about authorship [327, §2.1.9], and who acts as respondent and coordinator for enquiries from readers once the paper is published [185]. The corresponding author is also the author who would cooperate with the journal in any investigation of the integrity of the published work [303] [327].

The corresponding author is customarily designated on the title page of the article and is often the author occupying the first or last position in the byline [185], but not always so [173]. In fact, any author with a reasonably stable address can act as corresponding author.

12.6.5 Conflicts of Interest

All the contributors to a paper need to consider potential conflicts of interest [327, §2.1.3], that is, where professional judgement concerning the conduct and reporting of the research may be influenced by secondary interests, such as financial gain, work commitments, and personal relationships and beliefs [303] [339, ch. 5]. Conflicts of interest, perceived or actual, can take many forms. These examples are adapted from [12]:

- a researcher investigates the environmental effects of oil drilling but owns stock in an oil company;
- a researcher develops a novel graphics display and is also a consultant for a financial services firm that assesses investments in new technologies;
- a researcher submits a paper to a journal whose editor-in-chief is an academic colleague;
- a medical doctor writes a paper on emerging medical technologies but believes in traditional healing methods.

The presence of a conflict of interest does not necessarily mean that authorship is denied, merely that transparency should be maintained; that is, authors need to disclose their potential conflicts of interest on submission of the manuscript to the journal or conference proceedings [303] [327, §2.1.4].

In some cases, a degree of judgement may be required. In the examples, is the interest in the company or firm sufficient to influence the judgement of the investigator or developer? Is the editor-in-chief a member of the same institution or a larger regional grouping? Is the doctor's belief in traditional healing methods a matter of principle or expediency?

The IEEE Code of Ethics, however, is unambiguous: it requires both real and perceived conflicts of interest to be avoided where possible and for any such conflicts to be disclosed where they exist [181]. Yet a perceived conflict of interest may lead to the arbitrary subjective exclusion of a contributor when an actual conflict of interest does not exist. A more reasonable position may be that consideration of a conflict of interest should be limited to judgements founded on objective evidence [75, p. 52].

Whether perceived or actual, conflicts of interest are best determined before a study begins. Significant conflicts may need a potential author's contribution to be managed by an appropriate committee or administrator [339, §5d].

12.6.6 Declarations

The way in which conflicts of interest are disclosed varies across research fields. In some, especially biomedicine, journals ask authors to declare, either in a footnote or near the end of the article, any conflicts of interest. These declarations may comprise personal statements, the sources of support for the work, the role of the sponsors in the research (from design to interpretation and writing), and any constraints imposed on authors' access to data [303].

The conflict of interest statement in Example 12.18 is from one of the 15 authors of an article on stem cells [667]. It identifies the author's involvement in two businesses. A declaration of this kind does not necessarily invalidate the research in which the author participates. It merely provides the information that a reader needs in order to assess the significance of the work [60].

> Conflict of interest statement: W.L.M. is a founder and stockholder for Stem Pharm, Inc., and Tissue Regeneration Systems, Inc.
>
> (Biosciences or medical devices companies)

Example 12.18. *Conflict of interest statement, from a neuroscience article [667].*

In other fields, simply listing sources of financial support, for example, in the acknowledgements section of a paper (§3.9), is more usual. This information can still be helpful in signalling potential reporting biases [25].

12.7 Data Management

The practicalities of data description and statistical inference are discussed in Chapter 7. Ethical issues can arise in obvious ways with data analysis. It may be susceptible to researcher bias or wishful thinking, either of which could lead to the application of inappropriate procedures [58].

Ethical issues can also arise with data storage and access, for data consist of more than just the results of measurement and analysis. Data include the materials, methods, analyses, summaries, and the inputs and outputs of simulations and models, as well as the commonly understood results of that activity [301], indeed, all recorded factual material necessary to validate research findings [170].

Funders may expect a plan to manage data to be in place from the beginning of a research project. The plan aids good research conduct and compliance with regulations governing confidential data and risk management [170] [301].

12.7.1 Data Analysis and Reporting

Ethical guidelines for data analysis are sometimes contentious [58] [103]. The following list of unethical and questionable practices is drawn from several sources, including those in [137] [209] [355]:

- discarding unfavourable data: this practice should be distinguished from the systematic treatment of outliers [360], for which there are well-established procedures [21] [20, ch. 3], and from the use of robust statistics, such as the trimmed mean and median, which avoid the biasing effects of extreme values (§7.2);
- verifying post hoc hypotheses: hypotheses should be formulated before the data are collected [137, ch. 1]; to formulate them afterwards and verify them against the data is to commit what is called the Texas sharpshooter fallacy: the sharpshooter shoots at the side of a barn and then draws a bull's-eye target around the bullet hole [130];
- multiple testing for significance: if enough statistical tests are performed, eventually one will prove to be statistically significant (§7.9.2); the p-value found with multiple tests needs to be corrected [69, §8.5];
- overfitting: the goodness of fit of a model to a set of data can be improved by increasing the number of independent variables in the model; overfitting occurs when the model accounts not only for systematic effects but also random effects (§7.6.7); overfitting can be tested by cross-validation [159, §7.10];
- biasing: a linear regression fit to a set of data may suggest a pattern to the data, for example, a linear dependence, which hides a more accurate description (§7.6.4); this biasing effect can be countered with nonparametric methods (§7.6.5);
- significance and importance: a statistically significant effect may have little real importance [137, p. 105]; the size of the effect should always be reported (§7.9.3).

12.7.2 Data Storage and Access

Research funders, journals, publishers, learned and professional societies, and academic institutions expect research data to be made widely and freely accessible, and intelligible [318], subject to reasonable ethical and commercial constraints. The justification for transparency is set out comprehensively in [57] [98] [170] [266] [301] [339]. The main reasons are these:

- Research data obtained from publicly funded research should be regarded as a public good and should therefore be publicly accessible.
- Experimental findings may be validated by other researchers against the source data.
- Data may be analysed and exploited in ways that were not anticipated by the authors of the published article.
- The likelihood of research being duplicated is reduced.

The reuse of data from research that required informed consent is complicated. It depends on the nature of the consent, the type of data, and whether the data were anonymized. A balance may need to be struck between the benefits of data reuse, the limitations of consent, and data security. A judgement would normally be made by a research ethics committee or institutional research board [328].

All data are expected to be protected and stored safely, and special precautions taken with confidential data, including controlled access [339, ch. 6].

12.8 Submission and Publication

On submission of a paper to a journal, an author enters into an agreement with the publisher. Among other things, the agreement covers the conditions under which the journal considers the paper for publication and what can be done with the paper once it is accepted and published. Flouting this agreement is viewed as unethical and in some conditions may represent a breach of contract.

12.8.1 Duplicate Submission

As the name suggests, duplicate or simultaneous submission occurs when the same paper is submitted to two or more journals or conference proceedings at the same time. It is viewed as unethical because the duplicate submissions draw disproportionately on editorial resources in reviewing, editing, and publishing [303] [339]. Moreover, when published, duplicate articles distort the literature by inflating the apparent support for the work [339, §9c]. They can similarly distort meta-analyses, which combine the results from multiple studies and which assume that the data come from independent sources [314]. Duplicate submission, without explanation, breaches the usual undertaking given by authors on submission.

If, as a consequence of duplicate submission, duplicate publication takes place, then unless there is adequate justification, a statement may be attached to the first publication pointing out the duplication and the other publication may be retracted [367]. More extensive duplication may lead to more severe penalties, similar to those imposed for any other kind of plagiarism (§12.3).

Unless there is an agreement with the publishers involved, submitting different-language versions of the same manuscript is also viewed as duplicate submission [303] [314].

Submitting an expanded version of a conference paper need not, in itself, constitute duplicate submission, although its status depends on the degree of expansion (§12.2.5).

If a paper is rejected by a journal, then the author is naturally free to submit it to another. But it should accommodate the reasons for rejection (§9.3.9). Repeatedly submitting unrevised versions to successive journals has the same effect on editorial resources as duplicate submission.

12.8.2 Salami Slicing

Salami slicing or salami publication, also known as data fragmentation, is the practice of dividing a substantial single piece of research into several smaller pieces, often overlapping in content, to increase the total number of papers from the work [314] [339]. It is viewed as unethical much as duplicate publication is viewed as unethical (§12.8.1). It draws disproportionately on editorial resources and distorts the literature by inflating apparent support through studies that are not independent.

A similar practice, again viewed as unethical, is data augmentation. Here an author collects new data, adds them to previously published data, and presents the combination in a new publication [314]. The effect is again to distort the literature by inflating the apparent support.

From the author's viewpoint, salami slicing may also be counterproductive. Papers that make incremental advances are less likely to be accepted by reputable journals [303] [314]. If they are published, their total impact may be less than that of a single article that provides a comprehensive account of the research, in a digestible format.

12.8.3 Sharing Papers

The extent to which a paper can be shared depends on whether it is the manuscript or published article, and whether publication is in a subscription-only journal or an open access (OA) one (§9.2.2). What follows is a rough guide to sharing, although the policies of the journal or conference proceedings must be consulted [9] [13] [271] [330] [341]. What may be acceptable to one publisher may not be acceptable to another.

The submitted manuscript version of a paper involves no direct contribution from an editor, journal, or publisher, and it can generally be shared without constraint, except for a few elite journals where pre-exposure would damage the potential value of the article.

The accepted manuscript version of a paper has a different status from the submitted version. It incorporates author changes made in response to reviewers' comments and editorial advice, but none of the formatting and other changes required for publication. This pre-publication version may be shared privately with colleagues and students for their personal use, and after publication and possibly an embargo period, posted on your personal website and deposited in institutional and funding agencies' repositories. It may also be uploaded to preprint servers such as arXiv [10].

The final published version of a paper has a different status again. Among other operations by the publisher, it has undergone copy-editing to a greater or lesser extent, formatting, and citation and reference linking. It has been proofread, corrections made where necessary, pagination and indexing applied, and a digital object identifier (DOI) assigned. It is this version that appears on the journal website and in electronic and print copies. It constitutes the version of record. If the article is published in a subscription-only journal, then this version can be shared only privately with colleagues and students for personal use (§9.4.1). Some publishers do permit the published version to be placed in institutional and funding agencies' repositories after an embargo period, as well as posted on authors' personal websites [9]. Some learned-society publishers may waive the embargo period [376].

If the work is published as an OA article, it can generally be posted immediately on non-commercial sites together with a copyright notice and DOI linking it to the version of record on the publisher's website. Depending on the precise form of OA agreement, posting may also be permitted on commercial websites (§9.3.10). With some publishers, however, OA options may be available only for articles published in journals, not conference proceedings [13].

If you enter into an agreement with a publisher not to post the final published version on an unauthorized website and subsequently do post it, the publisher may issue a takedown notice, which requires the article to be removed.

12.9 Checklist for Ethical Issues

Plagiarism

1. Others' published and unpublished ideas and work acknowledged?
2. Your own published ideas and work acknowledged (including conference material)?
3. Any section of text (direct, interleaved, or paraphrased) from a previous publication?
4. Essential duplication indicated, e.g. 'Reproduced from . . .'?
5. Permission granted to reproduce copyrighted material, especially figures? Permission statement included?

Research ethics approval

6. For research with human beings and animals, name of the research ethics committee?
7. Statement of compliance with national and international standards?
8. For research with human subjects, statement about informed consent?
9. For clinical research, details of preregistration in a public database?

Citing and referencing

10. Citations adequately recognize the priority of previously published work?
11. Each cited reference read and honestly cited?
12. Citations match references?
13. Reference details verified against original sources?
14. Your own work proportionately citied? Any other potential citation bias?

Acknowledgements and declarations

15. Funding, technical support, and facilities identified?
16. Contributing individuals named, agreement with naming and format?
17. Conflicts of interest declared?

Authorship

18. Authorship reflects international guidance?
19. All authors agree with the byline ordering?
20. All authors agree on the identity of the corresponding author?

Data management

21. Methods of analysis and reporting of data justifiable?
22. Data stored safely and confidential data protected?
23. Subject to constraints, data freely accessible and intelligible?

Submission and publication

24. Overlap identified with any other submission to a journal or conference proceedings?
25. No salami slicing?
26. Plans for sharing adhere to the publisher agreement on submission?

13

Final Thoughts

Whatever you are working on—a paper, thesis, dissertation, report, or presentation—the challenge is to know when to stop. Unfortunately, there are competing forces that make it difficult: the temptation to add more and more detail, the need to reflect and revise, and, above all, the desire to keep refining. Halmos put it thus [150, p. 47]:

> There is always something left undone, always either something more to say, or a better way to say something, or, at the very least, a disturbing vague sense that the perfect addition or improvement is just around the corner.

So when to stop? Once feedback has been secured from colleagues, revisions made, and more feedback obtained; once the work has been left to one side for another inspection and revision; once these and any other checks have been completed and corrections made; once revisions have begun to repeat themselves—once you have reached this point, the time has come.

The work may be imperfect, but as Turabian observed [357], it has to be finished. Still, submission is only one step on the path to dissemination. When your work is returned for correction, you will have another chance for reflection and revision, another chance to make that perfect addition or improvement. In due course, your effort will be recognized, and your work will establish its rightful place in the community.

Notes

1. In the text examples with annotations, individual words and phrases of interest are usually underlined.
2. This version of the quotation attributed to Paul Dirac is taken from Farmelo [116, p. 121], and appears also in Monk [256, p. 130], who notes that there are many other versions, including one due to Bernstein [418, p. 23]: 'In physics we try to tell people things in such a way that they understand something that nobody knew before. Whereas in poetry . . .'.
3. A. Reeves, personal communication, 30 November 2015, reproduced with permission.
4. A callout is a short excerpt of text set away from the body and in larger or distinctive type designed to attract attention or highlight key points.
5. Reproduced with permission of the author.
6. The small discontinuity between the subject 'a new source separation technique' and the restrictive clause 'that exploits . . .' is conventionally acceptable; see Section 5.3.2.
7. Reproduced with permission of the author.
8. A fuller list of verbose words and phrases and their shorter equivalents is given in Section 5.6.25.
9. Word frequency was estimated from an analysis I made of 1036 journal articles published between 1995 and 2015. The articles were from 14 research areas: physics, chemistry, mathematics, computer science, general engineering, electrical engineering, materials science, geology, plant biology, molecular biology, neuroscience, physiology, psychology, and general and internal medicine. Across articles, the mean number of occurrences of 'might' was 4.1. The median was 2; the 95th and 99th centiles were 10 and 21, respectively.
10. An acronym is a word formed from the initial letters or parts of other words, but also used loosely, as here, for an abbreviation composed of initial letters [38]. Finer distinctions are described in [49] [311].
11. Problems arising with 'using' are discussed in Section 5.4.1.
12. Title statistics were estimated from an analysis I made of 4000 journal articles with a publication date of 2012 in Web of Science [375]. The articles were from eight research areas: physics, chemistry, mathematics, cell biology, computer science, general engineering, electrical and electronic engineering, and general and internal medicine (some articles had dual classifications, e.g. neuroscience and psychology). Across all articles, the mean title length was 12.3 words. The median was 12 words; the 5th and 95th centiles were 5 and 21 words, respectively; the 1st and 99th centiles were 3 and 26 words, respectively. The median varied little across research areas, except for chemistry, where it was 15 words. Generally, smaller values were obtained from a similar set of articles published in 2002. The mean was 10.9 words. The median was 10 words; the 5th and 95th centiles were 4 and 20 words, respectively; and the 1st and 99th centiles were 2 and 26 words, respectively. The median varied little across research areas, except again for chemistry, with 13 words.

13. From the same set of articles as in note 12, compound titles represented about 16% of the total in 2012 and 14% in 2002. Titles forming a question represented about 1% of the total in 2012 and 2002. Similar proportions were reported in [193] for a 2007 analysis of 2172 article titles in six PLOS (Public Library of Science) journals.
14. The claim that if a title is a question with a 'yes/no' answer, then the answer is 'no' has an uncertain origin, but it is known variously as Hinchliffe's rule and Betteridge's law.
15. A. Pierce, personal communication, 2 February 2015.
16. A search of Web of Science [375] yielded 586 titles containing 'celiac disease' in the year in which the review article was published.
17. The relative frequency of p-value statements in abstracts was estimated from the same set of articles as in note 12. The proportion was about 2%. For general and internal medicine the frequency was an order of magnitude larger; see also [46].
18. Abstract statistics were estimated from the same set of articles as in note 12, although not all articles contained abstracts. Across abstracts, the mean number of words was 160. The median was 153; the 5th and 95th centiles were 52 and 288, respectively; and the 1st and 99th centiles were 29 and 386, respectively. The median number of words varied across research areas, from 116 for mathematics to 274 for general and internal medicine. Generally smaller values were obtained with articles published in 2002. Across abstracts, the mean number of words was 139. The median was 129; the 5th and 95th centiles were 43 and 267, respectively; and the 1st and 99th centiles were 22 and 346, respectively. The median number of words varied across research areas, from 95 for mathematics to 212 for general and internal medicine.
19. According to Web of Science [375], this article received no citations in the ten years following its publication.
20. According to Web of Science [375], this article received no citations in the ten years following its publication.
21. According to Web of Science [375], this article received one citation in the ten years following its publication.
22. The frequencies of different kinds of article introductions were estimated from an analysis I made of 550 articles published in 2012. They were from cell biology, chemistry, mathematics, computer science, general engineering, civil engineering, electrical and electronic engineering, general and internal medicine, physics, and psychology. Not all introductions could be uniquely classified, but 43% began with a generalization about the field, 23% described the research question or problem, and the remainder, with descending frequency, reported recent or historic research, presented a key definition or the purpose of the work, or began with a question. The last was used in fewer than 1% of articles.
23. Hilbert's 1900 communication on 'Mathematical Problems' [165] is probably the most successful and long-lasting example of a set of challenges that have helped shape a discipline.
24. According to Web of Science [375], this article received over 100 citations in the ten years following its publication.
25. Famous exceptions containing no references are Einstein's 1905 article on special relativity [493] and Feynman's 1939 article on forces in molecules [501].
26. Reference management software systems can automatically order and format citations and references according to a particular journal style and then reformat them for another journal style. They can import references from online bibliographic databases, journal websites, and from the PDF of the article itself. They also allow annotations and other

bibliographic details to be inserted, which can be useful in grouping and sorting references for a given research topic. For reviews of some popular reference management systems, see [135] [233].

27. The original Vancouver system evolved into 'Recommendations for the Conduct, Reporting, Editing, and Publication of Scholarly Work in Medical Journals' [303].
28. Reproduced with permission of the author.
29. Reproduced with permission of the author.
30. For example, the article in Example 3.112 appears in the *Annual Review of Neuroscience*, 37(1), 17–38, with DOI 10.1146/annurev-neuro-071013-014100, and URL http://www.annualreviews.org/doi/abs/10.1146/annurev-neuro-071013-014100.
31. Crossref is a digital object identifier agency that allows cross-publisher citation linking of online journals [72].
32. Reproduced with permission of the author.
33. The phrase 'STEM tomography revealed' is an example of grammatical metaphor in which the instrument acts as the agent [92, §21.1].
34. For discussion of the use of nominalization to avoid a human agent, see [92, §21.2] [297, §17.52].
35. Some nouns can be considered singular or plural, for example, 'series' and 'species'. The interpretation depends on context, but once the interpretation is made, the verb agrees by grammar.
36. Traditional and more modern approaches to grammar differ in their treatment of tense. English has no verb form that marks the future, which needs to be signalled in other ways [32, §6.2] [92, §41.1]. For a practical overview, see [224, ch. 4] [340, pp. 235–66].
37. It is sometimes a source of confusion that the past tense of some verbs such as 'conducted' is identical in form to the past participle [32, p. 58] [297, §3.2].
38. The *THE: Times Higher Education* is a weekly magazine reporting on higher education, published by TES Global, London.
39. More accurately, top-heavy sentences have a long subject and a short predicate. In general, the part of the sentence before the verb should be shorter than the part of the sentence after it. This is the principle of end weight.
40. From the same set of articles as in note 9, 'It' or 'There' began a sentence on average 10.7 times per article.
41. 'Dummy subject' is used here as an umbrella term for a more nuanced division of function [32, pp. 231–3] [44, p. 27].
42. These constructions are all elliptical [297, §11.43, §15.51].
43. More precisely, 'are' is a linking or copular verb [224, pp. 27–32] [297, §2.16].
44. Both 'little' and 'few' give a positive or negative bias depending on whether they are preceded by the indefinite article. Thus 'a few books' is some or several books whereas 'few books' is not many books [297, §5.23].
45. The definitions are based largely on practice recorded in the *New Shorter Oxford English Dictionary* [38].
46. In an asymmetry, 'above' can act as adverb, noun, adjective, and preposition, yet 'below' only as adverb and preposition [280].
47. From the same set of articles as in note 9, 'but' appeared on average 22.7 times per article, 'however' 6.8, 'although' 3.5, 'though' 0.9, 'in contrast' 1.2, 'yet' 0.9, 'despite' 0.7, 'nevertheless' 0.4, 'by contrast' 0.2, 'nonetheless' 0.1, 'in spite of' 0.1, and 'even so', 'notwithstanding', and 'for all that' each less than 0.1.

48. From the same set of articles as in note 9, 'effect' or 'effects' appeared on average 5.7 times per article.
49. Reproduced with permission of the author.
50. Reproduced with permission of the author.
51. A. Reeves, personal communication, 23 November 2015, reproduced with permission.
52. The differences between theorems, propositions, lemmas, corollaries, conjectures, and hypotheses are discussed in [163, §3.1].
53. LaTeX is pronounced with 'TeX' as in 'technology' [206].
54. Snellen acuities are usually expressed as fractions. A value of 6/6 means that a patient can identify a letter at 6 m as well as a normal observer can at that distance, whereas a value of 6/24 means that a patient can identify a letter at 6 m that a normal observer can identify at 24 m. The data in Example 7.1 were taken from Table 1, left eye, of [648]. The entries show just the denominators of the Snellen fractions and include a null value, which has here been omitted.
55. Some authors refer to a 10% trimmed mean as one in which the lowest 5% and highest 5% of values are discarded.
56. From the same set of articles as in note 9, 'mean', as a statistical quantity, appeared on average about eight times as often as 'median'.
57. From the same set of articles as in note 9, 'standard deviation' appeared on average about 20 times as often as 'mean absolute deviation', 'interquartile range', and 'median absolute deviation'.
58. The number of significant figures is the number of digits between the leftmost nonzero digit and the rightmost nonzero digit or rightmost digit including zero if there is a decimal point [26].
59. The correction by $n/(n-1)$ eliminates bias from the estimate of the population variance, but not the estimate of the population standard deviation, a fact that is often ignored [283, §5.1].
60. In MATLAB [245], code for the 95% percentage point `t` for the t-distribution with $n-1$ degrees of freedom, where $n = 10$, is as follows:

```
n = 10; % sample size
p = 0.975; % upper limit
t = icdf('T', p, n-1);
```

61. In MATLAB [245], code for a 95% confidence interval `ci` based on the t-distribution estimate of the standard error is as follows:

```
x = [6 12 24 4.5 7.5 7.5 7.5 6 7.5 4.5]; % data
n = length(x);
p = 0.975; % upper limit
t = icdf('T', p, n-1); % 100p% percentage point
se = std(x)/sqrt(n); % standard error
ci = mean(x) + [-1, 1]*t*se;
```

62. To see the difference between the two statements, consider this situation described in [19, §7.2.2]. The weight of an empty dish is measured as 25.30 g with SD 0.14 g (assume normally distributed errors). A sample of powder is added to the dish. The weight of the dish and powder is measured as 25.50 g with SD 0.14 g. Subtracting the two measurements gives the weight of the powder as 0.20 g with SD 0.20 g (the SD of the difference of two independent normal distributions is the square root of the sum of

the squared SDs). Yet it is not true to say that there is a 68% probability of the weight falling within one SD of the mean and therefore a 32% probability of it falling outside, in particular, a 16% probability of it being negative. For an introduction to Bayesian methods, see [220] [223].

63. In MATLAB [245], code for a bootstrap estimate `se_boot` of the standard error is as follows:

```
x = [6 12 24 4.5 7.5 7.5 7.5 6 7.5 4.5]; % data
nboot = 200; % no. of iterations
bootstat = bootstrp(nboot, @mean, x); % nboot means
se_boot = std(bootstat);
```

64. In MATLAB [245], code for a 95% bootstrap percentile confidence interval `ci_boot` for the population mean is as follows:

```
x = [6 12 24 4.5 7.5 7.5 7.5 6 7.5 4.5]; % data
nboot = 1000; % no. of iterations
ci_boot = bootci(nboot, {@mean, x}, 'type', 'percentile');
```

65. The code is the same as in note 64 except that the argument `'percentile'` is replaced by `'bca'`.
66. Some authors call the highest power of *x* the degree of the polynomial, but practice varies, even within some reference works. Usage in this guide is the same as in [93] [110].
67. In MATLAB [245], code for the simple linear regression `xfit, yfit` is as follows. The arrays `x` and `y` contain the *x* and *y* data, not shown here.

```
X = [ones(size(x)), x]; % format x data
[b, bint] = regress(y, X); % coefficients, 95% CIs
xfit = linspace(min(x), max(x)); % evenly spaced points
yfit = b(1) + b(2)*xfit;
a0_hat = b(1); % coefficient a0
a1_hat = b(2); % coefficient a1
CI0 = bint(1,:); % CI for a0
CI1 = bint(2,:); % CI for a1
```

68. The remainder of Section 7.6.3 is based on [737].
69. The package modelfree, which contains code for either MATLAB [245] or R [298], is available at [241] and http://www.oup.co.uk/companion/foster/.
70. Formal definitions of these other functions can be found in the statistics literature, e.g. [17].
71. In MATLAB [245], code for the cubic regression `xfit, yfit` is as follows. The arrays `x` and `y` contain the *x* and *y* data, not shown here.

```
X = [ones(size(x)), x, x.^2, x.^3]; % format x data
xfit = linspace(min(x), max(x)); % evenly spaced points
[b, bint] = regress(y, X); % coefficients, 95% CIs
yfit = b(1) + b(2)*xfit + b(3)*xfit.^2 + b(4)*xfit.^3;
```

72. In MATLAB [245], code for the local linear regression `xfit, yfit` is as follows. The routine `loess` is from the Data Visualization Toolbox, available at [195] [196]. The arrays `x` and `y` contain the *x* and *y* data, not shown here.

```
xfit = linspace(min(x), max(x)); % evenly spaced points
h = 0.19; % neighbourhood size as fraction of range
deg = 1; % order of polynomial
yfit = loess(x, y, xfit, h, deg);
```

73. The term degrees of freedom is used in both the singular and plural in the literature.
74. The hat matrix is returned in the package modelfree, which contains code in both MATLAB [245] and R [298], and is available at [241] and http://www.oup.co.uk/companion/foster/.
75. In MATLAB [245], code for the quadratic regression `xfit`, `yfit` is as follows. The arrays `x` and `y` contain the *x* and *y* data, not shown here.

```
X = [ones(size(x)), x, x.^2]; % format x data
[b, bint] = regress(y, X); % coefficients, 95% CIs
xfit = linspace(min(x), max(x)); % evenly spaced points
yfit = b(1) + b(2)*xfit + b(3)*xfit.^2;
a1_hat = b(2); % coefficient a1
a2_hat = b(3); % coefficient a2
CI1 = bint(2,:); % CI for a1
CI2 = bint(3,:); % CI for a2
```

76. In MATLAB [245], code for the local linear regression `xfit`, `yfit` is as follows. The routine `loess` is from the Data Visualization Toolbox, available at [195] [196]. The arrays `x` and `y` contain the *x* and *y* data, not shown here.

```
xfit = x; % same x-values for R2 calculation
h = 0.6; % neighbourhood size as fraction of range
deg = 1; % order of polynomial
yfit = loess(x, y, xfit, h, deg);
yfit = reshape(yfit, size(y)); % ensure both column or both row
res = y - yfit; % residuals
R2 = 1 - sum(res.^2)/sum((y - mean(y)).^2); % prop variance
explained
```

77. Reproduced with permission of the author.
78. The dot plots described here should not be confused with dot plots used to represent frequency histograms [360].
79. The average cost to a publisher of producing an article in 2011 was between $3,500 and $4,000, according to one report [362]. Some article processing charges to authors are very much less, but some journals are cross-subsidized and not all journals deliver the same quality of publication or post-publication support.
80. See [317] for comments on the Thomson Reuters Journal Impact Factor and for links to earlier correspondence.
81. R. A. Audisio, personal communication, 22 May 2016, reproduced with permission.
82. Anonymized samples were abstracted by colleagues from peer review reports in their own subject areas, including physical and life sciences, behavioural and clinical sciences, mathematics, computer science, and engineering, along with some samples from my own editorial work. I discarded duplicates and idiosyncratic samples, consolidated the remainder, and finally edited them for clarity and generality.
83. A. Reeves, personal communication, 30 November 2015, reproduced with permission.

84. Reproduced with permission of coauthors K. Amano and S. M. C. Nascimento.
85. An even more accommodating 'CC0' mark enables authors to waive their interests and place an article as completely as possible in the public domain [70].
86. A font size range of 24–36 pt is compatible with published data on the sensitivity of the eye to spatial detail [276] [312]. Sensitivity peaks at spatial frequencies of about 3–4 cycles per degree of visual angle. At a viewing distance of 2 m, this spatial frequency range corresponds to letter sizes of about 25–33 pt. Different viewing distances give different optimum font sizes.
87. Historically, a slide was a small transparency whose image was projected optically onto a plain screen.
88. A font size range of 18–22 pt is compatible with the data described in note 86. In a theatre in which the maximum viewing distance is 12 m and in which a nominal slide width of 25.4 cm is projected at 2.25 m, a spatial frequency range of about 3–4 cycles per degree of visual angle corresponds to letter sizes of about 17–22 pt. Different viewing distances and different projected slide sizes give different optimum font sizes.
89. Some guides argue that bullet points should be avoided altogether [7, p. 122].
90. The Physiological Society guidelines for oral communications require that 'No Oral Communications shall be read (Domestic Rule DRA9) either from written script, slides or PowerPoint transmissions' [286].
91. Coincidentally or not, articles with duplicated material receive fewer citations [50].
92. I asked subscribers to an international mailing list CVNet (cvnet@mail.ewind.com) which of five methods of author ordering they used. I posted the results from 108 respondents on CVNet on 12 July 2004, along with some individual comments. Preferences were as follows, expressed as percentages:

Alphabetic order	1
Descending order of contribution	42
Descending order of contribution but senior (most experienced) author last	51
Descending order of need (for employment, promotion, etc.)	6
Random order	0

A copy of the results is available at http://www.oup.co.uk/companion/foster/.

93. *The Journal of Physiology*, first published in 1878, introduced alphabetical ordering of author names in 1928 but restored author choice in 1990, owing largely to competition from other journals [35]. The transition was difficult for authors who had published jointly authored papers in the journal before and after the transition date.

Bibliography

1. Aksnes, D. W. A macro study of self-citation. *Scientometrics*. 2003; 56 (2): 235–46.
2. Albert, T., Wager, E. How to handle authorship disputes: A guide for new researchers. The COPE report 2003: COPE. 2003. http://publicationethics.org/ [accessed 22 October 2015].
3. Allen, L., Jones, C., Dolby, K., et al. Looking for landmarks: The role of expert review and bibliometric analysis in evaluating scientific publication outputs. *PLOS ONE*. 2009; 4 (6): e5910. DOI: 10.1371/journal.pone.0005910.
4. Alley, M. *The craft of scientific writing* (3rd ed.). New York: Springer-Verlag; 1996.
5. Altman, D. G., Bland, J. M. Standard deviations and standard errors. *BMJ*. 2005; 331 (7521): 903.
6. Alzahrani, S. M., Salim, N., Abraham, A. Understanding plagiarism linguistic patterns, textual features, and detection methods. *IEEE Transactions on Systems, Man, and Cybernetics, Part C: Applications and Reviews*. 2012; 42 (2): 133–49.
7. Anderson, C. J. *TED talks: The official TED guide to public speaking*. London: Headline Publishing Group; 2016.
8. Armitage, P., Berry, G., Matthews, J. N. S. *Statistical methods in medical research* (4th ed.). Oxford: Blackwell Science Ltd; 2002.
9. Article sharing. Elsevier, 2016. http://www.elsevier.com [accessed 5 August 2016].
10. arXiv. Cornell University Library. http://arxiv.org/ [accessed 31 October 2015].
11. Author insights 2015 survey. Nature Publishing Group (July 2015). http://figshare.com/ [accessed 10 November 2015].
12. Author responsibilities & rights. Elsevier B.V. http://jp.elsevier.com/__data/assets/pdf_file/0019/175312/paul3_nsr.pdf [accessed 23 October 2015].
13. Author rights and responsibilities. IEEE, 2015. http://www.ieee.org [accessed 5 August 2016].
14. Authors: General information. *Annual Reviews*. Palo Alto, CA: Annual Reviews; 2016. http://www.annualreviews.org/ [accessed 17 March 2016].
15. Babyak, M. A. What you see may not be what you get: A brief, nontechnical introduction to overfitting in regression-type models. *Psychosomatic Medicine*. 2004; 66 (3): 411–21. DOI: 10.1097/01.psy.0000127692.23278.a9.
16. Baker, M. First results from psychology's largest reproducibility test. *Nature News*. 2015 (30 April 2015), DOI: 10.1038/nature.2015.17433.
17. Balakrishnan, N., Nevzorov, V. B. *A primer on statistical distributions*. Hoboken, NJ: John Wiley & Sons; 2003.
18. Bar-Ilan, J. Which h-index?—A comparison of WoS, Scopus and Google Scholar. *Scientometrics*. 2008; 74(2): 257–71. DOI: 10.1007/s11192-008–0216-y.
19. Barlow, R. J. *Statistics: A guide to the use of statistical methods in the physical sciences*. Chichester, West Sussex: John Wiley & Sons; 1989.
20. Barnett, V. *Environmental statistics: Methods and applications*. Chichester, West Sussex: John Wiley & Sons; 2004.

21. Barnett, V., Lewis, T. *Outliers in statistical data* (3rd ed.). Chichester, West Sussex: John Wiley & Sons; 1994.
22. Beall, J. List of publishers: Beall's list, 2016. http://scholarlyoa.com/ [archived at https://archive.org/web/, accessed 25 March 2017].
23. Bell Labs, Lucent Technologies. *Report of the investigation committee on the possibility of scientific misconduct in the work of Hendrik Schön and coauthors.* September 2002. American Physical Society.
24. Bem, D. J. Writing the empirical journal article. In: Zanna, M. P., Darley, J. M. (eds.). *The compleat academic: A practical guide for the beginning social scientist.* New York, NY: McGraw-Hill, Inc.; 1987; pp. 171–201.
25. Bes-Rastrollo, M., Schulze, M. B., Ruiz-Canela, M., et al. Financial conflicts of interest and reporting bias regarding the association between sugar-sweetened beverages and weight gain: A systematic review of systematic reviews. *PLOS Medicine.* 2013; 10 (12): e1001578. DOI: 10.1371/journal.pmed.1001578.
26. Bevington, P. R., Robinson, D. K. *Data reduction and error analysis for the physical sciences* (3rd ed.). New York, NY: McGraw-Hill; 2003.
27. bioRxiv. Cold Spring Harbor Laboratory. http://biorxiv.org/ [accessed 24 May 2016].
28. Björk, B.-C., Roos, A., Lauri, M. Scientific journal publishing: Yearly volume and open access availability. *Information Research.* 2009; 14 (1), 391. http://InformationR.net/ [accessed 31 July 2016].
29. Björk, B.-C., Solomon, D. Open access versus subscription journals: A comparison of scientific impact. *BMC Medicine.* 2012; 10 (1): 73.
30. Bland, J. M., Altman, D. G. Statistical methods for assessing agreement between two methods of clinical measurement. *Lancet.* 1986; 1 (8476): 307–10.
31. Booth, V. *Writing a scientific paper.* Colnbrook, Bucks, England: Koch-Light Laboratories Ltd; 1971.
32. Börjars, K., Burridge, K. *Introducing English grammar* (2nd ed.). London: Hodder Education; 2010.
33. Borowick, J. N. *Technical communication and its applications* (2nd ed.). Upper Saddle River, NJ: Prentice Hall; 2000.
34. Bould, M. D., Hladkowicz, E. S., Pigford, A.-A. E., et al. References that anyone can edit: Review of Wikipedia citations in peer reviewed health science literature. *BMJ.* 2014; 348: g1585.
35. Boyd, R. Physiology news. 2015. (100), Autumn 2015. The Physiological Society. pp. 20–3.
36. Breidthardt, A. Ally's resignation spares German chancellor drawn-out scandal. Reuters News Agency (10 February 2013). http://www.reuters.com/ [accessed 16 October 2015].
37. Broglio, K. R., Connor, J. T., Berry, S. M. Not too big, not too small: a Goldilocks approach to sample size selection. *Journal of Biopharmaceutical Statistics.* 2014; 24: 685–705.
38. Brown, L. (ed.). *New shorter Oxford English dictionary* (5th ed.). Oxford: Oxford University Press; 2002.
39. Brumfiel, G. Physicist found guilty of misconduct. *Nature* (26 September 2002), DOI: 10.1038/news020923-9
40. Burchfield, R. W. *The new Fowler's modern English usage* (3rd ed.). Oxford: Oxford University Press; 1996.

41. Cairncross, A. When share-dabbling cost Keynes his shirt. 1983. *The Guardian* (8 June 1983). London and Manchester.
42. Calcagno, V., Demoinet, E., Gollner, K., et al. Flows of research manuscripts among scientific journals reveal hidden submission patterns. *Science*. 2012; 338: 1065–9. DOI: 10.1126/science.1227833.
43. CAS source index (CASSI). American Chemical Society. 2016. http://cassi.cas.org [accessed 1 August 2016].
44. Chalker, S., Weiner, E. *The Oxford dictionary of English grammar*. Oxford: Oxford University Press; 1994.
45. Chambers, J. M., Cleveland, W. S., Kleiner, B., et al. *Graphical methods for data analysis*. Boca Raton: Chapman & Hall/CRC Press; 1998.
46. Chavalarias, D., Wallach, J. D., Li, A. H. T., et al. Evolution of reporting *P* values in the biomedical literature, 1990–2015. *JAMA*. 2016; 315 (11): 1141–8. DOI: 10.1001/jama.2016.1952.
47. Chernick, M. R. *Bootstrap methods: A practitioner's guide*. New York: John Wiley & Sons; 1999.
48. Chernin, E. The 'Harvard system': a mystery dispelled. *BMJ*. 1988; 297: 1062–3.
49. *The Chicago manual of style* (16th ed.). Chicago: University of Chicago Press; 2010.
50. Citron, D. T., Ginsparg, P. Patterns of text reuse in a scientific corpus. *Proceedings of the National Academy of Sciences of the United States of America*. 2015; 112 (1): 25–30.
51. Cleveland, W. S. Robust locally weighted regression and smoothing scatterplots. *Journal of the American Statistical Association*. 1979; 74 (368): 829–36.
52. Cleveland, W. S. LOWESS: A program for smoothing scatterplots by robust locally weighted regression. *American Statistician*. 1981; 35 (1): 54.
53. Cleveland, W. S. *Visualizing data*. Summit, NJ: Hobart Press; 1993.
54. Cleveland, W. S. *The elements of graphing data*. Summit, NJ: Hobart Press; 1994.
55. Coghill, A. M., Garson, L. R. (eds.). *The ACS style guide*. Washington, DC: American Chemical Society; 2006.
56. Cole, D. The Chinese room argument. In: Zalta, E. N. (ed.) *The Stanford encyclopedia of philosophy*, 2014. http://plato.stanford.edu/ [accessed 7 August 2016].
57. Commission recommendation on access to and preservation of scientific information. Brussels: European Commission (16 May 2013). http://ec.europa.eu/ [accessed 15 August 2016].
58. Committee on Professional Ethics of the American Statistical Association. Ethical guidelines for statistical practice: American Statistical Association (April 2016). http://www.amstat.org/ [accessed 1 August 2016].
59. Committee on Science, Engineering, and Public Policy, National Academy of Sciences, National Academy of Engineering, and Institute of Medicine. *On being a scientist: A guide to responsible conduct in research* (3rd ed.). Washington, DC: The National Academies Press; 2009.
60. Conflict of interest policy. Proceedings of the National Academy of Sciences of the United States of America (December 2014). http://www.pnas.org/site/authors/coi.xhtml [accessed 24 October 2015].
61. COPE. Text recycling guidelines: Committee on Publication Ethics (COPE) (22 February 2013). http://publicationethics.org/ [accessed 9 October 2015].
62. COPE. Committee on Publication Ethics. http://publicationethics.org/ [accessed 9 October 2015].

63. COPE. What constitutes authorship? COPE Discussion document. Committee on Publication Ethics (COPE) (9 June 2014). http://publicationethics.org/ [accessed 27 January 2016].
64. COPE. Compromised peer review system in published papers. Committee on Publication Ethics (COPE), 2012. http://publicationethics.org/ [accessed 17 November 2015].
65. COPE. Reprimanded author plagiarizes again. Committee on Publication Ethics (COPE) 2012. http://publicationethics.org/ [accessed 13 October 2015].
66. Copyright: Journal author rights. Elsevier, 2016. http://www.elsevier.com/ [accessed 1 August 2016].
67. Costas, R., Zahedi, Z., Wouters, P. Do 'altmetrics' correlate with citations? Extensive comparison of altmetric indicators with citations from a multidisciplinary perspective. *Journal of the Association for Information Science and Technology*. 2015; 66 (10): 2003–19. DOI: 10.1002/asi.23309.
68. Council of Science Editors. *Scientific style and format: The CSE manual for authors, editors, and publishers* (8th ed.). Chicago, IL: University of Chicago Press; 2014.
69. Cox, D. R., Donnelly, C. A. *Principles of applied statistics*. Cambridge: Cambridge University Press; 2011.
70. Creative Commons. About the licenses. Mountain View, CA: Creative Commons. http://creativecommons.org/ [accessed 18 December 2015].
71. Crompton, P. Hedging in academic writing: Some theoretical problems. *English for Specific Purposes*. 1997; 16 (4): 271–87.
72. Crossref. Lynnfield, MA: Publishers International Linking Association, Inc. http://www.crossref.org/ [accessed 23 December 2015].
73. Crystal, D. *The gift of the gab: How eloquence works*. New Haven, CT: Yale University Press; 2016.
74. Cumming, G., Fidler, F., Vaux, D. L. Error bars in experimental biology. *Journal of Cell Biology*. 2007; 177 (1): 7–11.
75. Davis, M., Johnston, J. Appendix C. Conflict of interest in four professions: A comparative analysis. In: Lo, B., Field, M. J. (eds.). *Conflict of interest in medical research, education, and practice*. Washington, DC: National Academies Press; 2009. http://www.ncbi.nlm.nih.gov/books/NBK22946/ [accessed 10 April 2016].
76. Davis, P. M., Lewenstein, B. V., Simon, D. H., et al. Open access publishing, article downloads, and citations: Randomised controlled trial. *BMJ*. 2008; 337. DOI: 10.1136/bmj.a568.
77. Davison, A. C., Hinkley, D. V. *Bootstrap methods and their application*. Cambridge: Cambridge University Press; 1997.
78. Day, R. A. *How to write & publish a scientific paper*. (5th ed.). Westport, CT: Oryx Press; 1998.
79. Day, R. A., Gastel, B. *How to write and publish a scientific paper* (7th ed.). Santa Barbara, CA: Greenwood; 2011.
80. Day, R. A., Sakaduski, N. *Scientific English: A guide for scientists and other professionals* (3rd ed.). Santa Barbara, CA: Greenwood; 2011.
81. Deakin, L., Docking, M., Graf, C., et al. Best practice guidelines on publishing ethics: A publisher's perspective. John Wiley & Sons, Ltd., 2014. http://exchanges.wiley.com/ethicsguidelines [accessed 1 August 2016].
82. Deng, B. Papers with shorter titles get more citations. *Nature News*. 2015 (26 August 2015), DOI: 10.1038/nature.2015.18246.

83. Desbiens, N. A. Let medicine lead the way beyond the registration of clinical trials: Registering hypotheses. *American Journal of the Medical Sciences*. 2008; 335 (2): 137–40.
84. Dickens, C. All the year round: A weekly journal. 1860; 3: 189. London.
85. Dobozi, P. University senate revokes Pál Schmitt's doctoral (dr. Univ.) title: Semmelweis University. 2012 (29 March 2012). http://semmelweis.hu/ [accessed 17 October 2015].
86. Dodge, Y. (ed.). *The Oxford dictionary of statistical terms*. Oxford: Oxford University Press; 2006.
87. The DOI system. International DOI Foundation. http://www.doi.org/ [accessed 14 July 2016].
88. Donovan, S. K. The importance of resubmitting rejected papers. *Journal of Scholarly Publishing*. 2007; 38 (3): 151–5. DOI: 10.3138/w706-5884-kg84-j282.
89. Doumont, J. English communication for scientists: Unit 2: Writing scientific papers. Nature Education, Nature Publishing Group. http://www.nature.com/scitable/ [accessed 25 June 2016].
90. Doumont, J. English communication for scientists: Unit 4: Giving oral presentations. Nature Education, Nature Publishing Group (17 January 2014). http://www.nature.com/scitable/ [accessed 6 January 2016].
91. Doumont, J. English communication for scientists: Unit 5: Giving poster presentations. Nature Education, Nature Publishing Group (17 January 2014). http://www.nature.com/scitable/ [accessed 10 January 2016].
92. Downing, A., Locke, P. *English grammar: A university course* (2nd ed.). London: Routledge; 2006.
93. Draper, N. R., Smith, H. *Applied regression analysis* (3rd ed.). New York: John Wiley & Sons, Inc.; 1998.
94. Durrani, M. Bell Labs physicist fired for misconduct. *Physics World*: Institute of Physics (25 September 2002). http://physicsworld.com/ [accessed 14 October 2015].
95. Eco, U. *How to write a thesis*. Cambridge, MA: MIT Press; 2015.
96. Editorial. How robust are your data? *Nature Cell Biology*. 2009; 11 (6): 667.
97. Editorial. With transparency comes trust. *Nature*. 2012; 485: 147.
98. Editorial. Sharing clinical trial data: A proposal from the International Committee of Medical Journal Editors. *Annals of Internal Medicine*. 2016; 164 (7): 505–6. DOI: 10.7326/M15-2928.
99. Efron, B., Tibshirani, R. J. *An Introduction to the bootstrap*. New York, NY: Chapman & Hall; 1993.
100. Egghe, L. Theory and practise of the *g*-index. *Scientometrics*. 2006; 69 (1): 131–52. DOI: 10.1007/s11192-006–0144-7.
101. Ehrenberg, A. S. C. *A primer in data reduction*. Chichester: John Wiley & Sons; 1994.
102. Eisen, J. A., MacCallum, C. J., Neylon, C. Expert failure: Re-evaluating research assessment. *PLOS Biology*. 2013; 11 (10): e1001677. DOI: 10.1371/journal.pbio.1001677.
103. Ellenberg, J. H. Ethical guidelines for statistical practice: A historical perspective. *American Statistician*. 1983; 37 (1): 1–4.
104. Elsevier Publishing Campus. Quick guides: Research and publishing ethics. Elsevier B.V. http://www.publishingcampus.elsevier.com/ [accessed 23 October 2015].

105. Ethical standards and procedures for research with human beings. World Health Organisation, 2016. http://www.who.int/ [accessed 25 October 2015].
106. Eubank, R. L., Speckman, P. Curve fitting by polynomial-trigonometric regression. *Biometrika*. 1990; 77 (1): 1–9.
107. Everitt, B. S., Skrondal, A. *The Cambridge dictionary of statistics* (4th ed.). Cambridge: Cambridge University Press; 2010.
108. Ewing, J. Paul Halmos: in his own words—On exposition. *Notices of the American Mathematical Society*. 2007; 54 (9): 1139.
109. Eyre-Walker, A., Stoletzki, N. The assessment of science: The relative merits of post-publication review, the impact factor, and the number of citations. *PLOS Biology*. 2013; 11 (10): e1001675. DOI: 10.1371/journal.pbio.1001675.
110. Fan, J., Gijbels, I. *Local polynomial modelling and its applications*. London: Chapman & Hall; 1996.
111. Fan, J., Heckman, N. E., Wand, M. P. Local polynomial kernel regression for generalized linear models and quasi-likelihood functions. *Journal of the American Statistical Association*. 1995; 90 (429): 141–50.
112. Fang, F. C., Casadevall, A. Retracted science and the retraction index. *Infection and Immunity*. 2011; 79 (10): 3855–9.
113. Fang, F. C., Steen, R. G., Casadevall, A. Misconduct accounts for the majority of retracted scientific publications. *Proceedings of the National Academy of Sciences of the United States of America*. 2012; 109 (42): 17028–33.
114. An FAQ on intellectual property rights for IEEE authors. IEEE. http://www.ieee.org/ [accessed 3 December 2015].
115. Faraway, J. J., Jhun, M. Bootstrap choice of bandwidth for density estimation. *Journal of the American Statistical Association*. 1990; 85 (412): 1119–22.
116. Farmelo, G. *The strangest man: The hidden life of Paul Dirac, quantum genius*. London: Faber and Faber; 2009.
117. Federal research misconduct policy. Rockville MD: Office of Research Integrity, U.S. Department of Health and Human Services (6 December 2000). http://ori.hhs.gov/ [accessed 10 April 2016].
118 Fisher, N. I. *Statistical analysis of circular data*. Cambridge: Cambridge University Press; 1993.
119. Fisher, R. A. On the mathematical foundations of theoretical statistics. *Philosophical Transactions of the Royal Society A*. 1922; 222: 309–68.
120. Former Kyoto U. chief to take helm at scandal-plagued Riken. *Japan Times*. 2015 (24 March 2015). http://www.japantimes.co.jp/ [accessed 10 April 2016].
121. Forsyth, M. *The elements of eloquence: How to turn the perfect English phrase*. London: Icon Books Ltd; 2013.
122. Fowler, H. W. *A dictionary of modern English usage* (2nd ed., revised by E. Gowers). Oxford: Oxford University Press; 1968.
123. Fowler, H. W., Fowler, F. G. *The King's English* (3rd ed.). London: Oxford University Press; 1973.
124. Fowler, J. H., Aksnes, D. W. Does self-citation pay? *Scientometrics*. 2007; 72 (3): 427–37. DOI: 10.1007/s11192-007–1777-2.
125. Freyne, J., Coyle, L., Smyth, B., et al. Relative status of journal and conference publications in computer science. *Communications of the ACM*. 2010; 53 (11): 124–32.

126. Fritz, C. O., Morris, P. E., Richler, J. J. Effect size estimates: Current use, calculations, and interpretation. *Journal of Experimental Psychology: General.* 2012; 141 (1): 2–18.
127. Gardner, M. J., Altman, D. G. Confidence intervals rather than P values: Estimation rather than hypothesis testing. *British Medical Journal.* 1986; 292: 746–50.
128. Garner, J. K., Alley, M. P. How the design of presentation slides affects audience comprehension: A case for the assertion-evidence approach. *International Journal of Engineering Education.* 2013; 29 (6): 1564–79.
129. Gasser, T., Sroka, L., Jennen-Steinmetz, C. Residual variance and residual pattern in nonlinear regression. *Biometrika.* 1986; 73 (3): 625–33.
130. Gawande, A. The cancer-cluster myth. 1999. New Yorker, 8 February 1999: Condé Nast.
131. Get noticed: Promoting your article for maximum impact. Elsevier, 2015. http://www.elsevier.com/ [accessed 20 December 2015].
132. Getting published in *Nature*: The editorial process. *Nature*: Macmillan Publishers Limited. 2015. http://www.nature.com/nature/authors/ [accessed 5 November 2015].
133. Gibney, E. I'm no plagiarist, I moved a comma. 2013. *THE: Times Higher Education*, 6 June 2013. London: TES Global Ltd, p. 9.
134. Gillman, L. *Writing mathematics well: A manual for authors.* The Mathematical Association of America; 1987.
135. Gilmour, R., Cobus-Kuo, L. Reference management software: A comparative analysis of four products. *Issues in Science and Technology Librarianship.* Association of College & Research Libraries. 2011, DOI: 10.5062/F4Z60KZF.
136. Global publishing: Changes in submission trends and the impact on scholarly publishers. Thomson Reuters. 2012 (April 2012). http://scholarone.com/media/pdf/GlobalPublishing_WP.pdf [accessed 30 November 2015].
137. Good, P. I., Hardin, J. W. *Common errors in statistics (and how to avoid them).* Hoboken, NJ: John Wiley & Sons, Inc.; 2003.
138. Google Scholar. Mountain View, CA: Google. http://scholar.google.co.uk/ [accessed 22 December 2015].
139. Gosling, P., Noordam, B. Mastering your PhD: Survival and success in the doctoral years and beyond (2nd ed.). Berlin: Springer, 2011, DOI: 10.1007/978–73-642–15847-6.
140. Goss, D. Some hints on mathematical style. Department of Mathematics, The Ohio State University. http://people.math.osu.edu/ [accessed 1 January 2015].
141. Gowers, E. *The complete plain words.* London: Her Majesty's Stationery Office; 1954.
142. Gowers, E. *The complete plain words* (3rd ed., revised by S. Greenbaum and J. Whitcut). London: Penguin Books Ltd; 1986.
143. Graphical abstracts. Elsevier, 2016. http://www.elsevier.com/ [accessed 7 June 2016].
144. Green, P. J., Silverman, B. W. *Nonparametric regression and generalized linear models: A roughness penalty approach.* Boca Raton, FL: Chapman & Hall/CRC; 1994.
145. Greenberg, S. A. How citation distortions create unfounded authority: Analysis of a citation network. *BMJ.* 2009; 339: b2680.
146. Grey Literature International Steering Committee. Guidelines for the production of scientific and technical reports: How to write and distribute grey literature. Version 1.0. 2006 (March 2006). http://eprints.rclis.org/ [accessed 10 August 2016].

147. Grieneisen, M. L., Zhang, M. A comprehensive survey of retracted articles from the scholarly literature. *PLOS ONE*. 2012; 7 (10): e44118. DOI: 10.1371/journal.pone.0044118.
148. Guidelines for responsible data management in scientific research. Rockville MD: Office of Research Integrity, U.S. Department of Health and Human Services. 2006. http://ori.hhs.gov/ [accessed 24 October 2015].
149. Halmos, P. R. 'Nicolas Bourbaki'. *Scientific American*. 1957; 196 (5): 88–99.
150. Halmos, P. R. How to write mathematics. In: Steenrod, N. E., Halmos, P. R., Schiffer, M. M., et al. (eds.). How to write mathematics: American Mathematical Society; 1973; pp. 19–48.
151. Hamby, S. On scientific writing in the information era: Tailoring papers for internet searching and other 21st century realities. *Psychology of Violence*. 2015; 5 (2): 103–11.
152. Hames, I. COPE ethical guidelines for peer reviewers. COPE: Committee on Publication Ethics (March 2013). http://publicationethics.org/ [accessed 17 November 2015].
153. Handoll, H. H. G., Atkinson, G. Snowballing citations. *BMJ*. 2015; 351: h6309.
154. Härdle, W. *Applied nonparametric regression*. Cambridge: Cambridge University Press; 1990.
155. Härdle, W., Werwatz, A., Müller, M., et al. *Nonparametric and semiparametric models*. Berlin: Springer; 2004.
156. Hart, J. D. *Nonparametric smoothing and lack-of-fit tests*. New York: Springer; 1997.
157. Hartley, J. *Academic writing and publishing: A practical handbook*. Abingdon, Oxon: Routledge; 2008.
158. Harvard guide to using sources: Avoiding plagiarism. Harvard College Writing Program, 2015. http://harvard.edu/ [accessed 9 October 2015].
159. Hastie, T., Tibshirani, R., Friedman, J. *The elements of statistical learning* (2nd ed.). New York: Springer-Verlag; 2009.
160. Hastie, T. J., Tibshirani, R. J. *Generalized additive models*. London: Chapman & Hall; 1990.
161. Hicks, W. *English for journalists* (2nd ed.). London: Routledge; 1998.
162. Higgins, J. P. T., Green, S. (eds.). *Cochrane handbook for systematic reviews of interventions, version 5.1.0 [updated March 2011]*. The Cochrane Collaboration, 2011. Available from http://www.handbook.cochrane.org.
163. Higham, N. J. *Handbook of writing for the mathematical sciences* (2nd ed.). Philadelphia: SIAM: Society for Industrial and Applied Mathematics; 1998.
164. Highlights. Elsevier, 2016. http://www.elsevier.com/ [accessed 24 February 2016].
165. Hilbert, D. Mathematical problems. Lecture delivered before the International Congress of Mathematicians at Paris in 1900. *Bulletin of the American Mathematical Society*. 1900/2000; 37 (4): 407–36. Originally published as *Mathematische Probleme. Vortrag, gehalten auf dem internationalen, Mathematike-Congress zu Paris 1900, Gött. Nachr. 1900*, 253–97, Vandenhoeck & Ruprecht, Göttingen.
166. Hirsch, J. E. An index to quantify an individual's scientific research output. *Proceedings of the National Academy of Sciences of the United States of America*. 2005; 102 (46): 16569–72.
167. Hitchings, H. *The language wars: A history of proper English*. London: John Murray; 2011.

168. Hoel, P. G. *Introduction to mathematical statistics* (5th ed.). New York: John Wiley & Sons; 1984.
169. Hofmann, A. H. *Scientific writing and communication: Papers, proposals, and presentations*. New York: Oxford University Press; 2010.
170. Holdren, J. P. Increasing access to the results of federally funded scientific research. Washington, DC: Office of Science and Technology Policy (22 February 2013). http://www.whitehouse.gov [accessed 24 October 2015].
171. Holladay, J. T. Proper method for calculating average visual acuity. *Journal of Refractive Surgery*. 1997; 13: 388–91.
172. How to deal with text recycling. BioMed Central, 2014. http://www.biomedcentral.com/ [accessed 14 August 2016].
173. How to publish in scholarly journals. Elsevier, 2015. http://www.elsevier.com [accessed 28 October 2015].
174. How to write for technical periodicals & conferences. Piscataway, NJ: IEEE, 2012. http://www.ieee.org [accessed 5 August 2016].
175. Howell, D. C. *Statistical methods for psychology* (8th ed.). Wadsworth, International Edition; 2013.
176. Hyland, K. Writing without conviction? Hedging in science research articles. *Applied Linguistics*. 1996; 17 (4): 433–54.
177. Hyland, K. Academic publishing and the myth of linguistic injustice. *Journal of Second Language Writing*. 2016; 31: 58–69.
178. IEEE Computer Society. Journal peer review: Author information. IEEE, 2015. http://www.computer.org/ [accessed 14 November 2015].
179. IEEE editorial style manual. Piscataway, NJ: IEEE, 2014. www.ieee.org [accessed 5 August 2016].
180. IEEE PES Transactions and Journals. IEEE Power and Energy Society, 2015. http://www.ieee-pes.org/ [accessed 18 November 2015].
181. IEEE policies: Section 7.8. IEEE Code of Ethics. IEEE, 2016. http://www.ieee.org/ [accessed 10 April 2016].
182. IEEE principles of scholarly publishing. IEEE, 2008. http://www.ieee.org [accessed 5 August 2016].
183. IEEE PSPB operations manual: 8.1.9 electronic information dissemination. IEEE (June 2015). http://www.ieee.org/ [accessed 25 May 2016].
184. IEEE Transactions on Industrial Informatics. Rejected manuscripts. IEEE Industrial Electronics Society. http://tii.ieee-ies.org/ [accessed 13 November 2016]
185. IFCC Ethics Task Force. Ethics in science: Background and resources on publication ethics. International Federation of Clinical Chemistry and Laboratory Medicine (IFCC). 2012 (20 March 2012). http://www.ifcc.org/ [accessed 10 March 2016].
186. Information for authors: Editorial policies. Proceedings of the National Academy of Sciences of the United States of America (October 2015). http://www.pnas.org/ [accessed 11 August 2016].
187. Information for authors: Procedures for submitting manuscripts. Proceedings of the National Academy of Sciences of the United States of America. http://www.pnas.org/ [accessed 11 August 2016].
188. Institutional Review Board. Rockville, MD: Office for Human Research Protections, U.S. Department of Health & Human Services. http://www.hhs.gov/ [accessed 8 October 2015].

189. International copyright basics. Amsterdam, Netherlands: RightsDirect, 2016. http://www.rightsdirect.com/ [accessed 13 August 2016].
190. The International System of Units (SI) (8th ed. and Supplement 2014). International Committee for Weights and Measures (CIPM). 2006/2014.
191. iThenticate. Oakland, CA: Turnitin, LLC. http://www.ithenticate.com [accessed 9 October 2015].
192. Jacques, T. S., Sebire, N. J. The impact of article titles on citation hits: An analysis of general and specialist medical journals. *Journal of the Royal Society of Medicine Short Reports*. 2010; 1: 2.
193. Jamali, H. R., Nikzad, M. Article title type and its relation with the number of downloads and citations. *Scientometrics*. 2011; 88: 653–61.
194. Janson, L., Fithian, W., Hastie, T. J. Effective degrees of freedom: a flawed metaphor. *Biometrika*. 2015; 102 (2): 479–85.
195. Johnson, R. Data visualization toolbox for MATLAB. Datatool. http://datatool.com/resources.html [accessed 1 June 2016].
196. Johnson, R. Data visualization toolbox. MATLAB Central: MathWorks; 2016. www.mathworks.com/matlabcentral/ [accessed 1 June 2016].
197. Johnson, V. E. Revised standards for statistical evidence. *Proceedings of the National Academy of Sciences of the United States of America*. 2013; 110 (48): 19313–17.
198. Journal Citation Reports: Journal Impact Factor. Thomson Reuters, 2012. http://admin-apps.webofknowledge.com/JCR/help/h_impfact.htm#impact_factor [accessed 13 November 2015].
199. Journal Citation Reports 2015 Thomson Reuters. 2015. http://apps.webofknowledge.com/ [accessed 11 November 2015].
200. Journal metrics. Elsevier B.V., 2016. http://www.journalmetrics.com/ [accessed 7 April 2017].
201. Keeping, E. S. *Introduction to statistical inference*. Mineola, NY: Dover Publications; originally published Princeton, NJ: Van Nostrand Company; 1962/1995.
202. Kelleher, C., Wagener, T. Ten guidelines for effective data visualization in scientific publications. *Environmental Modelling & Software*. 2011; 26: 822–7.
203. Kerans, M. E., de Jager, M. Handling plagiarism at the manuscript editor's desk. *European Science Editing*. 2010; 36 (3): 62–6.
204. Kilkenny, C., Browne, W. J., Cuthill, I. C., et al. Improving bioscience research reporting: The ARRIVE guidelines for reporting animal research. *PLOS Biology*. 2010; 8 (6): e1000412. DOI: 10.1371/journal.pbio.1000412.
205. Kirschbaum, E. German defense minister quits in plagiarism row. Reuters News Agency (1 March 2011). http://www.reuters.com/ [accessed 16 October 2015].
206. Knuth, D. E. *The TeXbook*. Reading, MA: Addison-Wesley; 1996.
207. Knuth, D. E., Larrabee, T., Roberts, P. M. *Mathematical writing*. MAA Notes Number 14. Mathematical Association of America; 1989.
208. Kosslyn, S. M. *Elements of graph design*. New York: W. H. Freeman and Co.; 1994.
209. Kromrey, J. D. Ethics and data analysis. *Educational Researcher*. 1993; 22: 24–7.
210. Krzywinski, M., Altman, N. Points of significance: error bars. *Nature Methods*. 2013; 10 (10): 921–2.
211. Kuehne, L. M., Olden, J. D. Opinion: Lay summaries needed to enhance science communication. *Proceedings of the National Academy of Sciences of the United States of America*. 2015; 112 (12): 3585–6.

212. Lamb, E. 5 sigma what's that? *Scientific American, Observations*. 2012 (17 July 2012). http://www.scientificamerican.com/ [accessed 14 August 2016].
213. Lamport, L. *LaTeX: A document preparation system*. Reading, MA: Addison-Wesley Publishing Company; 1994.
214. Larivière, V., Kiermer, V., MacCallum, C. J., et al. A simple proposal for the publication of journal citation distributions. *bioRxiv*. 2016, DOI: 10.1101/062109. http://biorxiv.org/ [accessed 14 August 2016].
215. Larkin, M. How to give a dynamic scientific presentation. Elsevier (4 August 2015). http://www.elsevier.com/ [accessed 6 January 2016].
216. Larsen, P. O., von Ins, M. The rate of growth in scientific publication and the decline in coverage provided by science citation index. *Scientometrics*. 2010; 84: 575–603. DOI: 10.1007/s11192-010–0202-z.
217. LaTeX: A document preparation system. LaTeX3 Project. http://www.latex-project.org/ [accessed 9 January 2016].
218. Law, J., Rennie, R. (eds.). *Dictionary of physics* (7th ed.). Oxford: Oxford University Press; 2015.
219. Lebrun, J.-L. *Scientific writing: A reader and writer's guide*. Singapore: World Scientific Publishing Co. Pte. Ltd.; 2008.
220. Lee, P. M. *Bayesian statistics: An introduction*. London: Edward Arnold; 1994.
221. Leek, J. T., Peng, R. D. What is the question? *Science*. 2015; 347 (6228): 1314–15. DOI: 10.1126/science.aaa6146.
222. Legislation for the protection of animals used for scientific purposes. Directive 2010/63/EU. Brussels, Belgium: European Commission, Directorate-General for Environment, 2010. http://ec.europa.eu/ [accessed 8 April 2016].
223. Leonard, T., Hsu, J. S. *Bayesian methods. An analysis for statisticians and interdisciplinary researchers*. Cambridge: Cambridge University Press; 1999.
224. Lester, M., Beason, L. *The McGraw-Hill handbook of English grammar and usage* (2nd ed.). New York: McGraw-Hill; 2013.
225. Letchford, A., Moat, H. S., Preis, T. The advantage of short paper titles. *Royal Society Open Science*. 2015; 2 (8), 150266. DOI: 10.1098/rsos.150266.
226. Letchford, A., Preis, T., Moat, H. S. The advantage of simple paper abstracts. *Journal of Informetrics*. 2016; 10 (1): 1–8. DOI: 10.1016/j.joi.2015.11.001.
227. Levine, T. R., Hullett, C. R. Eta squared, partial eta squared, and misreporting of effect size in communication research. *Human Communication Research*. 2002; 28 (4): 612–25.
228. Levis, A. W., Leentjens, A. F. G., Levenson, J. L., et al. Comparison of self-citation by peer reviewers in a journal with single-blind peer review versus a journal with open peer review. *Journal of Psychosomatic Research*. 2015; 79: 561–5. DOI: 10.1016/j.jpsychores.2015.08.004
229. Lichtfouse, E. *Scientific writing for impact factor journals*. New York: Nova Science Publishers, Inc.; 2013.
230. Lindman, H. R. *Analysis of variance in experimental design*. New York: Springer-Verlag; 1992.
231. List of title word abbreviations (LTWA). Paris: ISSN International Centre. 2016 (27 July 2015). http://www.issn.org/ [accessed 10 August 2016].
232. Loader, C. *Local regression and likelihood*. New York: Springer; 1999.

233. Lorenzetti, D. L., Ghali, W. A. Reference management software for systematic reviews and meta-analyses: An exploration of usage and usability. *BMC Medical Research Methodology*. 2013; 13: 141.
234. Maltoni, M., Schwetz, T. Testing the statistical compatibility of independent data sets. *Physical Review D*. 2003; 68: 033020.
235. Manage my author profile. Elsevier, 2015. http://www.elsevier.com/ [accessed 19 December 2015].
236. Manly, B. F. J. *Randomization, bootstrap and Monte Carlo methods in biology* (2nd ed.). London: Chapman & Hall; 1997.
237. A manual for authors of mathematical papers. *Bulletin of the American Mathematical Society*. 1943; 49 (3, Part 2): 1–16.
238. A manual for authors of mathematical papers. Providence, RI: American Mathematical Society; 1984.
239. Marcus, A. Authors get away with throwing quotation marks around plagiarized passages. Again. Retraction Watch (21 January 2015). http://retractionwatch.com/ [accessed 10 October 2015].
240. Mardia, K. V., Jupp, P. E. *Directional statistics*. Chichester: John Wiley & Sons; 2000.
241. Marín-Franch, I., Żychaluk, K., Foster, D. H. Modelfree. CRAN; 2012. http://cran.r-project.org/web/packages/modelfree/ [accessed 1 June 2016].
242. Martin, B. R. Editors' JIF-boosting stratagems—Which are appropriate and which not? *Research Policy*. 2016; 45: 1–7.
243. Martin, S., Darnley, L. *The teaching voice* (2nd ed.). London: Whurr Publishers Ltd; 2004.
244. Mason, R. L., Gunst, R. F., Hess, J. L. *Statistical design and analysis of experiments: With applications to engineering and science* (2nd ed.). Hoboken, NJ: John Wiley & Sons; 2003.
245. MATLAB with Statistics and Machine Learning Toolbox (R2015a). Natick, MA: The MathWorks, Inc. http://www.mathworks.com/ [accessed 10 August 2016].
246. Matta, C. F., Massa, L., Gubskaya, A. V., et al. Can one take the logarithm or the sine of a dimensioned quantity or a unit? Dimensional analysis involving transcendental functions. *Journal of Chemical Education*. 2011; 88 (1): 67–70.
247. McCartney, M. Margaret McCartney: The 'breakthrough' drug that's not been shown to help in Alzheimer's disease. *BMJ*. 2015; 351: h4064.
248. McCullagh, P., Nelder, J. A. *Generalized linear models* (2nd ed.). London: Chapman & Hall; 1989.
249. McDougall-Waters, J., Moxham, N., Fyfe, A. Philosophical Transactions: 350 years of publishing at the Royal Society (Exhibition Brochure): The Royal Society, 2015. http://royalsociety.org/ [accessed 7 August 2016].
250. Medawar, P. Is the scientific paper a fraud? In: Pyke, D. (ed.) *The threat and the glory: Reflections on science and scientists*. Oxford: Oxford University Press; 1990; pp. 228–33.
251. Medawar, P. B. Anglo-Saxon attitudes. 1965. *Encounter*; 25 (2) (August): 52–8.
252. Merton, R. K. Priorities in scientific discovery: A chapter in the sociology of science. *American Sociological Review*. 1957; 22 (6): 635–59.
253. Microsoft Word. Microsoft Corporation. http://products.office.com/en-gb/word [accessed 9 January 2016].
254. Miller, R. G., Jr. *Simultaneous statistical inference*. New York: Springer-Verlag; 1981.

255. Moed, H. F. Statistical relationships between downloads and citations at the level of individual documents within a single journal. *Journal of the American Society for Information Science and Technology*. 2005; 56 (10): 1088–97.
256. Monk, R. *Inside the centre: The life of J. Robert Oppenheimer*. London: Jonathan Cape; 2012.
257. Montgomery, D. B., Morrison, D. G. A note on adjusting R^2. *Journal of Finance*. 1973; 28 (4): 1009–13.
258. Montgomery, S. L. *The Chicago guide to communicating science*. Chicago: University of Chicago Press; 2003.
259. Mulligan, A., Taylor, M., Newsum, L. The challenges around defining authorship—You have your say: Elsevier (22 October 2014). http://www.elsevier.com/ [accessed 4 August 2016].
260. Murai, S. Riken serves Obokata, others with penalties; charges could follow. *Japan Times*. 2015 (10 February 2015). http://www.japantimes.co.jp/ [accessed 10 April 2016].
261. *Nature*: Manuscript formatting guide. Nature Publishing Group. http://www.nature.com/nature/authors [accessed 5 August 2016].
262. Neville, C. *The complete guide to referencing and avoiding plagiarism* (2nd ed.): Open University Press Educational and Professional Book Publishers. McGraw-Hill Education Europe; 2010.
263. *New Hart's rules: The Oxford style guide*. Oxford: Oxford University Press; 2014.
264. Nieuwenhuis, S., Forstmann, B. U., Wagenmakers, E.-J. Erroneous analyses of interactions in neuroscience: a problem of significance. *Nature Neuroscience*. 2011; 14 (9): 1105–7.
265. Nuzzo, R. Statistical errors. *Nature*. 2014; 506 (7487): 150–2.
266. OECD principles and guidelines for access to research data from public funding. Organisation for Economic Co-operation and Development (OECD), 2007. http://www.oecd.org/ [accessed 24 October 2015].
267. Office of Human Research Protections (OHRP). Rockville, MD: US Department of Health & Human Services. http://www.hhs.gov/ohrp/ [accessed 8 April 2016].
268. Office of Research Integrity. Findings of research misconduct [re Ryousuke Fujita, PhD, Columbia University]. *Federal Register*. Rockville MD: Office of Research Integrity, U.S. Department of Health and Human Services (7 April 2015). http://www.federalregister.gov/ [accessed 8 October 2015].
269. Okoli, C., Mehdi, M., Mesgari, M., et al. The people's encyclopedia under the gaze of the sages: A systematic review of scholarly research on Wikipedia. Social Sciences Research Network. 2012 (Revised: 17 May 2014), DOI: 10.2139/ssrn.2021326.
270. Olejnik, S., Algina, J. Generalized eta and omega squared statistics: Measures of effect size for some common research designs. *Psychological Methods*. 2003; 8 (4): 434–47.
271. Open access: Self-archiving. John Wiley & Sons, Inc., 2015. http://olabout.wiley.com/ [accessed 31 October 2015].
272. Oransky, I. Boeing engineer has two papers retracted for duplication. Retraction Watch (21 July 2014). http://retractionwatch.com/ [accessed 4 August 2016].
273. ORCID. Bethesda, MD: ORCID, Inc. http://orcid.org/ [accessed 19 December 2015].
274. Orwell, G. Politics and the English language. *Horizon*. April 1946; 252–64. http://www.unz.org/ [accessed 9 August 2016].

275. OUP statement on publication ethics. Oxford: Oxford University Press, 2015. http://oxfordjournals.org/en/authors [accessed 8 October 2015].
276. Owsley, C., Sekuler, R., Siemsen, D. Contrast sensitivity throughout adulthood. *Vision Research*. 1983; 23 (7): 689–99.
277. Oxford dictionaries online. Language matters: avoiding clichés. Oxford: Oxford University Press, 2016. http://blog.oxforddictionaries.com/avoiding-cliches/ [accessed 11 March 2016].
278. Park, I.-U., Peacey, M. W., Munafo, M. R. Modelling the effects of subjective and objective decision making in scientific peer review. *Nature*. 2014; 506: 93–6. DOI: 10.1038/nature12786.
279. Partridge, E. *You have a point there: A guide to punctuation and its allies*. London: Hamish Hamilton; republished Abingdon, Oxon: Routledge; 1953/2005.
280. Partridge, E. *Usage and abusage: A guide to good English* (new ed.; ed. J. Whitcut). New York, NY: W. W. Norton & Co., Inc.; 1997.
281. Pasterkamp, G., Rotmans, J. I., de Kleijn, D. V. P., et al. Citation frequency: A biased measure of research impact significantly influenced by the geographical origin of research articles. *Scientometrics*. 2007; 70 (1): 153–65. DOI: 10.1007/s11192-007–0109-5.
282. Patrias, K. Citing medicine: The NLM style guide for authors, editors, and publishers (2nd ed.). Bethesda, MD: National Library of Medicine (US); Wendling D. L., technical ed., 2011. www.nlm.nih.gov/citingmedicine/ [accessed 15 August 2016].
283. Pawitan, Y. *In all likelihood: Statistical modelling and inference using likelihood*. Oxford: Clarendon Press; 2001.
284. Peters, P. *The Cambridge guide to English usage*. Cambridge: Cambridge University Press; 2004.
285. Physical Measurement Laboratory of NIST. SI unit rules and style conventions: Check list for reviewing manuscripts. NIST: National Institute of Standards and Technology, US Department of Commerce, 2004. http://www.nist.gov/ [accessed 11 August 2016].
286. Physiology 2015: Guidelines for oral communications. London: Physiological Society, 2014.
287. Policies: Article sharing policy. Wiley, 2016. http://olabout.wiley.com/ [accessed 15 August 2016].
288. Poster and lecture presentation guidelines. IEEE Communications Society, ICC 2012, 2012. http://icc2012.ieee-icc.org/posterguideline.html [accessed 16 January 2016].
289. Poster preparation guidelines. London: Physiological Society, 2012. http://www.physoc.org/ [accessed 10 August 2016].
290. Preparation of papers for IEEE Transactions and Journals (December 2013). IEEE, 2013. http://www.ieee.org/ [accessed 5 August 2016].
291. Presenter resources: Poster presentations. Society for Neuroscience, Neuroscience 2015. http://www.sfn.org/ [accessed 16 January 2016].
292. 'Printed publications' as prior art [R-11.2013] (9th ed.). *Manual of Patent Examining Procedure, Chapter 2100, Section 2128*. United States Patent and Trademark Office (November 2015). http://www.uspto.gov/ [accessed 10 August 2016].
293. *Publication manual of the American Psychological Association*. (6th ed.). Washington, DC: American Psychological Association; 2010

294. Publishing ethics and policies: Authorship. London: Royal Society. 2015 (13 July 2015). http://royalsociety.org/ [accessed 8 October 2015].
295. Publishing ethics resource kit (PERK). Elsevier, 2015. http://www.elsevier.com/ [accessed 8 October 2015].
296. Puhan, M. A., ter Riet, G., Eichler, K., et al. More medical journals should inform their contributors about three key principles of graph construction. *Journal of Clinical Epidemiology*. 2006; 59: 1017–22.
297. Quirk, R., Greenbaum, S., Leech, G., et al. *A comprehensive grammar of the English language*. London: Longman; 1985.
298. R Core Team. R: a language and environment for statistical computing. Vienna, Austria: R Foundation for Statistical Computing; 2015. http://www.R-project.org/ [accessed 11 August 2016].
299. Radicchi, F., Fortunato, S., Castellano, C. Universality of citation distributions: Toward an objective measure of scientific impact. *Proceedings of the National Academy of Sciences of the United States of America*. 2008; 105 (45): 17268–72.
300. Rasberry, L. Citing Wikipedia: Don't do it—Wikipedians wouldn't. *BMJ*. 2014; 348: g1819.
301. RCUK common principles on data policy. Research Councils UK. 2011 (revised July 2015). http://www.rcuk.ac.uk/research/datapolicy/ [accessed 24 October 2015].
302. re3data.org: Registry of research data repositories. Hannover, Germany: DataCite. http://re3data.org/ [accessed 24 October 2015].
303. Recommendations for the conduct, reporting, editing, and publication of scholarly work in medical journals. ICMJE International Committee of Medical Journal Editors (December 2015). http://www.icmje.org/ [accessed 23 February 2016].
304. Reich, E. S. *Plastic fantastic: How the biggest fraud in physics shook the scientific world*. New York: Palgrave Macmillan; 2009.
305. Research Publication Investigative Committee. Report on STAP cell research paper investigation. Japan: RIKEN (25 December 2014). http://www.riken.jp/ [accessed 10 April 2016].
306. ResearcherID. Thomson Reuters, 2015. http://wokinfo.com/researcherid/ [accessed 19 December 2015].
307. Responsible Conduct Working Group. Guidelines for responsible conduct regarding scientific communication. Washington, DC: Society for Neuroscience (20 July 2010). http://www.sfn.org/ [accessed 29 October 2015]
308. Retraction Watch. http://retractionwatch.com/ [accessed 16 October 2015].
309. Rice, J. A. *Mathematical statistics and data analysis* (3rd ed.). Belmont, CA: Duxbury, Thomson Corp.; 2007.
310. Rightsdirect. Amsterdam: Copyright Clearance Center. http://www.rightsdirect.com/ [accessed 1 August 2016].
311. Ritter, R. M. *The Oxford guide to style*. Oxford: Oxford University Press; 2002.
312. Robson, J. G. Spatial and temporal contrast-sensitivity functions of the visual system. *Journal of the Optical Society of America*. 1966; 56 (8): 1141–2. DOI: 10.1364/JOSA.56.001141.
313. Rodriguez, N. Infographic: Tips for designing better research posters. Elsevier (7 May 2015). http://www.elsevier.com/ [accessed 12 January 2016].
314. Roig, M. Avoiding plagiarism, self-plagiarism, and other questionable writing practices: A guide to ethical writing. Rockville, MD: Office of Research Integrity,

U.S. Department of Health and Human Services (15 May 2013). http://ori.hhs.gov/ [accessed 10 October 2015].

315. Rosner, J. L. Reflections of science as a product. *Nature*. 1990; 345: 108.
316. Ross, D. Big science: sizing up a colossus. 2015. *THE: Times Higher Education*: World University Rankings, 1 October 2015. London: TES Global Ltd. pp. 64–5.
317. Rossner, M., Van Epps, H., Hill, E. Irreproducible results: a response to Thomson Scientific. *Journal of Cell Biology*. 2008; 180 (2): 254–5.
318. Royal Society Policy Centre. *Science as an open enterprise*. London: Royal Society; 2012.
319. Rüger, S. How to write a good PhD thesis and survive the viva. Open University, UK. 2013 (12 July 2013). http://people.kmi.open.ac.uk/stefan/thesis-writing.pdf [accessed 10 February 2016].
320. Ruppert, D., Sheather, S. J., Wand, M. P. An effective bandwidth selector for local least squares regression. *Journal of the American Statistical Association*. 1995; 90 (432): 1257–70.
321. Ruppert, D., Wand, M. P., Carroll, R. J. *Semiparametric regression*. Cambridge: Cambridge University Press; 2003.
322. Russey, W. E., Ebel, H. F., Bliefert, C. *How to write a successful science thesis*. Weinheim: Wiley-VCH Verlag GmbH & Co. KGaA; 2006.
323. San Francisco declaration on research assessment DORA. 2012. http://www.ascb.org/dora/ [accessed 12 November 2015].
324. Schiermeier, Q. Jan Hendrik Schön loses his PhD. *Nature News*. 2011; 477 (19 September 2011), DOI: 10.1038/477374a.
325. *Scientific and technical reports—preparation, presentation, and preservation*. Baltimore, MD: National Information Standards Organization; 2010 (ANSI/NISO Z39.18–2005 (R2010)). http://www.niso.org/ [accessed 14 March 16].
326. Scopus. Elsevier B.V. http://www.elsevier.com/solutions/scopus/ [accessed 11 November 2015].
327. Scott-Lichter, D. and the Editorial Policy Committee, Council of Science Editors. CSE's white paper on promoting integrity in scientific journal publications, 2012 update (3rd rev. ed.). Wheat Ridge, CO: Council of Science Editors. 2012. http://www.councilscienceeditors.org/ [accessed 30 July 2016].
328. Secretary's Advisory Committee on Human Research Protections (SACHRP). Attachment A: Human subjects research implications of 'big data' studies. Rockville, MD: Office for Human Research Protections, U.S. Department of Health & Human Services (25 April 2015). http://www.hhs.gov/ [accessed 4 August 2016].
329. Şekercioğlu, Ç. H. Citation opportunity cost of the high impact factor obsession. *Current Biology*. 2013; 23 (17): R701–2. DOI: 10.1016/j.cub.2013.07.065.
330. Self-archiving policy Springer, 2015. http://www.springer.com/gp/open-access/authors-rights [accessed 31 October 2015].
331. Shao, J., Tu, D. *The jackknife and bootstrap*. New York: Springer-Verlag; 1995.
332. Sheskin, D. J. *Handbook of parametric and nonparametric statistical procedures* (3rd ed.). Boca Raton: Chapman & Hall/CRC; 2004.
333. Sicilia, M.-A., Sánchez-Alonso, S., García-Barriocanal, E. Comparing impact factors from two different citation databases: The case of computer science. *Journal of Informetrics*. 2011; 5: 698–704.

334. Siler, K., Lee, K., Bero, L. Measuring the effectiveness of scientific gatekeeping. *Proceedings of the National Academy of Sciences of the United States of America*. 2015; 112 (2): 360–5.
335. Silverman, B. W. *Density estimation for statistics and data analysis*. London: Chapman & Hall; 1986.
336. Simonoff, J. S. *Smoothing methods in statistics*. New York: Springer; 1996.
337. Sonin, A. A. *The physical basis of dimensional analysis* (2nd ed.). Cambridge, MA: Department of Mechanical Engineering, MIT; 2001.
338. Standards and operational guidance for ethics review of health-related research with human participants. World Health Organization. 2011. http://www.who.int/ [accessed 25 October 2015].
339. Steneck, N. H. ORI: Introduction to the responsible conduct of research. Rockville MD: Office of Research Integrity, U.S. Department of Health and Human Services. 2007 (August 2007). http://www.ori.dhhs.gov/ [accessed 21 October 2015].
340. Stilman, A. *Grammatically correct: The essential guide to spelling, style, usage, grammar, and punctuation* (2nd ed.). Cincinnati, OH: Writer's Digest Books; 2010.
341. STM consultation on article sharing. International Association of Scientific, Technical and Medical Publishers (STM). 2015. http://www.stm-assoc.org/stm-consultations/scn-consultation-2015/ [accessed 27 January 2016].
342. Strevens, M. The role of the priority rule in science. *Journal of Philosophy*. 2003; 100 (2): 55–79.
343. Student [W. S. Gosset]. The probable error of a mean. *Biometrika*. 1908; 6 (1): 1–25.
344. Sun, X.-L., Zhou, J. English versions of Chinese authors' names in biomedical journals: Observations and recommendations. *Science Editor*. 2002; 25 (1): 3–4.
345. Suppe, F. The structure of a scientific paper. *Philosophy of Science*. 1998; 65: 381–405. DOI: 10.2307/188275.
346. Swan, M. *Practical English usage* (3rd ed.). Oxford: Oxford University Press; 2007.
347. Szakacs, G., Than, K. Hungarian president quits after plagiarism row. Reuters News Agency (2 April 2012). http://www.reuters.com/ [accessed 1 August 2016].
348. Technical Committee ISO/TC 46. Information and documentation—Guidelines for bibliographic references and citations to information resources (ISO 690:2010(E)) (3rd ed.): ISO: International Organization for Standardization. 2010 (15 June 2010).
349. Teixeira da Silva, J. A., Dobránszki, J. Multiple authorship in scientific manuscripts: Ethical challenges, ghost and guest/gift authorship, and the cultural/disciplinary perspective. *Science and Engineering Ethics*. 2016; 22: 1457–1472. DOI: 10.1007/s11948-015-9716-3.
350. The European Association of Science Editors (EASE). EASE statement on inappropriate use of impact factors 2007. http://www.ease.org.uk/publications/ [accessed 31 July 2016].
351. The Symbols Committee of the Royal Society. *Quantities, units, and symbols* (2nd ed.). London: The Royal Society; 1975.
352. Thomson, P. Aims and objectives—What's the difference? 2014 (9 June 2014). http://patthomson.net/2014/06/09/ [accessed 5 August 2016].
353. Tufte, E. R. *The visual display of quantitative information*. Cheshire, CT: Graphics Press; 1983.
354. Tufte, E. R. *Visual explanations: Images and quantities, evidence and narrative*. Cheshire, CT: Graphics Press; 1997.

355. Tukey, J. W. Analyzing data: Sanctification or detective work? *American Psychologist*. 1969; 24: 83–91.
356. Tukey, J. W., McLaughlin, D. H. Less vulnerable confidence and significance procedures for location based on a single sample: Trimming/Winsorization 1. *Sankhyā: Indian Journal of Statistics, Series A*. 1963; 25 (3): 331–52.
357. Turabian, K. L. *A manual for writers of research papers, theses, and dissertations* (7th ed.). Chicago: University of Chicago Press; 2007.
358. Typefaces for symbols in scientific manuscripts. National Institute of Standards and Technology (NIST), U.S. Department of Commerce, 1998. http://physics.nist.gov/ [accessed 26 December 2015].
359. Unicode 9.0 character code charts. Mountain View, CA: The Unicode Consortium. 2016. http://www.unicode.org/charts/ [accessed 11 August 2016].
360. Upton, G., Cook, I. *A dictionary of statistics* (3rd ed.). Oxford: Oxford University Press; 2014.
361. Van Noorden, R. Who is the best scientist of them all? Online ranking compares h-index metrics across disciplines. *Nature News*: Macmillan Publishers Ltd. 2013 (6 November 2013), DOI: 10.1038/nature.2013.14108.
362. Van Noorden, R. The true cost of science publishing. *Nature*. 2013; 495: 426–9.
363. Vanclay, J. K. Impact factor: Outdated artefact or stepping-stone to journal certification? *Scientometrics*. 2012; 92: 211–38. DOI: 10.1007/s11192-011–0561-0.
364. Venables, W. N., Ripley, B. D. *Modern applied statistics with S* (4th ed.). New York: Springer-Verlag; 2002.
365. Vinkers, C. H., Tijdink, J. K., Otte, W. M. Use of positive and negative words in scientific PubMed abstracts between 1974 and 2014: Retrospective analysis. *BMJ*. 2015; 351: h6467.
366. Wager, E. How should editors respond to plagiarism? COPE Discussion paper. Committee on Publication Ethics, COPE. 2011 (26 April 2011). http://publicationethics.org/ [accessed 14 October 2016].
367. Wager, E., Barbour, V., Yentis, S., et al. Retraction guidelines. Committee on Publication Ethics (COPE) 2009 (1 September 2009). http://publicationethics.org/ [accessed 9 October 2015].
368. Wager, E., Godlee, F., Jefferson, T. How to survive peer review. London: BMJ Books; 2002.
369. Wager, E., Kleinert, S. Responsible research publication: International standards for authors. A position statement developed at the 2nd World Conference on Research Integrity, Singapore, 22–4 July 2010. In: Mayer, T., Steneck, N. (eds.). *Promoting research integrity in a global environment*. Singapore: Imperial College Press/World Scientific Publishing; 2011; pp. 309–16.
370. Waldron, A., Judd, P., Miller, V. *Physical Review style and notation guide*. Ridge, NY: American Physical Society; 2011.
371. Walter, S. D. Methods of reporting statistical results from medical research studies. *American Journal of Epidemiology*. 1995; 141 (10): 896–906.
372. Ware, M., Mabe, M. The STM report: An overview of scientific and scholarly journal publishing. Oxford: International Association of Scientific, Technical and Medical Publishers, 2009.
373. Warton, D. I., Wright, I. J., Falster, D. S., et al. Bivariate line-fitting methods for allometry. *Biological Reviews*. 2006; 81 (2): 259–91. DOI: 10.1017/s1464793106007007.

374. Wasserstein, R. L., Lazar, N. A. The ASA's statement on *p*-values: Context, process, and purpose. *American Statistician*. 2016; 70 (2): 129–33.
375. Web of Science. Thomson Reuters, later Clarivate Analytics. http://wokinfo.com/ [accessed 19 December 2015].
376. Web posting guidelines for papers in AIP journals and proceedings. American Institute of Physics, 2015. http://publishing.aip.org/authors/ [accessed 31 October 2015].
377. Weinberger, C. J., Evans, J. A., Allesina, S. Ten simple (empirical) rules for writing science. *PLOS Computational Biology*. 2015; 11 (4): e1004205. DOI: 10.1371/journal.pcbi.1004205.
378. Wikipedia, The Free Encyclopedia. Wikimedia Foundation, Inc. http://en.wikipedia.org/ [accessed 31 July 2016].
379. Wilcox, R. R., Keselman, H. J. Modern robust data analysis methods: Measures of central tendency. *Psychological Methods*. 2003; 8 (3): 254–74.
380. Wilkinson, L. *The grammar of graphics*. New York: Springer-Verlag; 1999.
381. Winer, B. J., Brown, D. R., Michels, K. M. *Statistical principles in experimental design* (3rd ed.). Boston, MA: McGraw-Hill, Inc.; 1991.
382. Wislar, J. S., Flanagin, A., Fontanarosa, P. B., et al. Honorary and ghost authorship in high impact biomedical journals: A cross sectional survey. *BMJ*. 2011; 343: d6128.
383. WMA Declaration of Helsinki—Ethical principles for medical research involving human subjects. World Medical Association 2016. http://www.wma.net/ [accessed 4 August 2016].
384. Wolfowitz, J. Additive partition functions and a class of statistical hypotheses. *Annals of Mathematical Statistics*. 1942; 13 (3): 247–79.
385. Wong, B. B. M., Kokko, H. Is science as global as we think? *Trends in Ecology and Evolution*. 2005; 20 (9): 475–6.
386. World Medical Association. Ferney-Voltaire, France: WMA The World Medical Association, 2016. http://www.wma.net/ [accessed 8 April 2016].
387. Wyatt, P. J. Too many authors, too few creators. *Physics Today*. 2012; 65 (4): 9–10.
388. Xia, Y., Li, W. K. Asymptotic behavior of bandwidth selected by the cross-validation method for local polynomial fitting. *Journal of Multivariate Analysis*. 2002; 83: 265–87.
389. Young, M. *The technical writer's handbook: Writing with style and clarity*. Mill Valley, CA: University Science Books; 1989.
390. Zigmond, M. J., Fischer, B. A. Beyond fabrication and plagiarism: The little murders of everyday science—Commentary on 'Six domains of research ethics' (K.D. Pimple). *Science and Engineering Ethics*. 2002; 8 (2): 229–34.
391. Zinsser, W. *On writing well* (7th ed.). New York: HarperCollins; 2006.

Research Works

392. Aad, G., Abbott, B., Abdallah, J., et al. Combined measurement of the Higgs boson mass in *pp* collisions at $\sqrt{s}$ = 7 and 8 TeV with the ATLAS and CMS experiments. *Physical Review Letters*. 2015; 114 (19): 191803.
393. Abey-Wickrama, I., a'Brook, M. F., Gattoni, F. E. G., et al. Mental-hospital admissions and aircraft noise. *Lancet*. 1969; 294 (7633): 1275–7.
394. Adelberger, E. G., Heckel, B. R., Nelson, A. E. Tests of the gravitational inverse-square law. *Annual Review of Nuclear and Particle Science*. 2003; 53: 77–121.
395. Agarwal, R. P., O'Regan, D., Sambandham, M. Random degree and essentiality for countably condensing maps. *Stochastic Analysis and Applications*. 2002; 20 (6): 1169–76.
396. Ahmed, I., Sutton, A. J., Riley, R. D. Assessment of publication bias, selection bias, and unavailable data in meta-analyses using individual participant data: A database survey. *BMJ*. 2012; 344: d7762.
397. Al-Khayat, H. A., Kensler, R. W., Squire, J. M., et al. Atomic model of the human cardiac muscle myosin filament. *Proceedings of the National Academy of Sciences of the United States of America*. 2013; 110 (1): 318–323. DOI: 10.1073/pnas.1212708110.
398. Ali, B. S. A safety assessment framework for automatic dependent surveillance broadcast (ADS-B) and its potential impact on aviation safety. PhD thesis. Imperial College London, London, Centre for Transport Studies, Department of Civil and Environmental Engineering; 2013.
399. Archer, G., Titterington, D. M. On some Bayesian/regularization methods for image-restoration. *IEEE Transactions on Image Processing*. 1995; 4 (7): 989–95.
400. Arik, S. New criteria for global robust stability of delayed neural networks with norm-bounded uncertainties. *IEEE Transactions on Neural Networks and Learning Systems*. 2014; 25 (6): 1045–52.
401. Arkwright, J. W., Blenman, N. G., Underhill, I. D., et al. Measurement of muscular activity associated with peristalsis in the human gut using fiber Bragg grating arrays. *IEEE Sensors Journal*. 2012; 12 (1): 113–17.
402. Arlow, T., Scott, K., Wagenseller, A., et al. Proteasome inhibition rescues clinically significant unstable variants of the mismatch repair protein Msh2. *Proceedings of the National Academy of Sciences of the United States of America*. 2013; 110 (1): 246–51. DOI: 10.1073/pnas.1215510110.
403. Arribas, S. M., Hillier, C., González, C., et al. Cellular aspects of vascular remodeling in hypertension revealed by confocal microscopy. *Hypertension*. 1997; 30: 1455–64.
404. Atkinson, J. On the lattice-geometry and birational group of the six-point multi-ratio equation. *Proceedings of the Royal Society A*. 2015; 471: 20140612.
405. ATLAS Collaboration. Observation of a new particle in the search for the Standard Model Higgs boson with the ATLAS detector at the LHC. *Physics Letters B*. 2012; 716: 1–29.

406. Axelsson, P., Gustafsson, F. Discrete-time solutions to the continuous-time differential Lyapunov equation with applications to Kalman filtering. *IEEE Transactions on Automatic Control*. 2015; 60 (3): 632–43.
407. Backfisch, B., Veedin Rajan, V. B., Fischer, R. M., et al. Stable transgenesis in the marine annelid *Platynereis dumerilii* sheds new light on photoreceptor evolution. *Proceedings of the National Academy of Sciences of the United States of America*. 2013; 110 (1): 193–8.
408. Badro, J., Côté, A. S., Brodholt, J. P. A seismologically consistent compositional model of Earth's core. *Proceedings of the National Academy of Sciences of the United States of America*. 2014; 111 (21): 7542–5.
409. Baglio, S., Bloemer, M., Savalli, N., et al. Development of novel optoelectromechanical systems based on 'transparent metals' PBG structures. *IEEE Sensors Journal*. 2001; 1 (4): 288–95.
410. Balazs, P., Feichtinger, H. G., Hampejs, M., et al. Double preconditioning for Gabor frames. *IEEE Transactions on Signal Processing*. 2006; 54 (12): 4597–4610.
411. Balter, M. 'Hellboy' dino was a close relative of *Triceratops*. *Science: News, Paleontology*. 2015 (4 June 2015), DOI: 10.1126/science.aac4672.
412. Banerjee, K., Utture, S., Dasgupta, S., et al. Multiresidue determination of 375 organic contaminants including pesticides, polychlorinated biphenyls and polyaromatic hydrocarbons in fruits and vegetables by gas chromatography-triple quadrupole mass spectrometry with introduction of semi-quantification approach. *Journal of Chromatography A*. 2012; 1270: 283–95.
413. Barber, A. C., Hippert, C., Duran, Y., et al. Repair of the degenerate retina by photoreceptor transplantation. *Proceedings of the National Academy of Sciences of the United States of America*. 2013; 110 (1): 354–9.
414. Barber, J. A., Thompson, S. G. Analysis and interpretation of cost data in randomised controlled trials: Review of published studies. *BMJ*. 1998; 317: 1195–1200.
415. Beckett, N., Peters, R., Tuomilehto, J., et al. Immediate and late benefits of treating very elderly people with hypertension: Results from active treatment extension to hypertension in the very elderly randomised controlled trial. *BMJ*. 2012; 344: d7541. DOI: 10.1136/bmj.d7541.
416. Belouchrani, A., Abed-Meraim, K., Cardoso, J.-F., et al. A blind source separation technique using second-order statistics. *IEEE Transactions on Signal Processing*. 1997; 45 (2): 434–44.
417. Bengtson, P. Open nomenclature. *Palaeontology*. 1988; 31 (1): 223–7.
418. Bernstein, J. *Oppenheimer: Portrait of an enigma*. London: Duckworth; 2004.
419. Björnson, E., Hoydis, J., Kountouris, M., et al. Massive MIMO systems with non-ideal hardware: Energy efficiency, estimation, and capacity limits. *IEEE Transactions on Information Theory*. 2014; 60 (11): 7112–39.
420. Bloch, I., Dalibard, J., Nascimbène, S. Quantum simulations with ultracold quantum gases. *Nature Physics*. 2012; 8: 267–76.
421. Boudraa, A.-O., Dehak, S. M. R., Zhu, Y.-M., et al. Automated segmentation of multiple sclerosis lesions in multispectral MR imaging using fuzzy clustering. *Computers in Biology and Medicine*. 2000; 30: 23–40.
422. Bourbaki, N. *Elements of mathematics. General topology. Part 1*: Paris: Hermann; Reading, MA: Addison-Wesley; 1966.
423. Bourbaki, N. *Elements of mathematics. Algebra I*: Paris: Hermann; Reading, MA: Addison-Wesley; 1974.

424. Bourgain, J. On the Cauchy problem for the Kadomstev-Petviashvili equation. *Geometric and Functional Analysis*. 1993; 3 (4): 315–41.
425. Boyd, G. E., Adamson, A. W., Myers, L. S., Jr. The exchange adsorption of ions from aqueous solutions by organic zeolites. II. Kinetics. *Journal of the American Chemical Society*. 1947; 69 (11): 2836–48.
426. Brell, C. G. Generalized color codes supporting non-Abelian anyons. *Physical Review A*. 2015; 91: 042333.
427. Brewin, C. R. Episodic memory, perceptual memory, and their interaction: Foundations for a theory of posttraumatic stress disorder. *Psychological Bulletin*. 2014; 140 (1): 69–97.
428. Brindley, G. S. *Physiology of the retina and visual pathway* (2nd ed.). London: Edward Arnold (Publishers) Ltd; 1970.
429. Brod, D. J. Complexity of simulating constant-depth BosonSampling. *Physical Review A*. 2015; 91: 042316.
430. Brown, C. M., Henderson, D. M. A new horned dinosaur reveals convergent evolution in cranial ornamentation in ceratopsidae. *Current Biology*. 2015; 25: 1641–8. DOI: 10.1016/j.cub.2015.04.041.
431. Brugman, B., Tijssen, T., van Oosterom, P. Validating a 3D topological structure of a 3D space partition. In: Geertman, S., Reinhardt, W., Toppen, F. (eds.). *Advancing geoinformation science for a changing world*. Berlin: Springer; 2011; pp. 359–78.
432. Brzezinski, D., Stefanowski, J. Reacting to different types of concept drift: The accuracy updated ensemble algorithm. *IEEE Transactions on Neural Networks and Learning Systems*. 2014; 25 (1): 81–94.
433. Buckley, S. M., Aranda-Orgilles, B., Strikoudis, A., et al. Regulation of pluripotency and cellular reprogramming by the ubiquitin-proteasome system. *Cell Stem Cell*. 2012; 11: 783–98.
434. Buffington, S. A., Huang, W., Costa-Mattioli, M. Translational control in synaptic plasticity and cognitive dysfunction. *Annual Review of Neuroscience*. 2014; 37: 17–38.
435. Cabral, R., De la Torre, F., Costeira, J. P., et al. Matrix completion for weakly-supervised multi-label image classification. *IEEE Transactions on Pattern Analysis and Machine Intelligence*. 2015; 37 (1): 121–35.
436. Carter, C. S., Braver, T. S., Barch, D. M., et al. Anterior cingulate cortex, error detection, and the online monitoring of performance. *Science*. 1998; 280: 747–9. DOI: 10.1126/science.280.5364.747.
437. Carter, G. A., Knapp, A. K. Leaf optical properties in higher plants: Linking spectral characteristics to stress and chlorophyll concentration. *American Journal of Botany*. 2001; 88 (4): 677–84. DOI: 10.2307/2657068.
438. Carullo, A., Parvis, M. An ultrasonic sensor for distance measurement in automotive applications. *IEEE Sensors Journal*. 2001; 1 (2): 143–7.
439. Cevenini, G., Borzelli, G., Rubegni, P., et al. Modified Karhunen-Loéve expansion for evaluating skin-colour-associated melanoma risk factors. *Computers in Biology and Medicine*. 2000; 30: 171–89.
440. Chatterjee, S., Bandyopadhyay, A., Sarkar, K. Effect of iron oxide and gold nanoparticles on bacterial growth leading towards biological application. *Journal of Nanobiotechnology*. 2011; 9: 34.

441. Chatterjee, S., Sarkar, K. Retracted article: Surface-functionalized gold nanoparticles mediate bacterial transformation: A nanobiotechnological approach. *Biotechnology Letters*. 2014; 36 (2): 265–71.
442. Chen, G., King, J. A., Burgess, N., et al. How vision and movement combine in the hippocampal place code. *Proceedings of the National Academy of Sciences of the United States of America*. 2013; 110 (1): 378–83.
443. Chen, Q., Freymueller, J. T., Yang, Z., et al. Spatially variable extension in southern Tibet based on GPS measurements. *Journal of Geophysical Research*. 2004; 109: B09401.
444. Chen, T.-Y. Structure and function of CLC channels. *Annual Review of Physiology*. 2005; 67: 809–839. DOI: 10.1146/annurev.physiol.67.032003.153012.
445. Chi, M., Gay-Balmaz, F., Putkaradze, V., et al. Dynamics and optimal control of flexible solar updraft towers. *Proceedings of the Royal Society A*. 2015; 471: 20140539.
446. Chiu, T.-F. A proposed IPC-based clustering method for exploiting expert knowledge and its application to strategic planning. *Journal of Information Science*. 2014; 40 (1): 50–66.
447. Christmas, J., Everson, R. Robust autoregression: Student-t innovations using variational Bayes. *IEEE Transactions on Signal Processing*. 2011; 59 (1): 48–57.
448. Clamp, M., Cuff, J., Searle, S. M., et al. The Jalview Java alignment editor. *Bioinformatics*. 2004; 20 (3): 426–7. DOI: 10.1093/bioinformatics/btg430.
449. Cleland, T. A. Early transformations in odor representation. *Trends in Neurosciences*. 2010; 33 (3): 130–9.
450. Clinical trials results and new data analyses in amyloid-related therapies from the Alzheimer's Association International Conference 2015. Washington, DC (22 July 2015). http://www.alz.org/aaic/_downloads/Wed-7am.pdf [accessed 10 August 2016].
451. Codocedo, J. F., Allard, C., Godoy, J. A., et al. Sirt1 regulates dendritic development in hippocampal neurons. *PLOS ONE*. 2012; 7 (10): e47073. DOI: 10.1371/journal.pone.0047073.
452. Cohen, L. Probability distributions with given multivariate marginals. *Journal of Mathematical Physics*. 1984; 25 (8): 2402–3.
453. Comaniciu, D., Ramesh, V., Meer, P. Kernel-based object tracking. *IEEE Transactions on Pattern Analysis and Machine Intelligence*. 2003; 25 (5): 564–77.
454. Committee on Surface Temperature Reconstructions for the Last 2,000 Years, National Research Council. *Surface temperature reconstructions for the last 2,000 years*. Washington, DC: National Research Council; 2006.
455. Cooper, R., Strand, B. H., Hardy, R., et al. Physical capability in mid-life and survival over 13 years of follow-up: British birth cohort study. *BMJ*. 2014; 348: g2219. DOI: 10.1136/bmj.g2219.
456. Cordeira, J. W., Felsted, J. A., Teillon, S., et al. Hypothalamic dysfunction of the thrombospondin receptor α2δ-1 underlies the overeating and obesity triggered by brain-derived neurotrophic factor deficiency. *Journal of Neuroscience*. 2014; 34 (2): 554–65.
457. Courtney, A. J., Chan, H. S. Visual lobe dimensions and search performance for targets on a competing homogeneous background. *Perception & Psychophysics*. 1986; 40 (1): 39–44.

458. Cumpson, P. J., Seah, M. P. *Stability of reference masses III: Mechanism and long term effects of mercury contamination on platinum-iridium mass standards*. Teddington, Middlesex, UK: National Physical Laboratory; 1994 (NPL Report DMM(A)125).
459. Daina, N. Modelling electric vehicle use and charging behaviour. PhD thesis. Imperial College London, London, Centre for Transport Studies, Department of Civil and Environmental Engineering; 2014.
460. Dalton, L. A., Dougherty, E. R. Bayesian minimum mean-square error estimation for classification error—Part II: linear classification of Gaussian models. *IEEE Transactions on Signal Processing*. 2011; 59 (1): 130–44.
461. Dalzell, C. J., Han, T. P. J., Ruddock, I. S., et al. Two-photon excited fluorescence in rare-earth doped optical fiber for applications in distributed sensing of temperature. *IEEE Sensors Journal*. 2012; 12 (1): 51–4.
462. Damjanovic, D., Muralt, P., Setter, N. Ferroelectric sensors. *IEEE Sensors Journal*. 2001; 1 (3): 191–206.
463. Danev, R., Buijsse, B., Khoshouei, M., et al. Volta potential phase plate for in-focus phase contrast transmission electron microscopy. *Proceedings of the National Academy of Sciences of the United States of America*. 2014; 111 (44): 15635–40. DOI: 10.1073/pnas.1418377111.
464. Dani, A. P., Chung, S.-J., Hutchinson, S. Observer design for stochastic nonlinear systems via contraction-based incremental stability. *IEEE Transactions on Automatic Control*. 2015; 60 (3): 700–14.
465. Dastidar, P., Mäenpää, J., Heinonen, T., et al. Magnetic resonance imaging based volume estimation of ovarian tumours: Use of a segmentation and 3D reformation software. *Computers in Biology and Medicine*. 2000; 30: 329–40.
466. Davis, J. C. S., Lee, D.-H. Concepts relating magnetic interactions, intertwined electronic orders, and strongly correlated superconductivity. *Proceedings of the National Academy of Sciences of the United States of America*. 2013; 110 (44): 17623–30.
467. Davis, S. W., Dennis, N. A., Buchler, N. G., et al. Assessing the effects of age on long white matter tracts using diffusion tensor tractography. *Neuroimage*. 2009; 46: 530–41. DOI: 10.1016/j.neuroimage.2009.01.068.
468. Day, B. L., Marsden, C. D. Accurate repositioning of the human thumb against unpredictable dynamic loads is dependent upon peripheral feed-back. *Journal of Physiology*. 1982; 327: 393–407.
469. De Cesarei, A., Loftus, G. R. Global and local vision in natural scene identification. *Psychonomic Bulletin & Review*. 2011; 18: 840–7.
470. de Pablo-Galán, L., de Pablo, J. J., Chávez-García, M. Diagenesis and rheology of a recent-Pleistocene volcanogenic sedimentary sequence, Mexican Basin. *Clays and Clay Minerals*. 2002; 50 (6): 807–23.
471. Deary, I. J. Intelligence. *Annual Review of Psychology*. 2012; 63: 453–82.
472. Debnath, S., Qin, J., Bahrani, B., et al. Operation, control, and applications of the modular multilevel converter: A review. *IEEE Transactions on Power Electronics*. 2015; 30 (1): 37–53.
473. DECC/Ofgem Smart Grid Forum Workstream 3. *Developing networks for low carbon*. Energy Networks Association and DECC/Ofgem Smart Grid Forum; 2011. http://www.ofgem.gov.uk/ [accessed 10 August 2016].

474. Deng, Z., Yin, Y., Yang, S., et al. Fabrication of superhydrophobic surface from binary micro-/nano-structure of mullite-whisk-based films. *Applied Physics A: Materials Science & Processing*. 2013; 113: 591–6.
475. DiCarlo, J. J., Zoccolan, D., Rust, N. C. How does the brain solve visual object recognition? *Neuron*. 2012; 73: 415–434. DOI: 10.1016/j.neuron.2012.01.010.
476. Diffie, W., Hellman, M. E. New directions in cryptography. *IEEE Transactions on Information Theory*. 1976; 22 (6): 644–54.
477. Diokh, T., Le-Roux, E., Jeannot, S., et al. Investigation of the impact of the oxide thickness and RESET conditions on disturb in HfO2-RRAM integrated in a 65nm CMOS technology. Presented at 2013 IEEE International Reliability Physics Symposium (IRPS); Monterey, CA. 14–18 April 2013.
478. Dollár, P., Wojek, C., Schiele, B., et al. Pedestrian detection: An evaluation of the state of the art. *IEEE Transactions on Pattern Analysis and Machine Intelligence*. 2012; 34 (4): 743–61.
479. Donaldson, K., Stone, V., Tran, C. L., et al. Nanotoxicology. *Occupational & Environmental Medicine*. 2004; 61: 727–8.
480. Dong, H., Wang, Z., Gao, H. Distributed filtering for a class of time-varying systems over sensor networks with quantization errors and successive packet dropouts. *IEEE Transactions on Signal Processing*. 2012; 60 (6): 3164–73.
481. Dorfman, D., Fernandez, D. C., Chianelli, M., et al. Post-ischemic environmental enrichment protects the retina from ischemic damage in adult rats. *Experimental Neurology*. 2013; 240: 146–56.
482. Doyle, J. C., Glover, K., Khargonekar, P. P., et al. State-space solutions to standard H_2 and H_∞ control problems. *IEEE Transactions on Automatic Control*. 1989; 34 (8): 831–47.
483. Dreaden, E. C., Alkilany, A. M., Huang, X., et al. The golden age: Gold nanoparticles for biomedicine. *Chemical Society Reviews*. 2012; 41: 2740–79.
484. Dréo, J., Siarry, P. A new ant colony algorithm using the heterarchical concept aimed at optimization of multiminima continuous functions. In: Dorigo, M., Di Caro, G., Sampels, M. (eds.). *Ant algorithms*. Berlin: Springer; 2002; pp. 216–21.
485. Du, P., Eisenberg, R. Catalysts made of earth-abundant elements (Co, Ni, Fe) for water splitting: Recent progress and future challenges. *Energy & Environmental Science*. 2012; 5: 6012–21. DOI: 10.1039/c2ee03250c.
486. Dubois, D., Prade, H. Gradualness, uncertainty and bipolarity: Making sense of fuzzy sets. *Fuzzy Sets and Systems*. 2012; 192: 3–24.
487. Dunning, T. H., Jr. Gaussian basis functions for use in molecular calculations. I. Contraction of (9s5p) atomic basis sets for first-row atoms. *Journal of Chemical Physics*. 1970; 53 (7): 2823–33.
488. Durfee, C. G., III, Backus, S., Murnane, M. M., et al. Design and implementation of a TW-class high-average power laser system. *IEEE Journal of Selected Topics in Quantum Electronics*. 1998; 4 (2): 395–406.
489. Dutta, B., Raghavan, G. S. V., Ngadi, M. Surface characterization and classification of slow and fast pyrolyzed biochar using novel methods of pycnometry and hyperspectral imaging. *Journal of Wood Chemistry and Technology*. 2012; 32: 105–20.
490. Edelman, S., Butler, J., Hershatter, B. W., et al. The effects of androgen deprivation therapy on cardiac function and heart failure: Implications for management of prostate cancer. *Clinical Genitourinary Cancer*. 2014; 12 (6): 399–407.

491. Editors, PLOS ONE. Retraction: *MicroRNA-302a* functions as a putative tumor suppressor in colon cancer by targeting *Akt*. *PLOS ONE*. 2015; 10 (9): e0139497. DOI: 10.1371/journal.pone.0139497.
492. EerNisse, E. P., Wiggins, R. B. Review of thickness-shear mode quartz resonator sensors for temperature and pressure. *IEEE Sensors Journal*. 2001; 1 (1): 79–87.
493. Einstein, A. Zur Elektrodynamik bewegter Körper. *Annalen der Physik*. 1905; 17: 891–921.
494. El-Dossoki, F. I. Refractive index and density measurements for selected binary protic-protic, aprotic-aprotic, and aprotic-protic systems at temperatures from 298.15 K to 308.15 K. *Journal of the Chinese Chemical Society*. 2007; 54: 1129–37.
495. El Aissati, A., McQueen, J. M., Cutler, A. Finding words in a language that allows words without vowels. *Cognition*. 2012; 124: 79–84.
496. Executive summaries for the hydrogen storage materials centers of excellence. Office of Energy Efficiency and Renewable Energy, U. S. Department of Energy (April 2012). http://energy.gov/ [accessed 18 March 2016].
497. Faraway, J. J. Bootstrap selection of bandwidth and confidence bands for nonparametric regression. *Journal of Statistical Computation and Simulation*. 1990; 37 (1): 37–44.
498. Farooque, M., Coughlin, R. W. Electrochemical gasification of coal (investigation of operating conditions and variables). *Fuel*. 1979; 58: 705–12.
499. Fasano, A., Catassi, C. Celiac disease. *New England Journal of Medicine*. 2012; 367 (25): 2419–26.
500. Femia, N., Petrone, G., Spagnuolo, G., et al. Optimization of perturb and observe maximum power point tracking method. *IEEE Transactions on Power Electronics*. 2005; 20 (4): 963–73.
501. Feynman, R. P. Forces in molecules. *Physical Review*. 1939; 56: 340–3.
502. Fibich, G., Gavious, A., Solan, E. Averaging principle for second-order approximation of heterogeneous models with homogeneous models. *Proceedings of the National Academy of Sciences of the United States of America*. 2012; 109 (48): 19545–50. DOI: 10.1073/pnas.1206867109.
503. Finocchiaro, P., Pappalardo, A., Cosentino, L., et al. Characterization of a novel 100-channel silicon photomultiplier—Part 1: Noise. *IEEE Transactions on Electron Devices*. 2008; 55 (10): 2757–64. DOI: 10.1109/ted.2008.2003996.
504. Fisher, R. A. On the dominance ratio. *Proceedings of the Royal Society of Edinburgh*. 1922; 42: 321–341.
505. Fleurence, A., Friedlein, R., Ozaki, T., et al. Experimental evidence for epitaxial silicene on diboride thin films. *Physical Review Letters*. 2012; 108: 245501. DOI: 10.1103/PhysRevLett.108.245501.
506. Foster, D. H. Color constancy. *Vision Research*. 2011; 51: 674–700.
507. Foster, D. H., Amano, K., Nascimento, S. M. C. Time-lapse ratios of cone excitations in natural scenes. *Vision Research*. 2016; 120: 45–60.
508. Foster, D. H., Amano, K., Nascimento, S. M. C., et al. Frequency of metamerism in natural scenes. *Journal of the Optical Society of America A*. 2006; 23 (10): 2359–72. DOI: 10.1364/JOSAA.23.002359.
509. Frankland, P. W., Köhler, S., Josselyn, S. A. Hippocampal neurogenesis and forgetting. *Trends in Neurosciences*. 2013; 36 (9): 497–503.
510. French, A. S., Dick, S. H., Rasmusson, D. D. Postsynaptic dorsal column and cuneate correlations in the raccoon: A re-evaluation by parallel-cascade analysis. *Journal of Neurophysiology*. 2002; 88: 3372–6.

511. Fussell, S. R., Setlock, L. D., Yang, J., et al. Gestures over video streams to support remote collaboration on physical tasks. *Human-Computer Interaction*. 2004; 19: 273–309.
512. Galuppi, L., Royer-Carfagni, G. F. Effective thickness of laminated glass beams: New expression *via* a variational approach. *Engineering Structures*. 2012; 38: 53–67.
513. Gandhi, M. S., Redd, C. B., Tuckett, R. P., et al. A novel device to evaluate the vibrotactile threshold. *Journal of Medical Devices*. 2012; 6: 031001.
514. Gandhi, S., Lyon, R. E. *Health hazards of combustion products from aircraft composite materials*. Washington, DC: Office of Aviation Research; 1998 (DOT/FAA/AR-98/34).
515. García, S., Derrac, J., Cano, J. R., et al. Prototype selection for nearest neighbor classification: Taxonomy and empirical study. *IEEE Transactions on Pattern Analysis and Machine Intelligence*. 2012; 34 (3): 417–35.
516. Genzer, J. Surface-bound gradients for studies of soft materials behavior. *Annual Review of Materials Research*. 2012; 42: 435–68.
517. George, R. E., Robledo, L. M., Maroney, O. J. E., et al. Opening up three quantum boxes causes classically undetectable wavefunction collapse. *Proceedings of the National Academy of Sciences of the United States of America*. 2013; 110 (10): 3777–81.
518. Giddey, S., Kulkarni, A., Badwal, S. P. S. Low emission hydrogen generation through carbon assisted electrolysis. *International Journal of Hydrogen Energy*. 2015; 40: 70–4.
519. Gilbey, J. To see with new eyes. 2015. *THE: Times Higher Education*, 30 July 2015. London: TES Global Ltd. p. 28.
520. Gomes, M. T. S. R. Application of piezoelectric quartz crystals to the analysis of trace metals in solution: A review. *IEEE Sensors Journal*. 2001; 1 (2): 109–18.
521. Goodwin, J. E., Feng, Y., Velazquez, H., et al. Endothelial glucocorticoid receptor is required for protection against sepsis. *Proceedings of the National Academy of Sciences of the United States of America*. 2013; 110 (1): 306–11.
522. Granlund, S., Lindqvist, P., Martio, O. *F*-harmonic measure in space. *Annales Academiae Scientiarum Fennicae, Series A I Mathematica*. 1982; 7: 233–47.
523. Groppa, S., Herzog, J., Falk, D., et al. Physiological and anatomical decomposition of subthalamic neurostimulation effects in essential tremor. *Brain*. 2014; 137: 109–21.
524. Gruzinov, A. V., Diamond, P. H. Self-consistent theory of mean-field electrodynamics. *Physical Review Letters*. 1994; 72 (11): 1651–3.
525. Gunathillake, A., Samarasinghe, K. An unequal clustering algorithm for an emergency response wireless sensor network. In: *2013 IEEE Ninth International Conference on Mobile Ad-Hoc and Sensor Networks (MSN)*, Dalian, Liaoning, China, 11–13 December 2013, pp. 383–8. IEEE Computer Society.
526. Guo, Y. Aircraft flap side edge noise modeling and prediction. Presented at 17th AIAA/CEAS Aeroacoustics Conference (32nd AIAA Aeroacoustics Conference); Portland, OR. 5–8 June 2011.
527. Guo, Y. Flap side edge noise modeling and prediction. *Journal of Sound and Vibration*. 2013; 332: 3846–68.
528. Guo, Y. Retraction notice to: 'Flap side edge noise modeling and prediction' [J. Sound Vibr. 332 (2013) 3846–3868]. *Journal of Sound and Vibration*. 2014; 333: 4403.

529. Gustafsson, J.- Å. Estrogen receptor β—A new dimension in estrogen mechanism of action. *Journal of Endocrinology*. 1999; 163: 379–83.
530. Haïssaguerre, M., Jaïs, P., Shah, D. C., et al. Spontaneous initiation of atrial fibrillation by ectopic beats originating in the pulmonary veins. *New England Journal of Medicine*. 1998; 339 (10): 659–66.
531. Halliday, I. G. Lattice field theories. *Reports on Progress in Physics*. 1984; 47: 987–1057.
532. Halmos, P. R. Random alms. *Annals of Mathematical Statistics*. 1944; 15: 182–9.
533. Hama, T., Ueta, H., Kouchi, A., et al. Quantum tunneling observed without its characteristic large kinetic isotope effects. *Proceedings of the National Academy of Sciences of the United States of America*. 2015; 112 (24): 7438–43.
534. Han, L., Islam, A., Chen, H., et al. High-efficiency dye-sensitized solar cell with a novel co-adsorbent. *Energy & Environmental Science*. 2012; 5: 6057–60.
535. Hannibal, S., Kettmann, P., Croitoru, M. D., et al. Quench dynamics of an ultracold Fermi gas in the BCS regime: Spectral properties and confinement-induced breakdown of the Higgs mode. *Physical Review A*. 2015; 91: 043630.
536. Hardy, J., Selkoe, D. J. The amyloid hypothesis of Alzheimer's disease: Progress and problems on the road to therapeutics. *Science*. 2002; 297 (5580): 353–6.
537. Hasegawa, P. M., Bressan, R. A., Zhu, J.-K., et al. Plant cellular and molecular responses to high salinity. *Annual Review of Plant Physiology and Plant Molecular Biology*. 2000; 51: 463–99.
538. He, J.-J., Koteles, E. S., Lamontagne, B., et al. Integrated polarization compensator for WDM waveguide demultiplexers. *IEEE Photonics Technology Letters*. 1999; 11 (2): 224–6.
539. He, J., Li, Y. W., Munir, M. S. A flexible harmonic control approach through voltage-controlled DG-grid interfacing converters. *IEEE Transactions on Industrial Electronics*. 2012; 59 (1): 444–55.
540. Hinde, E., Digman, M. A., Hahn, K. M., et al. Millisecond spatiotemporal dynamics of FRET biosensors by the pair correlation function and the phasor approach to FLIM. *Proceedings of the National Academy of Sciences of the United States of America*. 2013; 110 (1): 135–40.
541. Hirai, Y., Harada, S., Isaka, S., et al. Nano-imprint lithography using replicated mold by Ni electroforming. *Japanese Journal of Applied Physics*. 2002; 41 (6B): 4186–9.
542. Hobbs, N. Z., Farmer, R. E., Rees, E. M., et al. Short-interval observational data to inform clinical trial design in Huntington's disease. *Journal of Neurology, Neurosurgery & Psychiatry*. 2015; 86: 1291–8.
543. Holmes-Cerfon, M., Gortler, S. J., Brenner, M. P. A geometrical approach to computing free-energy landscapes from short-ranged potentials. *Proceedings of the National Academy of Sciences of the United States of America*. 2013; 110 (1): E5–E14.
544. Houghton, R. L., Benson, D. R., Reynolds, L. D., et al. A multi-epitope synthetic peptide and recombinant protein for the detection of antibodies to *Trypanosoma cruzi* in radioimmunoprecipitation-confirmed and consensus-positive sera. *Journal of Infectious Diseases*. 1999; 179: 1226–34.
545. Huffman, W. E. Kahneman's psychology of value: The sixth T. W. Schultz lecture. *American Journal of Agricultural Economics*. 2012; 94 (2): 285–90.
546. Hwang, S. S., Kim, Y. U., Lee, S., et al. Transcription factor YY1 is essential for regulation of the Th2 cytokine locus and for Th2 cell differentiation. *Proceedings*

of the National Academy of Sciences of the United States of America. 2013; 110 (1): 276–81.

547. Iacoboni, M., Dapretto, M. The mirror neuron system and the consequences of its dysfunction. *Nature Reviews Neuroscience*. 2006; 7: 942–51.
548. Irwin, J. J., Sterling, T., Mysinger, M. M., et al. ZINC: A free tool to discover chemistry for biology. *Journal of Chemical Information and Modeling*. 2012; 52: 1757–68.
549. Isegawa, M., Peverati, R., Truhlar, D. G. Performance of recent and high-performance approximate density functionals for time-dependent density functional theory calculations of valence and Rydberg electronic transition energies. *Journal of Chemical Physics*. 2012; 137: 244104.
550. Ismail, R., Azuara-Blanco, A., Ramsay, C. R. Variation of clinical outcomes used in glaucoma randomised controlled trials: A systematic review. *British Journal of Ophthalmology*. 2014; 98: 464–8.
551. Jadbabaie, A., Lin, J., Morse, A. S. Coordination of groups of mobile autonomous agents using nearest neighbor rules. *IEEE Transactions on Automatic Control*. 2003; 48 (6): 988–1001.
552. Jiang, X., Ma, Z. J., Ren, W.-X. Crack detection from the slope of the mode shape using complex continuous wavelet transform. *Computer-Aided Civil and Infrastructure Engineering*. 2012; 27: 187–201.
553. Jimbo, M. A q-difference analog of U($\mathfrak{g}$) and the Yang-Baxter equation. *Letters in Mathematical Physics*. 1985; 10: 63–9.
554. Jones-Rhoades, M. W., Bartel, D. P., Bartel, B. MicroRNAs and their regulatory roles in plants. *Annual Review of Plant Biology*. 2006; 57: 19–53.
555. Kalal, Z., Mikolajczyk, K., Matas, J. Tracking-learning-detection. *IEEE Transactions on Pattern Analysis and Machine Intelligence*. 2012; 34 (7): 1409–22.
556. Kales, H. C., Gitlin, L. N., Lyketsos, C. G. Assessment and management of behavioral and psychological symptoms of dementia. *BMJ*. 2015; 350: h369.
557. Kang, D., Yu, D., Min, K., et al. A highly efficient and linear Class-AB/F power amplifier for multimode operation. *IEEE Transactions on Microwave Theory and Techniques*. 2008; 56 (1): 77–87.
558. Kanungo, J., Basu, S., Sarkar, C. K. Fabrication and characterization of ZnO/p-Si and TiO_2/p-Si heterojunctions for hydrogen detection—influence of Pd functionalization. *IEEE Sensors Journal*. 2015; 15 (12): 6954–61.
559. Kaplunovsky, V. S. One-loop threshold effects in string unification. *Nuclear Physics B*. 1988; 307: 145–56.
560. Kawahara, A. Y., Barber, J. R. Tempo and mode of antibat ultrasound production and sonar jamming in the diverse hawkmoth radiation. *Proceedings of the National Academy of Sciences of the United States of America*. 2015; 112 (20): 6407–12.
561. Kim, Y., Evans, R. G., Iversen, W. M. Remote sensing and control of an irrigation system using a distributed wireless sensor network. *IEEE Transactions on Instrumentation and Measurement*. 2008; 57 (7): 1379–87.
562. King, I. R. The structure of star clusters. III. Some simple dynamical models. *Astronomical Journal*. 1966; 71 (1): 64–75.
563. Kitahara, T., Kajita, Y., Nishimoto, Y., et al. Collision test between steel bars with shock absorbing rubber for bridge restrainer system using frictionless impact testing apparatus. In: *Proceedings of the First International Conference on Advances in Experimental Structural Engineering: AESE 2005, Volume 2*, Nagoya, Japan, 19–21 July 2005, pp. 901–7. Ichiryusha, Handa, Japan.

564. Kokiopoulou, E., Frossard, P. Distributed classification of multiple observation sets by consensus. *IEEE Transactions on Signal Processing*. 2011; 59 (1): 104–14.
565. Kotov, V. N., Uchoa, B., Pereira, V. M., et al. Electron-electron interactions in graphene: Current status and perspectives. *Reviews of Modern Physics*. 2012; 84: 1067–1125.
566. Kounavis, M. E., Berry, F. L. A systematic approach to building high performance software-based CRC generators. In: *Proceedings of the 10th IEEE Symposium on Computers and Communications (ISCC'05)*, Cartagena, Murcia, Spain, 27–30 June 2005, pp. 855–62. IEEE.
567. Kourkoutis, L. F., Plitzko, J. M., Baumeister, W. Electron microscopy of biological materials at the nanometer scale. *Annual Review of Materials Research*. 2012; 42: 33–58.
568. Krebs, H. I., Volpe, B. T., Williams, D., et al. Robot-aided neurorehabilitation: A robot for wrist rehabilitation. *IEEE Transactions on Neural Systems and Rehabilitation Engineering*. 2007; 15 (3): 327–35.
569. Kress, J. P., Hall, J. B. ICU-acquired weakness and recovery from critical illness. *New England Journal of Medicine*. 2014; 370: 1626–35.
570. Kuczewski, N., Fourcaud-Trocmé, N., Savigner, A., et al. Insulin modulates network activity in olfactory bulb slices: Impact on odour processing. *Journal of Physiology*. 2014; 592 (13): 2751–69.
571. Kumar, A., Vlachos, I., Aertsen, A., et al. Challenges of understanding brain function by selective modulation of neuronal subpopulations. *Trends in Neurosciences*. 2013; 36 (10): 579–86.
572. Kumar, S., Parks, D. M. On the hyperelastic softening and elastic instabilities in graphene. *Proceedings of the Royal Society A*. 2014; 471: 20140567.
573. Laferrère, B., Teixeira, J., McGinty, J., et al. Effect of weight loss by gastric bypass surgery *versus* hypocaloric diet on glucose and incretin levels in patients with type 2 diabetes. *Journal of Clinical Endocrinology & Metabolism*. 2008; 93 (7): 2479–85.
574. Land, D. V., Campbell, A. M. A quick accurate method for measuring the microwave dielectric properties of small tissue samples. *Physics in Medicine and Biology*. 1992; 37: 183–92.
575. Lappin, J. S., Bell, H. H., Harm, O. J., et al. On the relation between time and space in the visual discrimination of velocity. *Journal of Experimental Psychology: Human Perception and Performance*. 1975; 1 (4): 383–94.
576. Larsson, L., Grimby, G., Karlsson, J. Muscle strength and speed of movement in relation to age and muscle morphology. *Journal of Applied Physiology*. 1979; 46 (3): 451–6.
577. Learned, J. G., Mannheim, K. High-energy neutrino astrophysics. *Annual Review of Nuclear and Particle Science*. 2000; 50: 679–749.
578. LeBeau, A. M., Lee, M., Murphy, S. T., et al. Imaging a functional tumorigenic biomarker in the transformed epithelium. *Proceedings of the National Academy of Sciences of the United States of America*. 2013; 110 (1): 93–8.
579. Lee, C.-H., Chen, C.-Y. Natural product-based therapeutics for the treatment of cancer stem cells: A patent review (2010–2013). *Expert Opinion on Therapeutic Patents*. 2015; 25 (6): 1–11.
580. Lee, C., Wei, X., Kysar, J. W., et al. Measurement of the elastic properties and intrinsic strength of monolayer graphene. *Science*. 2008; 321 (5887): 385–8.

581. Lee, K. H., Kim, D. H., Lee, S. W., et al. *In vivo* import experiments in protoplasts reveal the importance of the overall context but not specific amino acid residues of the transit peptide during import into chloroplasts. *Molecules and Cells*. 2002; 14 (3): 388–97.
582. Leela, J. K., Syeda, A. H., Anupama, K., et al. Rho-dependent transcription termination is essential to prevent excessive genome-wide R-loops in *Escherichia coli*. *Proceedings of the National Academy of Sciences of the United States of America*. 2013; 110 (1): 258–63.
583. Lennard, L., Van Loon, J. A., Lilleyman, J. S., et al. Thiopurine pharmacogenetics in leukemia: Correlation of erythrocyte thiopurine methyltransferase activity and 6-thioguanine nucleotide concentrations. *Clinical Pharmacology & Therapeutics*. 1987; 41 (1): 18–25.
584. Leung, W., Shaffer, C. D., Reed, L. K., et al. *Drosophila* Muller F elements maintain a distinct set of genomic properties over 40 million years of evolution. *G3: Genes, Genomes, Genetics*. 2015; 5: 719–40.
585. Li, L.-B., Bonini, N. M. Roles of trinucleotide-repeat RNA in neurological disease and degeneration. *Trends in Neurosciences*. 2010; 33: 292–8. DOI: 10.1016/j.tins.2010.03.004.
586. Link, S., El-Sayed, M. A. Optical properties and ultrafast dynamics of metallic nanocrystals. *Annual Review of Physical Chemistry*. 2003; 54: 331–66.
587. Liqun, H., Bergmann, N. W. Induction motor condition monitoring using industrial wireless sensor networks. In: *Proceedings of the 2010 Sixth International Conference on Intelligent Sensors, Sensor Networks and Information Processing (ISSNIP)*, Brisbane, Australia, 7–10 December 2010, pp. 49–54. IEEE.
588. Liu, C.-H., Vaidyanathan, P. P. MMSE DFE transceiver design over slowly time-varying MIMO channels using ST-GTD. *IEEE Transactions on Signal Processing*. 2011; 59 (1): 277–289. DOI: 10.1109/tsp.2010.2087329.
589. Liu, G., Zou, S., Fu, L., et al. Roles of chain conformation and interpenetration in the growth of a polyelectrolyte multilayer. *Journal of Physical Chemistry B*. 2008; 112 (14): 4167–71. DOI: 10.1021/jp077286f.
590. Liu, L., Johnson, H. L., Cousens, S., et al. Global, regional, and national causes of child mortality: An updated systematic analysis for 2010 with time trends since 2000. *Lancet*. 2012; 379: 2151–61.
591. Liu, X., Lee, D.-Y., Cai, S., et al. Regulation of the actin-activated MgATPase activity of *Acanthamoeba* myosin II by phosphorylation of serine 639 in motor domain loop 2. *Proceedings of the National Academy of Sciences of the United States of America*. 2013; 110 (1): E23–E32.
592. Loretz, M., Rosskopf, T., Boss, J. M., et al. Single-proton spin detection by diamond magnetometry. *Science*. 2014, DOI: 10.1126/science.1259464
593. Loretz, M., Rosskopf, T., Boss, J. M., et al. Retraction. *Science*. 2015; 347 (6218), DOI: 10.1126/science.347.6218.139-a.
594. Lu, J., Tang, T., Tang, H., et al. The accumulation of deleterious mutations in rice genomes: A hypothesis on the cost of domestication. *Trends in Genetics*. 2006; 22 (3): 126–31. DOI: 10.1016/j.tig.2006.01.004.
595. Lui, K. W., Murphy, O. H., Toumazou, C. 32-μW wirelessly-powered sensor platform with a 2-m range. *IEEE Sensors Journal*. 2012; 12 (6): 1919–24.
596. Mac Low, M.-M., Klessen, R. S. Control of star formation by supersonic turbulence. *Reviews of Modern Physics*. 2004; 76 (1): 125–94.

597. Magny, M., Joannin, S., Galop, D., et al. Holocene palaeohydrological changes in the northern Mediterranean borderlands as reflected by the lake-level record of Lake Ledro, northeastern Italy. *Quaternary Research*. 2012; 77: 382–96.
598. Maniyar, F. H., Sprenger, T., Monteith, T., et al. Brain activations in the premonitory phase of nitroglycerin-triggered migraine attacks. *Brain*. 2014; 137: 232–41.
599. Mann, M. E., Zhang, Z., Hughes, M. K., et al. Proxy-based reconstructions of hemispheric and global surface temperature variations over the past two millennia. *Proceedings of the National Academy of Sciences of the United States of America*. 2008; 105 (36): 13252–7.
600. Marinovic, W., Tresilian, J. R., de Rugy, A., et al. Corticospinal modulation induced by sounds depends on action preparedness. *Journal of Physiology*. 2014; 592 (1): 153–69.
601. Matei, I., Baras, J. S. The asymptotic consensus problem on convex metric spaces. *IEEE Transactions on Automatic Control*. 2015; 60 (4): 907–21.
602. McLeod, D. Emergent fibrinolysis for central artery occlusion is inappropriate after 2 hours' ischemia. *Canadian Journal of Ophthalmology: Journal Canadien d'Ophtalmologie*. 2015; 50 (3): 249.
603. McMains, S. A., Somers, D. C. Processing efficiency of divided spatial attention mechanisms in human visual cortex. *Journal of Neuroscience*. 2005; 25 (41): 9444–8.
604. Meijer, G. C. M., Wang, G., Fruett, F. Temperature sensors and voltage references implemented in CMOS technology. *IEEE Sensors Journal*. 2001; 1 (3): 225–34.
605. Mikolajczyk, K., Schmid, C. A performance evaluation of local descriptors. *IEEE Transactions on Pattern Analysis and Machine Intelligence*. 2005; 27 (10): 1615–30.
606. Mikropoulos, P. N., Tsovilis, T. E. Striking distance and interception probability. *IEEE Transactions on Power Delivery*. 2008; 23 (3): 1571–80.
607. Milnor, J. On the immersion of n-manifolds in $(n + 1)$-space. *Commentarii Mathematici Helvetici*. 1956; 30 (1): 275–84.
608. Milnor, J. On isometries of inner product spaces. *Inventiones Mathematicae*. 1969; 8 (2): 83–97.
609. Milnor, J. Fundamental groups of complete affinely flat manifolds. *Advances in Mathematics*. 1977; 25: 178–87.
610. Milnor, J. On the concept of attractor. *Communications in Mathematical Physics*. 1985; 99: 177–95.
611. Milnor, J., Tresser, C. On entropy and monotonicity for real cubic maps (with an appendix by Adrien Douady and Pierrette Sentenac). *Communications in Mathematical Physics*. 2000; 209: 123–78.
612. Milnor, J. W. *Topology from the differentiable viewpoint*. Princeton, NJ: Princeton University Press; 1965.
613. Minns, R. J., Craxford, A. D. Pressure under the forefoot in rheumatoid arthritis: A comparison of static and dynamic methods of assessment. *Clinical Orthopaedics and Related Research*. 1984; 187: 235–42.
614. Miranda, M. A., Henson, D. B. Perimetric sensitivity and response variability in glaucoma with single-stimulus automated perimetry and multiple-stimulus perimetry with verbal feedback. *Acta Ophthalmologica*. 2008; 86: 202–6.
615. Mishra, A., Bäuerle, P. Small molecule organic semiconductors on the move: Promises for future solar energy technology. *Angewandte Chemie—International Edition*. 2012; 51 (9): 2020–67.

616. Modi, K., Brodutch, A., Cable, H., et al. The classical-quantum boundary for correlations: Discord and related measures. *Reviews of Modern Physics*. 2012; 84 (4): 1655–1707.
617. Mondal, B., Heath, R. W., Jr. Channel adaptive quantization for limited feedback MIMO beamforming systems. *IEEE Transactions on Signal Processing*. 2006; 54 (12): 4717–29.
618. Moreau, T., Coles, A., Wing, M., et al. Transient increase in symptoms associated with cytokine release in patients with multiple sclerosis. *Brain*. 1996; 119: 225–37.
619. Morrison, A. P., Turkington, D., Pyle, M., et al. Cognitive therapy for people with schizophrenia spectrum disorders not taking antipsychotic drugs: A single-blind randomised controlled trial. *Lancet*. 2014; 383: 1395–1403.
620. Mostame, S., Castelnovo, C., Moessner, R., et al. Tunable nonequilibrium dynamics of field quenches in spin ice. *Proceedings of the National Academy of Sciences of the United States of America*. 2014; 111 (2): 640–5.
621. Muller, S. J. *Asymmetry: The foundation of information*. Berlin: Springer; 2007.
622. Murayama, H., Pierce, A. Realistic Dirac leptogenesis. *Physical Review Letters*. 2002; 89 (27): 271601.
623. Nelson, D. D., Shorter, J. H., McManus, J. B., et al. Sub-part-per-billion detection of nitric oxide in air using a thermoelectrically cooled mid-infrared quantum cascade laser spectrometer. *Applied Physics B: Lasers and Optics*. 2002; 75: 343–50.
624. Nelson, S. O., Trabelsi, S., Kraszewski, A. W. RF sensing of grain and seed moisture content. *IEEE Sensors Journal*. 2001; 1 (2): 119–26.
625. Nian, H., Cheng, P., Zhu, Z. Q. Independent operation of DFIG-based WECS using resonant feedback compensators under unbalanced grid voltage conditions. *IEEE Transactions on Power Electronics*. 2015; 30 (7): 3650–61.
626. Nik-Zainal, S., Alexandrov, L. B., Wedge, D. C., et al. Mutational processes molding the genomes of 21 breast cancers. *Cell*. 2012; 149: 979–93.
627. Nordin, M., Gutman, P.-O. Controlling mechanical systems with backlash—A survey. *Automatica*. 2002; 38: 1633–49.
628. Ntranos, V., Maddah-Ali, M. A., Caire, G. Cellular interference alignment. *IEEE Transactions on Information Theory*. 2015; 61 (3): 1194–1217.
629. Obokata, H., Wakayama, T., Sasai, Y., et al. Retraction: Stimulus-triggered fate conversion of somatic cells into pluripotency. *Nature*. 2014; 511 (7507). DOI: 10.1038/nature13598.
630. Obus, A., Wewers, S. Cyclic extensions and the local lifting problem. *Annals of Mathematics*. 2014; 180: 233–84.
631. Ohtsuka, H., Xu, Y., Wada, H. Alignment of ferrite grains during austenite to ferrite transformation in a high magnetic field. *Materials Transactions, JIM*. 2000; 41 (8): 907–10.
632. Olejniczak, S. H., La Rocca, G., Gruber, J. J., et al. Long-lived microRNA-Argonaute complexes in quiescent cells can be activated to regulate mitogenic responses. *Proceedings of the National Academy of Sciences of the United States of America*. 2013; 110 (1): 157–62. DOI: 10.1073/pnas.1219958110.
633. Omodei, A., Legnani, G., Adamini, R. Three methodologies for the calibration of industrial manipulators: Experimental results on a SCARA robot. *Journal of Robotic Systems*. 2000; 17 (6): 291–307.

634. Pan, W. W., Liu, Q. F., Han, R., et al. Microwave absorption properties of the Ni nanofibers fabricated by electrospinning. *Applied Physics A: Materials Science & Processing*. 2013; 113: 755–61.
635. Panzeri, S., Brunel, N., Logothetis, N. K., et al. Sensory neural codes using multiplexed temporal scales. *Trends in Neurosciences*. 2010; 33 (3): 111–20. DOI: 10.1016/j.tins.2009.12.001.
636. Pathak, P. N., Choppin, G. R. Sorption studies of europium(III) on hydrous silica. *Journal of Radioanalytical and Nuclear Chemistry*. 2006; 270 (2): 277–83.
637. Patzek, T. W., Male, F., Marder, M. Gas production in the Barnett Shale obeys a simple scaling theory. *Proceedings of the National Academy of Sciences of the United States of America*. 2013; 110 (49): 19731–6.
638. Peterson, J., Dixit, P. D., Dill, K. A. A maximum entropy framework for nonexponential distributions. *Proceedings of the National Academy of Sciences of the United States of America*. 2013; 110 (51): 20380–5.
639. Phedorin, M. A., Goldberg, E. L. Prediction of absolute concentrations of elements from SR XRF scan measurements of natural wet sediments. *Nuclear Instruments and Methods in Physics Research A*. 2005; 543: 274–9.
640. Podsakoff, P. M., MacKenzie, S. B., Podsakoff, N. P. Sources of method bias in social science research and recommendations on how to control it. *Annual Review of Psychology*. 2012; 63: 539–69.
641. Poggiale, J.-C., Auger, P., Cordoleani, F., et al. Study of a virus-bacteria interaction model in a chemostat: Application of geometrical singular perturbation theory. *Philosophical Transactions of the Royal Society A*. 2009; 367: 4685–97.
642. Popova, E. M. Processing schlieren-background patterns by constructing the direction field. *Journal of Optical Technology*. 2004; 71 (9): 572–4.
643. Postoyan, R., Tabuada, P., Nešić, D., et al. A framework for the event-triggered stabilization of nonlinear systems. *IEEE Transactions on Automatic Control*. 2015; 60 (4): 982–96.
644. Prajapati, C. S., Kushwaha, A., Sahay, P. P. Optoelectronics and formaldehyde sensing properties of tin-doped ZnO thin films. *Applied Physics A: Materials Science & Processing*. 2013; 113: 651–62.
645. Price, M. N., Dehal, P. S., Arkin, A. P. FastTree 2—Approximately maximum-likelihood trees for large alignments. *PLOS ONE*. 2010; 5 (3): e9490. DOI: 10.1371/journal.pone.0009490.
646. Ratches, J. A., Vollmerhausen, R. H., Driggers, R. G. Target acquisition performance modeling of infrared imaging systems: Past, present, and future. *IEEE Sensors Journal*. 2001; 1 (1): 31–40.
647. Ratledge, C., Editor-in-Chief. Retraction note to: Biotechnol Lett (2014) 36:265–271, DOI 10.1007/s10529-013-1360-x. *Biotechnology Letters*. 2015; 37: 1527–8.
648. Regan, D., Neima, D. Low-contrast letter charts as a test of visual function. *Ophthalmology*. 1983; 90 (10): 1192–1200.
649. Riotto, A., Trodden, M. Recent progress in baryogenesis. *Annual Review of Nuclear and Particle Science*. 1999; 49: 35–75.
650. Roberts, B. M., Dzuba, V. A., Flambaum, V. V. Parity and time-reversal violation in atomic systems. *Annual Review of Nuclear and Particle Science*. 2015; 65: 63–86. DOI: 10.1146/annurev-nucl-102014-022331.

651. Roldán, A., Martínez-Moreno, J., Roldán, C. Multidimensional fixed point theorems in partially ordered complete metric spaces. *Journal of Mathematical Analysis and Applications*. 2012; 396: 536–45.
652. Russo, S. J., Dietz, D. M., Dumitriu, D., et al. The addicted synapse: Mechanisms of synaptic and structural plasticity in nucleus accumbens. *Trends in Neurosciences*. 2010; 33: 267–76.
653. Rust, B. R. Depositional models for braided alluvium. *Fluvial Sedimentology*. 1977; 5: 605–25.
654. Sadeghi, M. M., Jo, I., Shi, L. Phonon-interface scattering in multilayer graphene on an amorphous support. *Proceedings of the National Academy of Sciences of the United States of America*. 2013; 110 (41): 16321–6.
655. Saegusa, R., Metta, G., Sandini, G., et al. Developmental perception of the self and action. *IEEE Transactions on Neural Networks and Learning Systems*. 2014; 25 (1): 183–202.
656. Saghi, Z., Midgley, P. A. Electron tomography in the (S)TEM: From nanoscale morphological analysis to 3D atomic imaging. *Annual Review of Materials Research*. 2012; 42: 59–79.
657. Saito, R., Fujita, M., Dresselhaus, G., et al. Electronic structure of chiral graphene tubules. *Applied Physics Letters*. 1992; 60 (18): 2204–6. DOI: 10.1063/1.107080.
658. Salje, E. K. H. Ferroelastic materials. *Annual Review of Materials Research*. 2012; 42: 265–83.
659. Samet, B., Vetro, C., Vetro, P. Fixed point theorems for α-ψ-contractive type mappings. *Nonlinear Analysis*. 2012; 75: 2154–65.
660. Sanderson, A. J. R., Ponman, T. J. X-ray modelling of galaxy cluster gas and mass profiles. *Monthly Notices of the Royal Astronomical Society*. 2010; 402: 65–72.
661. Schaller, M. J., Dziadosz, G. M. Individual-differences in adult foveal visual asymmetries. *Journal of Experimental Psychology: Human Perception and Performance*. 1975; 1 (4): 353–65.
662. Schamiloglu, E., Schoenbach, K. H., Vidmari, R. Basic research leading to compact, portable pulsed power: FY'01 MURI Consortium. University of New Mexico, Pulsed Power, Beams and Microwaves Laboratory, Department of Electrical & Computer Engineering. 2007 (31 March 2007).
663. Schirmer, M., Butler, B. J., Roy, J. W., et al. A relative-least-squares technique to determine unique Monod kinetic parameters of BTEX compounds using batch experiments. *Journal of Contaminant Hydrology*. 1999; 37: 69–86.
664. Schmidt-Mende, L., Fechtenkötter, A., Müllen, K., et al. Self-organized discotic liquid crystals for high-efficiency organic photovoltaics. *Science*. 2001; 293: 1119–22. DOI: 10.1126/science.293.5532.1119.
665. Schmogrow, R., Nebendahl, B., Winter, M., et al. Error vector magnitude as a performance measure for advanced modulation formats. *IEEE Photonics Technology Letters*. 2012; 24 (1): 61–3.
666. Schneeloch, J. A., Zhong, R. D., Xu, Z. J., et al. Dependence of superconductivity in $Cu_xBi_2Se_3$ on quenching conditions. *Physical Review B*. 2015; 91: 144506.
667. Schwartz, M. P., Hou, Z., Propson, N. E., et al. Human pluripotent stem cell-derived neural constructs for predicting neural toxicity. *Proceedings of the National Academy of Sciences of the United States of America*. 2015; 112 (40): 12516–21.

668. Sen, S., Nehorai, A. Adaptive OFDM radar for target detection in multipath scenarios. *IEEE Transactions on Signal Processing.* 2011; 59 (1): 78–90.
669. Shang, X., Shim, S.-H., de Hoop, M., et al. Multiple seismic reflectors in Earth's lowermost mantle. *Proceedings of the National Academy of Sciences of the United States of America.* 2014; 111 (7): 2442–6.
670. Sheen, D. M., McMakin, D. L., Hall, T. E. Three-dimensional millimeter-wave imaging for concealed weapon detection. *IEEE Transactions on Microwave Theory and Techniques.* 2001; 49 (9): 1581–92.
671. Shi, X., Xie, N., Fortune, K., et al. Durability of steel reinforced concrete in chloride environments: An overview. *Construction and Building Materials.* 2012; 30: 125–38.
672. Shi, Z., Wang, J., Genack, A. Z. Microwave conductance in random waveguides in the cross-over to Anderson localization and single-parameter scaling. *Proceedings of the National Academy of Sciences of the United States of America.* 2014; 111 (8): 2926–30.
673. Siffrin, V., Vogt, J., Radbruch, H., et al. Multiple sclerosis—Candidate mechanisms underlying CNS atrophy. *Trends in Neurosciences.* 2010; 33 (4): 202--10.
674. Singh, N., Yadavalli, V. S. S., Peiris, M. S. A note on the modelling and analysis of vector arma processes with nonstationary innovations. *Mathematical and Computer Modelling.* 2002; 36: 1409–24. DOI: 10.1016/s0895-7177(02)00297-2.
675. Sinopoli, B., Schenato, L., Franceschetti, M., et al. Kalman filtering with intermittent observations. *IEEE Transactions on Automatic Control.* 2004; 49 (9): 1453–64.
676. Slee, D., Stepan, J., Wei, W., et al. Introduction to printed circuit board failures. In: *Proceedings of the 2009 IEEE Symposium on Product Compliance Engineering*, Toronto, Canada, 26–28 October 2009, pp. 1–8. IEEE.
677. Smith, D. R., Pendry, J. B., Wiltshire, M. C. K. Metamaterials and negative refractive index. *Science.* 2004; 305: 788–92.
678. Sniderman, A. D., Toth, P. P., Thanassoulis, G., et al. Taking a longer term view of cardiovascular risk: The causal exposure paradigm. *BMJ.* 2014; 348: g3047.
679. Spivak, D. I. *Category theory for the sciences.* Cambridge, MA: MIT Press; 2014.
680. Stock, N., Biswas, S. Synthesis of metal-organic frameworks (MOFs): Routes to various MOF topologies, morphologies, and composites. *Chemical Reviews.* 2012; 112: 933–69.
681. Suhling, J. C., Jaeger, R. C. Silicon piezoresistive stress sensors and their application in electronic packaging. *IEEE Sensors Journal.* 2001; 1 (1): 14–30.
682. Sun, S., Zhang, G., Wu, Z., et al. *MicroRNA-302a* functions as a putative tumor suppressor in colon cancer by targeting *Akt. PLOS ONE.* 2014; 9 (12): e115980. DOI: 10.1371/journal.pone.0115980.
683. Sundari, S. T. Optical absorption in Ar^+ irradiated silicon studied through spectroscopic ellipsometry. *Nuclear Instruments and Methods in Physics Research B.* 2004; 215: 157–61.
684. Supuran, C. T., Knowles, E. Retracted article: Natural product-based therapeutics for the treatment of cancer stem cells: A patent review (2010–2013). *Expert Opinion on Therapeutic Patents.* 2015; 25 (6): 737. DOI: 10.1517/13543776.2015.1042242.
685. Sutor, A., Lerch, R., Hohe, H.-P., et al. New CMOS-compatible mechanical shear stress sensor. *IEEE Sensors Journal.* 2001; 1 (4): 345–51.
686. Swets, D. L., Weng, J. J. Using discriminant eigenfeatures for image retrieval. *IEEE Transactions on Pattern Analysis and Machine Intelligence.* 1996; 18 (8): 831–6.

687. Ta, N., Liu, J., Chenna, S., et al. Stabilized gold nanoparticles on ceria nanorods by strong interfacial anchoring. *Journal of the American Chemical Society*. 2012; 134: 20585–8.
688. Takahashi, S., Kakiuchi, S., Nanba, Y., et al. The perfusion index derived from a pulse oximeter for predicting low superior vena cava flow in very low birth weight infants. *Journal of Perinatology*. 2010; 30: 265–9. DOI: 10.1038/jp.2009.159.
689. Tappen, M. F., Freeman, W. T., Adelson, E. H. Recovering intrinsic images from a single image. *IEEE Transactions on Pattern Analysis and Machine Intelligence*. 2005; 27 (9): 1459–72. DOI: 10.1109/tpami.2005.185.
690. Tarasenko, Y. N., Fleming, S. T., Schoenberg, N. E. The relationship between perceived burden of chronic conditions and colorectal cancer screening among Appalachian residents. *Journal of Rural Health*. 2014; 30: 40–9.
691. Taylor, N. L., Heazlewood, J. L., Day, D. A., et al. Differential impact of environmental stresses on the pea mitochondrial proteome. *Molecular & Cellular Proteomics*. 2005; 4 (8): 1122–33.
692. Te Morenga, L., Mallard, S., Mann, J. Dietary sugars and body weight: Systematic review and meta-analyses of randomised controlled trials and cohort studies. *BMJ*. 2013; 346: e7492. DOI: 10.1136/bmj.e7492.
693. Techapalokul, S., Alleman, J. H., Chen, Y. Economics of standard: A survey and framework. In: *Proceedings from the 2nd IEEE Conference on Standardization and Innovation in Information Technology*, Boulder, CO, 3–6 October 2001, pp. 193–205. IEEE.
694. Teutschbein, C., Seibert, J. Bias correction of regional climate model simulations for hydrological climate-change impact studies: Review and evaluation of different methods. *Journal of Hydrology*. 2012; 456–457: 12–29. DOI: 10.1016/j.jhydrol.212.05.052.
695. THE Staff. Overseas briefing. 2008. *THE: Times Higher Education*, 6 November 2008. London: TES Global Ltd.
696. Thomalla, G., Jonas, M., Bäumer, T., et al. Costs of control: Decreased motor cortex engagement during a Go/NoGo task in Tourette's syndrome. *Brain*. 2014; 137: 122–36.
697. Thomashow, M. F. Plant cold acclimation: Freezing tolerance genes and regulatory mechanisms. *Annual Review of Plant Physiology and Plant Molecular Biology*. 1999; 50: 571–99.
698. Toberer, E. S., Baranowski, L. L., Dames, C. Advances in thermal conductivity. *Annual Review of Materials Research*. 2012; 42: 179–209.
699. Troshin, S. M., Tyurin, N. E. Unitarity constraints and role of geometrical effects. *Nuclear Physics A*. 2002; 711: 236c–239c.
700. Trotter, C. M. Characterising the topographic effect at red wavelengths using juvenile conifer canopies. *International Journal of Remote Sensing*. 1998; 19 (11): 2215–21.
701. Truby, J. M., Mueller, S. P., Llewellin, E. W., et al. The rheology of three-phase suspensions at low bubble capillary number. *Proceedings of the Royal Society A*. 2015; 471: 20140557.
702. van der Heijden, G. M. F., Powers, J. S., Schnitzer, S. A. Lianas reduce carbon accumulation and storage in tropical forests. *Proceedings of the National Academy of Sciences of the United States of America*. 2015; 112 (43): 13267–71.

703. van der Weele, C. M., Jiang, H. S., Palaniappan, K. K., et al. A new algorithm for computational image analysis of deformable motion at high spatial and temporal resolution applied to root growth. Roughly uniform elongation in the meristem and also, after an abrupt acceleration, in the elongation zone. *Plant Physiology.* 2003; 132 (3): 1138–48.
704. Vedel, S., Tay, S., Johnston, D. M., et al. Migration of cells in a social context. *Proceedings of the National Academy of Sciences of the United States of America.* 2013; 110 (1): 129–34.
705. Velosa, E., Gouveia, C., Frazão, O., et al. Digital control of a white light interrogation system for optical fiber interferometers. *IEEE Sensors Journal.* 2012; 12 (1): 201–6.
706. Vetterli, M., Herley, C. Wavelets and filter banks: Theory and design. *IEEE Transactions on Signal Processing.* 1992; 40 (9): 2207–32.
707. Vieira, M., Fernandes, M., Martins, J., et al. Laser-scanned p-i-n photodiode (LSP) for image detection. *IEEE Sensors Journal.* 2001; 1 (2): 158–67.
708. Villain, J. Insulating spin glasses. *Zeitschrift für Physik B.* 1979; 33: 31–42.
709. Wada, A., Tanaka, S., Takahashi, N. Optical fiber vibration sensor using FBG Fabry-Perot interferometer with wavelength scanning and Fourier analysis. *IEEE Sensors Journal.* 2012; 12 (1): 225–9.
710. Wang, C., Levin, M. Topological invariants for gauge theories and symmetry-protected topological phases. *Physical Review B.* 2015; 91: 165119.
711. Wang, M., Gu, D., Zhang, Z., et al. *XPD* polymorphisms, cigarette smoking, and bladder cancer risk: A meta-analysis. *Journal of Toxicology and Environmental Health, Part A.* 2009; 72: 698–705.
712. Wang, Y., Feng, C., Zhang, M., et al. Visible light active N-doped TiO_2 prepared from different precursors: Origin of the visible light absorption and photoactivity. *Applied Catalysis B: Environmental.* 2011; 104: 268–74.
713. Westheimer, G. Scaling of visual acuity measurements. *Archives of Ophthalmology.* 1979; 97 (2): 327–30.
714. Wilson, D. M., Hoyt, S., Janata, J., et al. Chemical sensors for portable, handheld field instruments. *IEEE Sensors Journal.* 2001; 1 (4): 256–74.
715. Wolff, U., Dickert, F. L., Fischerauer, G. K., et al. SAW sensors for harsh environments. *IEEE Sensors Journal.* 2001; 1 (1): 4–13.
716. Wu, B., Li, C., Yang, H., et al. Formation of polyelectrolyte multilayers by flexible and semiflexible chains. *Journal of Physical Chemistry B.* 2012; 116: 3106–14.
717. Wu, H., Moshkina, N., Min, J., et al. Correction. *Proceedings of the National Academy of Sciences of the United States of America.* 2013; 110 (45): 18339.
718. Wu, H., Moshkina, N., Min, J., et al. Correction. *Proceedings of the National Academy of Sciences of the United States of America.* 2015; 112 (3): E340.
719. Wu, J., Boström, P., Sparks, L. M., et al. Beige adipocytes are a distinct type of thermogenic fat cell in mouse and human. *Cell.* 2012; 150: 366–76.
720. Wu, R.-B., Brif, C., James, M. R., et al. Limits of optimal control yields achievable with quantum controllers. *Physical Review A.* 2015; 91: 042327.
721. Wu, Z. Stability criteria of random nonlinear systems and their applications. *IEEE Transactions on Automatic Control.* 2015; 60 (4): 1038–49.
722. Yablon, I. G., Heller, F. G., Shouse, L. The key role of the lateral malleolus in displaced fractures of the ankle. *Journal of Bone & Joint Surgery.* 1977; 59 (2): 169–73.

723. Yagi, K. A simple fluorometric assay for lipoperoxide in blood plasma. *Biochemical Medicine*. 1976; 15 (2): 212–16.
724. Yamada, Y., Kawabe, T. Localizing non-retinotopically moving objects. *PLOS ONE*. 2013; 8 (1): e53815. DOI: 10.1371/journal.pone.0053815.
725. Yang, J., Ulukus, S. Optimal packet scheduling in an energy harvesting communication system. *IEEE Transactions on Communications*. 2012; 60 (1): 220–30.
726. Yermolayeva, Y., Rakison, D. H. Connectionist modeling of developmental changes in infancy: Approaches, challenges, and contributions. *Psychological Bulletin*. 2014; 140 (1): 224–55.
727. Yeung, C. H., Saad, D., Wong, K. Y. M. From the physics of interacting polymers to optimizing routes on the London Underground. *Proceedings of the National Academy of Sciences of the United States of America*. 2013; 110 (34): 13717–22.
728. Zamarrón, C., Romero, P. V., Rodriguez, J. R., et al. Oximetry spectral analysis in the diagnosis of obstructive sleep apnoea. *Clinical Science*. 1999; 97: 467–73. DOI: 10.1042/cs0970467.
729. Zangmeister, C. D., Radney, J. G., Dockery, L. T., et al. Packing density of rigid aggregates is independent of scale. *Proceedings of the National Academy of Sciences of the United States of America*. 2014; 111 (25): 9037–41.
730. Zaridze, D., Lewington, S., Boroda, A., et al. Alcohol and mortality in Russia: Prospective observational study of 151 000 adults. *Lancet*. 2014; 383: 1465–73.
731. Zhang, G.-M., Bao, C.-Y., Wan, F.-N., et al. MicroRNA-302a suppresses tumor cell proliferation by inhibiting *AKT* in prostate cancer. *PLOS ONE*. 2015; 10 (4): e0124410. DOI: 10.1371/journal.pone.0124410.
732. Zhang, X., Liu, L., Wu, Y. The eigenvalue problem for a singular higher order fractional differential equation involving fractional derivatives. *Applied Mathematics and Computation*. 2012; 218: 8526–36.
733. Zhang, X., Rowland, S. M. Stability and energy of low current surface discharges on wet surfaces. *IEEE Transactions on Dielectrics and Electrical Insulation*. 2012; 19 (6): 2055–62.
734. Zhang, Y. M., Winkler, D., Claeson, T. Linewidth measurements of Josephson flux-flow oscillators in the band 280–330 GHz. *Applied Physics Letters*. 1993; 62 (24): 3195–7.
735. Zhao, X., Leotta, A., Kustanovich, V., et al. A unified genetic theory for sporadic and inherited autism. *Proceedings of the National Academy of Sciences of the United States of America*. 2007; 104 (31): 12831–6.
736. Zhu, X., Liu, X., Li, Y., et al. Molecular high-order-harmonic generation due to the recollision mechanism by a circularly polarized laser pulse. *Physical Review A*. 2015; 91: 043418.
737. Żychaluk, K., Foster, D. H. Model-free estimation of the psychometric function. *Attention, Perception, & Psychophysics*. 2009; 71 (6): 1414–25. DOI: 10.3758/APP.71.6.1414.

Index

Locators in *italics* refer to examples.

Z

The manufacturer's authorised representative in the EU for product safety is Oxford University Press España S.A. of el Parque Empresarial San Fernando de Henares, Avenida de Castilla, 2 – 28830 Madrid (www.oup.es/en or product.safety@oup.com). OUP España S.A. also acts as importer into Spain of products made by the manufacturer.

www.ingramcontent.com/pod-product-compliance
Ingram Content Group UK Ltd.
Pitfield, Milton Keynes, MK11 3LW, UK
UKHW021824120726
473206UK00013B/161

* 9 7 8 0 1 9 8 7 0 4 2 4 9 *